Design for Environment as a Tool for the Development of a Sustainable Supply Chain

Maurizio Bevilacqua · Filippo Emanuele Ciarapica
Giancarlo Giacchetta

Design for Environment as a Tool for the Development of a Sustainable Supply Chain

Springer

Prof. Maurizio Bevilacqua
Dipartimento di Ingegneria Industriale e Scienze Matematiche
Università Politecnica delle Marche
Via Brecce Bianche
Ancona
Italy
e-mail: m.bevilacqua@univpm.it

Prof. Giancarlo Giacchetta
Dipartimento di Ingegneria Industriale e Scienze Matematiche
Università Politecnica delle Marche
Via Brecce Bianche
Ancona
Italy
e-mail: g.giachetta@univpm.it

Prof. Filippo Emanuele Ciarapica
Facoltà di Scienze e Tecnologie
Libera Università di Bolzano
Piazza Università 5
Bolzano
Itlay
e-mail: filippoemanuele.ciarapica@unibz.it

ISBN 978-1-4471-2460-3 e-ISBN 978-1-4471-2461-0
DOI 10.1007/978-1-4471-2461-0
Springer London Dordrecht Heidelberg New York

British Library Cataloguing in Publication Data
A catalogue record for this book is available from the British Library

Library of Congress Control Number: 2011943354

Printed on acid-free paper

Springer is part of Springer Science+Business Media (www.springer.com)

Preface

This book was born from the idea that Design for Environment (DfE) methodologies could be used as a tool for the development of a more sustainable supply chain. This idea led the authors of this text to focus their research toward the development of new procedures and techniques that allowed the designers to integrate DfE methods and Environmental Supply Chain Management.

In the past decade, Environmental Design ('Eco Design' or 'Design for Environment') has become one of the most significant agendas for many manufacturing companies. Yet, there is a wide variety in their approaches, strategies and in their levels of execution. The initial reaction of many companies in the late 1980s to the challenge posed by environmental concern and the need to move toward sustainability was often a relatively superficial change to their products and their approaches to marketing communications. With the emergence of the enviropreneuring paradigm, and the acceptance of the business logic, which underpins the need to consider eco-performance, more substantive changes could be expected.

The development of innovative methodologies in Environmental Management fields and the application of these in several case studies allowed the authors to publish several papers in international journals and proceedings. These case studies have been used in this book in order to better explain the theoretical topics. The integration of case studies and basic concepts such as Design for Environmental, Supply Chain Management and Life Cycle Assessment could be useful in order to make these topics understandable for large numbers of readers.

This text is directed to several professional figures such as industrial designers, production process owners, heads of technical/marketing/sales departments, mechanical/managing/production engineers and to everyone involved in environmental issues. The aim is to provide to all of these professional figures some guidelines to develop, within their area of interest, Design for Environmental (DfE) programs with the objective of redesigning the entire supply chain.

Moreover the aims of this book are to set target specifications for the product and its life cycle, and to establish eco-design concepts.

In this text the Design for Environmental concepts have been enlarged. In fact we want to evaluate not only the way in which the product design phase could influence environmental performances of a supply chain but also how re-designing the supply chain could provide a decrease of environmental impact. While much early environmental new product development (ENPD) work employed a design-for-environment approach, which emphasized the reduction of the post-use environmental burden, more recently, there has been an increased emphasis on the "embodied" environmental burdens of the materials used. Suppliers have an important role in determining all aspects of product quality including eco-performance. ENPD requires a detailed understanding of the socio-environmental impacts of the whole supply chain, down to the simplest ingredient, which may previously have been perceived as standardized and unlikely to pose quality problems.

Contents

Chapter 1
Introduction

Recently there has been increasing emphasis on the need for industrialised countries to address issues of sustainable development, understood as "development which satisfies the needs of the present without compromising the possibility for future generations to satisfy their own needs". Industries and designers have therefore found themselves faced with the necessity to adopt opportunely new tools and reference parameters for production and design.

The "sustainable development" philosophy has been gaining a growing interest from public institution, customers and companies that, to consolidate and increase their position on the market, have to necessarily deal with environmental issues. Through the techniques and the instruments of the Eco-Efficiency, the economy and the ecology are not in contradiction, on the contrary they can guide the companies to a greater competitiveness. Promotion of eco-efficiency is based on a management approach that allows companies to introduce environmental improvement of the process and of the product, by a market-oriented point of view.

In order to improve environmental performance of firms, two business processes were analysed in this book: the *supply chain management* and the *new-product development process*.

- The supply chain consists of the sub-processes of procurement, production, distribution, and after-sales support. The extended supply chain is the cross-company supply chain resulting from linking suppliers and customers (and possibly suppliers' suppliers and customers' customers) to the supply chain of a particular manufacturing or service company.
- The new-product development process consists of the business processes, from research and development to product and process design to product launch activities that are required to design a new-product or service and to maximize the likelihood that the product is a success in the market.

Many papers and books have been written in the interim about both supply chain management and new-product development. By the mid-1990s, excellence

M. Bevilacqua et al., *Design for Environment as a Tool for the Development of a Sustainable Supply Chain*, DOI: 10.1007/978-1-4471-2461-0_1,

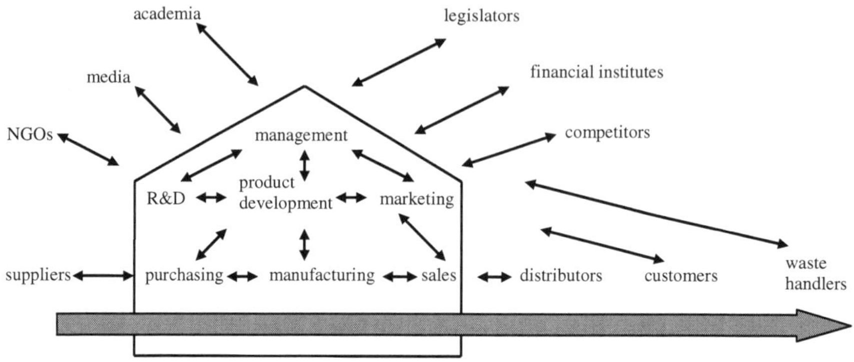

Fig 1.1 Product development in context (modified by Baumann et al. 2002)

in these two processes was viewed widely as a sine qua non for competitive success, with the speed, quality, customer focus, and cost of these processes being central aspects of both profitability and long-term strategy and competence.

At the same time as these revolutionary developments were occurring worldwide, a second and quieter revolution, sometimes referred to as *industrial ecology*, was also occurring.

To use the metaphor of the extended supply chain, the basic driver of industrial ecology was the notion that business itself exists within an extended supply chain of environmental and ecological resources, and waste and inefficiency in this eco-supply chain must be eliminated, just as economic waste must be eliminated in the narrower supply chain of business activities.

Indeed, in the industrial ecology framework, each company has a special role as a steward of the environment and the ecosystem within which it operates. Naturally, this role of product stewardship and environmental waste and risk management encompasses both suppliers and customers, just as extended value-chain analysis encompassed suppliers and customers in the traditional supply chain improvement process. And just as in the traditional model, it also has been found in industrial ecology that product and process design at the front end are better than end-of-pipe continuous improvement activities.

Ourframework(Fig. 1.1)isbasedontheideathatthedevelopmentofagreenproductisa processwithintheinternalprocessesofacompany,whichinturnareembeddedinaproduct supply chain (the other actors have some role in producing, consuming, recycling, and disposingtheproduct),aswellassociety(media,politics,technologicaldevelopments, etc.).

In Europe, one of the first attempts to integrate environmental considerations into product design was the term ecological functionalism (Moller 1982), which contained an ecological checklist for designers and manufacturers which formed the base for a working group for ecology and design in the "Verband Deutscher Industrie-Designer". In the Netherlands, the Advisory Council for Research on Nature and the Environment met already in 1984–1985 (Van Weenen et al. 1991).

In the literature, this was noticed a few years later with references such as "Green design" (Burall 1991), Design for the Environment (Mackenzie 1991), and Green design (Mackenzie 1991). Green consumerism arrived in 1988 with the publication of the Green Consumer Guide (Elkington and Hailes 1988) and the Green Consumer Week by the Friends of the Earth in 1988.

By the end of the 1980s, the focus of attention moved towards global consumption and distribution. The issues on the agenda were fundamental choices regarding lifestyles, patterns of production and consumer priorities. In the 1990s, the oppositional nature of ecological design is more apparent, since pragmatic attempts to apply ecological principals to design seem inevitably to challenge existing practices and ideologies Baumann et al. (2002).

Design for Environmental (DfE) is one of the possible approaches in the new-product development process. The concept of DfE can be summarised in the principle "do more, using less" which is the incentive for reducing the quantity of energy and material used to provide goods and services. In parallel with the general reduction in this type of consumption there are also problems regarding the use of scarce raw materials and the emission of substances which are harmful for the environment and for man.

Much research focuses on the environmental implications of design decisions and on methods to determine and influence products' environmental impact. Less attention is paid to the organizational consequences of such insight (Bakker et al. 2002). Moreover, Nielsen and Wenzel (2002) are of the opinion that some of the most important decisions concerning the environmental properties of a new-product are made during product development. Thus, significant environmental improvements can very often be achieved by integrating environmental properties as an optimization parameter during product development together with parameters such as function, production cost, ergonomics, etc.

Recently, more attention has been paid to environmental supply chain management, defined as "the set of supply chain management policies held, actions taken and relationships formed in response to concerns related to the natural environment with regard to the design, acquisition, production, distribution, use, reuse, and disposal of the firm's goods and services (Zsidisin and Siferd 2001).

Today many companies design and engineer their products to prevent pollution and conserve natural resources throughout the supply chain. The objective is to achieve environmental responsiveness in tandem with sound business management. It is necessary to reduce the environmental impact of products beginning at the design stage as cross-functional product design teams work to make thoughtful and effective decisions that will have positive environmental results throughout the equipment's life cycle. During the early design stages of a product even a small environmental consideration can have a large effect on the environmental impact of a product throughout its life cycle. Some companies developed an internal Design for Environment (DfE) program (see Box A and B) incorporating into product development environmental attributes such as reduction of environmentally sensitive materials, decreases in equipment energy consumption, extension of product life span and utilization of parts that can be reused, resold or recycled.

Box A: Design for Environmental program of Dell Computer

Dell's DfE approach works to achieve higher product quality and longevity, improved customer satisfaction and innovations in materials management. Greater efficiencies in manufacturing and service along with reduced costs associated with asset recovery also result from this program.

Environmentally Sensitive Materials Elimination

Dell directs its suppliers to restrict and/or eliminate environmentally sensitive materials in the components and products supplied to Dell.

Consistent with our Chemical Use Policy, periodic revisions of our restricted materials list will be made to incorporate legal requirements, relevant aspects of international treaties and conventions (such as the Convention for the Protection of the Marine Environment of the N.E. Atlantic (OSPAR) Chemicals for Priority Action), and specific market demands.

Product Content Restrictions

Dell's Restricted Materials Guidance Document lists substances that Dell bans or restricts. For each listed substance, a threshold limit has been established to account for unavoidable impurities along with exemptions consistent with regulatory requirements.

Dell has implemented robust compliance assurance processes to ensure Dell's precautionary chemical use policies are met throughout the entire supply chain.

Compliance Verification Process

Dell is planning to implement a software solution that will improve the collection of materials compliance declarations throughout the supply base. This new solution will integrate with Dell's product record system to enhance the accessibility of materials compliance data and enable compliance roll-up throughout each product's bill of materials. In addition, this solution will enable improved tracking of compliance throughout a product's life cycle. Part of the deployment effort includes training, both at Dell and through the supply chain. This solution is targeted to be in place by the middle of 2006.

Until this data collection solution is deployed, Dell is managing compliance verification of parts and products by way of a Supplier's Declaration of Conformity (SDoC), modeled after ISO/IEC 17050-1 and through an audit program. Dell is collecting SDoCs to support compliance verification efforts. To sign the SDoC, the supplier must ensure that the product meets the Dell Materials Restricted for Use specification and record any applicable exemptions. At Dell's request, the supplier must also be able to provide technical documentation in the form of internal design controls, supplier declarations, or analytical test data. Dell's goal is to collect supplier declarations on each part in a product bill of materials. This will ensure each product meets the legislated materials requirements.

Compliance Verification Audit Program

A second tier in Dell's compliance verification strategy is a supplier audit program. This program can be divided into two parts: a traditional audit and an in-depth supplier survey. A traditional audit is conducted on a quarterly basis in which Dell parts are selected at random and submitted for third-party analytical testing. Samples are tested for restricted materials including those highlighted by the Restriction of Hazardous Substances (RoHS) Directive. The audit allows Dell to investigate gray areas, enforce SDoC diligence, and assist in supplier compliance uncertainty. Dell also actively screens samples in-house by using an X-Ray Fluorescence (XRF) unit.

Box B: Hitachi Group Ecodesign Management Guidelines

The Hitachi Group Ecodesign Management Guidelines is based on the IEC 62430 international standard (environmentally conscious design for electrical and electronic products and systems). Using these guidelines, the company established a management system where divisions such as business planning, design, procurement, manufacturing, and quality control are required to be environmentally conscious and to keep records on processes and results. The first step is to analyze environmental requirements, both legal and those of stakeholders. Next, the company identify the environmental attributes of these requirements, such as energy efficiency, ease of disassembly and disposal, and the effects of these attributes. The analysis and the target setting (plan) leads to a PDCA (plan-do-check-act) cycle (Fig. 1.2), followed by environmentally conscious design and development (do), and review and ongoing improvement (check and act). Each of these processes uses analysis and evaluation tools, and involves communication and information management.

Moreover, the company incorporated the ecodesign concept into quality or environmental management systems at production facilities that have acquired ISO 9001 and ISO 14001 certification.

Products achieving a score above a certain level are designated as Eco-Products, and their environmental information is disclosed. DfE assessment is applied not only to hardware but also to systems, software, and other products.

Development of Super Eco-Products

To further develop environmentally conscious products, the company has established the Super Eco-Products category for those Eco-Products that meet even more demanding requirements. The global warming prevention efficiency or resource efficiency of these products must be at least ten times

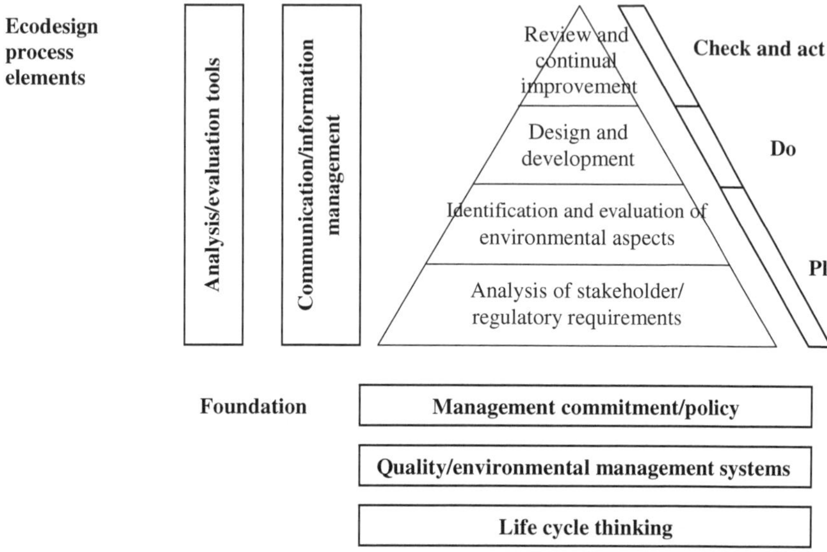

Fig 1.2 PDCA (plan-do-check-act) cycle

greater than reference products (products of fiscal 2000), or they must be leaders in their industry in the energy efficiency standard achievement rate and other factors, or must be highly rated outside the company. In fiscal 2008, 1,103 products and 6,961 models were registered as Eco-Products, for a registration rate of 98% (applicable product scope: information and telecommunication systems, digital media and consumer products) and 86% (applicable product scope: electronic devices, power and industrial systems, high functional materials and components, logistics, services and others). The ratio of Eco-Product sales to total sales of the Hitachi Group increased to 47%. The number of Super Eco-Products included in these totals was 129 products and 375 models, or 18% of all Eco-Products.

The concepts at the root of supply chain design and improvement as a means of promoting sustainable development are not new. Life-cycle analysis, as a tool to understanding a product's complete impact on the environment, has been known for some time (White et al. 1995). This understanding is also widespread in the business community. Product stewardship, as promoted by the Responsible Care Code of the Chemical Manufacturers Association (CMA), emphasizes the need to encompass a product's entire span, from cradle to grave, to ensure environmental responsibility. Reverse logistics (Council of Logistics Management 1993) is a related tool to reduce environmental impacts through recycling of packaging and products. Planning a product's prolonged life span, after normal use, also requires comprehending possible uses for used material. Finally, the approval in 1996 of the first four of the International Organization for Standardization's (ISO) 14000 standards for environmental management systems can be expected to add significant further impetus to internal and extended supply chain environmental stewardship activities.

Information technology (IT) could play an important role in furthering the environmental and revenue benefits of such initiatives. Some examples of potential IT uses include gathering information on inputs and outputs of different processes, surveying customers and using prototypes to understand environmental impacts during product design, tracking product movements, optimizing transportation policies, and analyzing recycling and reuse behavior. With data from these sources, companies may improve the environmental aspects of their products at three important levels:

(1) product and supply chain design to minimize environmental impacts,
(2) ongoing waste minimization and risk mitigation after the product has been deployed, and
(3) diagnostic feedback from supply chain participants to assess opportunities for new-products and processes and to spawn future environmental initiatives.

These environmental improvements, in turn, can lead to economic benefits for companies in several areas: in helping customers to improve their performance and regulatory compliance; in reducing risk and strategic vulnerabilities internally and for their customers; in improving the atmosphere between themselves and regulators and possibly reducing compliance costs; and in improving their reputation and reducing transactions costs in dealing with local communities, environmental groups, and other external stakeholders. In all of these areas, careful tracking of environmental information (cost, value, and performance) is essential in understanding, managing, and legitimizing investments in product stewardship activities in the supply chain.

1.1 Objectives and Outlines of this Book

Our aims in this book are essentially to:

- Define procedures and techniques that allow the companies to develop internal Design for Environment (DfE) programs in order to improve the environmental impact of the whole supply chain.
- Define new methods for designing the entire supply chain using business process re-engineering (BPR) techniques.

In this text the supply chain concepts, generally focused on suppliers/manufacturers/customers, have been enlarged taking into consideration other stakeholders such as

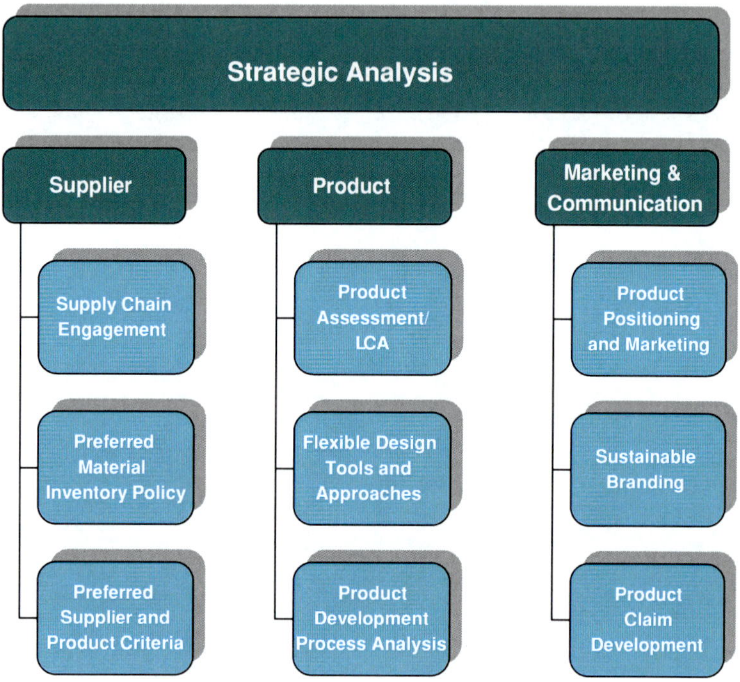

Fig 1.3 Environmental strategic analysis

government, recycling companies and authorities (legislation, regulation, policies). These stakeholders could influence the supply chain strategy, the design choices and the strategic analysis (see Fig. 1.3).

This analysis must regard:

- The development of a product life cycle directed towards the environmental impact minimization;
- Suppliers selection and management based on environmental criteria;
- Increase in value of environmental sensitive products using the best marketing and choices of communications.

The following chapters are organized according to this strategic analysis: Chaps. from 2 to 7 focused on sustainable product design. From strategic point of view design a sustainable supply chain means starting from a design of sustainable products. While the practical application varies among disciplines, some common principles of sustainable design are as follows:

- Low-impact materials: choose non-toxic, sustainably produced or recycled materials which require little energy to process
- Energy efficiency: use manufacturing processes and produce products which require less energy

- Quality and durability: longer-lasting and better-functioning products will have to be replaced less frequently, reducing the impacts of producing replacements
- Design for reuse and recycling.
- Design Impact Measures for total carbon footprint and life-cycle assessment for any resource used are increasingly required and available.
- Sustainable Design Standards and project design guides are also increasingly available and are vigorously being developed by a wide array of private organizations and individuals.

Chapters from 8 to 11 focused on methods for designing a sustainable supply chain. Many companies are limited to measuring the sustainability of their own business operations and are unable to extend this evaluation to their suppliers and customers. This makes determining their true environmental costs highly challenging and reduces their ability to remove waste from the supply chains. These chapters include a discussion on the meaning of sustainability when it comes to SCM as well as a suggestion of how to incorporate sustainability into the SCM concept. These chapters are conceptual and based on a literature review and secondary data analysis of illustrative case examples.

Based on the analysis of the previous sections, Chaps. 12 and 13 focused on DfE procedures for developing a more sustainable supply chain. In particular by combining LCA (Life Cycle Assessment) techniques and by using the QFD (Quality Function Deployment) multi-criteria matrices, an "environmental compromise" can be reached.

The last chapters focused on relation between Enviromental Marketing, New Product Development and Supply Chain Environmental Policy. Today the emphasis has shifted from the "green consumer" to the "responsible retailer" (Knight 2002) whereby the retailer/service provider and the brand owner assume responsibility for ensuring that consumers can buy products and services with confidence in their source and manufacture. Similarly the relationship between manufacturers and retailers has shifted. In the past manufacturers were the drivers of the supply chain, managing the pace at which products were manufactured and distributed. Today it is retailers that drive the agenda and successful manufacturers are those who can meet customer demands for options, styles or features as well as fulfill and deliver orders quickly.

References

Bakker FGA, Fisscher OAM, Brack AJP (2002) Brack AJP (2002) Organizing product-oriented environmental management from a firm's perspective. J Clean Prod 10:455–464

BaumannH,BoonsF,BragdA(2002)Mappingthegreenproductdevelopmentfield:engineering,policy andbusinessperspectives.JCleanProd10(5):409–425

Burall P (1991) Green design. London, Design Council

Council of Logistics Management (1993) Reuse and Recycling—Reverse Logistics Opportunities. Council of Logistics Management, Oak Brook, Ill

Elkington J, Hailes J (1988) The green consumer guide. Victor Gollancz, London
Knight A (2002) Head of Social Responsibility, Kingfisher: 'How green is my kitchen?' Report
Mackenzie D (1991) Green design: design for the environment, Laurence King Ltd., London
Mackenzie D (1991) Design for the environment. Rizzoli International Publishing Inc, New York
Moller E (1982) Design-Philosophie der 80er Jahre. Kommt reit dem Ende der wegwerfideologie ein okologischer functionalsimus? Form 98 und Unternehm Pro Umwelt. Munich Lexika
Nielsen PH, Wenzel H (2002) Integration of environmental aspects in product development: a stepwise procedure based on quantitative life cycle assessment. J Clean Prod 10:247–257
Van Weenen JC, Bakker CA, de Keijser IV (1991) Eco-design: an exploration of the environment. Univeristeit van Amsterdam, Milieukunde, The Netherlands
White PH, Franke M, Hindle P (1995) Integrated solid waste management: a lifecycle inventory. Blackie, London
Zsidisin G, Siferd SP (2001) Environmental purchasing: a framework for theory development. Eur J Purch Supply Manag 7:61–73

Chapter 2
Integration of Design for Environmental Concepts in Product Life Cycle

Several enterprises in various geographic regions recognized that good environmental performance is an important factor for their future success and does a lot to reduce environmental impacts by treatment actions, implementation of cleaner technologies, and by product modifications. Thus, significant environmental improvements can very often be achieved by integrating environmental properties as an optimization parameter during the product development together with parameters such as function, production costs, aesthetics, ergonomics etc.

In some markets such as the car industry, environmental considerations have had a fundamental impact on product development processes with the quest for energy-efficient, low/zero-emission vehicles (Thornton 1999). In particular markets, a large proportion of new product introductions involve products marketed at least partly on the basis of their environmental performance. In the USA, the proportion of green products among new product introductions rose from 1.1% in 1986 to a high of 13.4% in 1991 (Ottman 1998). Tighter regulations and consumer skepticism have since led to a reduction in such introductions, but in markets such as paper towels, dishwasher liquids, and diapers, steady market growth for green brands has continued (Speer 1997). In 1997, green products accounted for 10% of all new US product introductions, with the highest proportion in the ''household products'' category, accounting for 30% of product introductions (Fuller 1999). This phenomenon is also not limited to manufacturers and the design of physical products. Eco-tourism products and green and ethical investment products address fast-growing segments of the tourism and financial services markets.

Some of the most important decisions with respect to environmental properties of a new product are taken during the product development.

The growing interest in 'sustainable development' has led many companies to examine the ways in which they deal with environmental issues (Glazebrook 2000). To achieve sustainable industry, environmentally conscious design (eco-design) or design for environment (DfE) is becoming an increasingly important topic (Brezet and Van Hemel 1997). The introduction of Design for Environment (DfE) methodologies in manufacturing firms allows attention to be

M. Bevilacqua et al., *Design for Environment as a Tool for the Development of a Sustainable Supply Chain*, DOI: 10.1007/978-1-4471-2461-0_2,

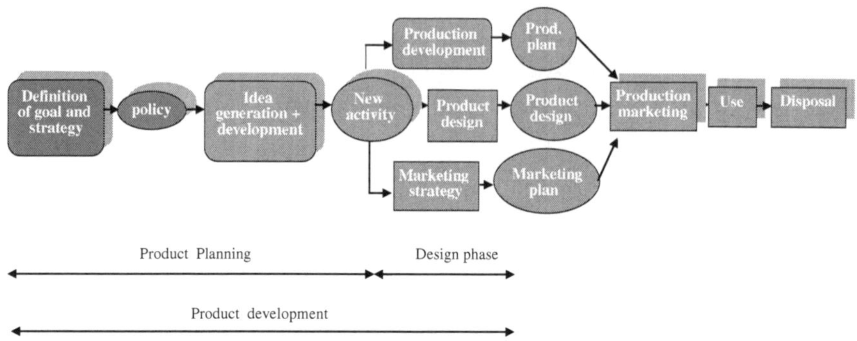

Fig. 2.1 Product development process in a company

paid to environmental aspects right from the start of the design stage leading to a reduction in the materials used and the waste products, avoiding any future weaknesses and inefficiencies.

DfE bears in mind the potential environmental impact throughout the life cycle of the product: emission of harmful substances, excessive use of energy or non-renewable energy sources. It also considers the life cycle of the materials from extraction to disposal. In this way the designers do not create just a product but a whole life cycle.

The use of DfE is also proof of a sense of responsibility toward the consumer and may improve the market position of the firm. Many firms have decided to develop DfE for different core products for several reasons, the main one being that customers are increasingly asking for information concerning the environmental performance of products. This is due to growing environmental awareness and the desire to compare products of different types and from different companies in an environmental context. The development and presentation of Environmental Product Declarations (EPDs) based on the International Standard ISO TR 14025—Environmental labels and declaration: Type III environmental declarations—are a logical way of achieving this (Fraser of Allander Institute 2001). The information incorporated in each EPD is based on life cycle assessments (LCAs), according to the international ISO 14040 standard—Environmental management, Life cycle assessment: principles and framework. The resulting EPDs can also serve as good sales arguments for environmentally friendly products.

2.1 Environmental Aspects in Product Life Cycle

Before speaking of environmental aspects involved in product design processes it is necessary to analyse some general concepts of this company's process.

Figure 2.1 shows the main phases of product development as outlined by several authors. During each phase, aspects such as technical properties,

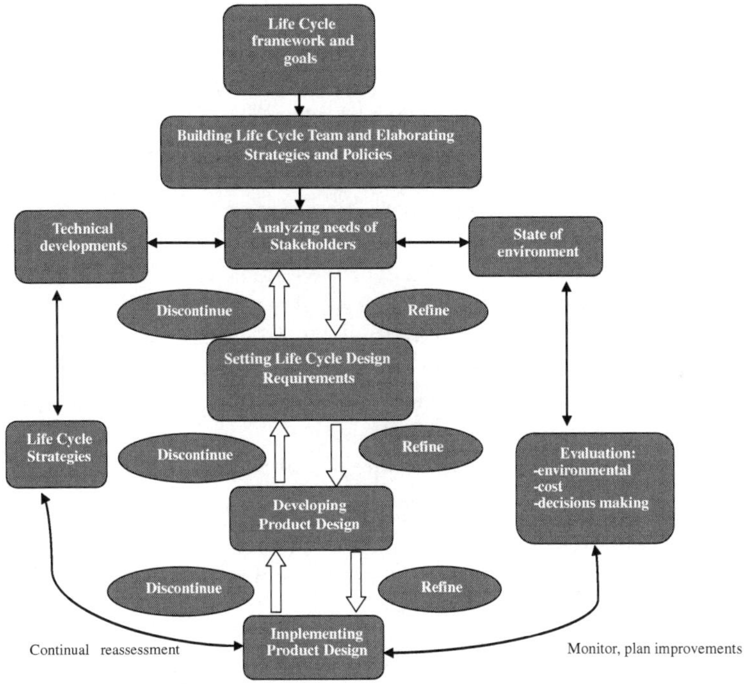

Fig. 2.2 Flow chart: life cycle design process. (Source: Keoleian and Mezerey 1993)

ergonomic properties, economic properties, health properties, and environmental properties of the product are taken more or less into account and the final product usually comes out as a compromise between the different priorities.

After building an interdisciplinary multi-stakeholder life cycle team, understanding, and integration of life cycle thinking, and development of life cycle design goals and principles, life cycle design process should be elaborated. Keoleian and Menerey (1995) have elaborated and offered a useful plan to build Life Cycle Design Process (see Fig. 2.2) that consists of six main steps: (1) Developing of Life Cycle Framework and Goals, (2) Building Life Cycle Team and Elaborating Strategies and Policies (Integration of Life Cycle Thinking idea), (3) Analyzing needs of Stakeholders (Stakeholders Management), (4) Setting Life Cycle Design Requirements, (5) Developing Product Design, and (6) Implementing Product Design.

As Fig. 2.2 shows, product development is a complex system. In the diagram, life cycle goals are very important, therefore they are located at the top. Management influences all the stages of life cycle design process. Concurrent design and life cycle quality provide models for life cycle design. Moreover, measures of success, life cycle team coordination, and policy, strategy are needed in order to support all life cycle design process with design projects.

According to Keoleian and Menerey (1995), a typical design project begins with a needs analysis that identifies customers' needs and ideas of a company. Here I also add that not only customers' needs and interests should be analyzed but also other stakeholders' needs and interests (suppliers, government, competitors, consumer, and environmental organizations). After identification and analysis of the needs, the project team formulates the requirements. Requirements can be set with a use of Requirements Matrix that allows project team to study the interactions between life cycle requirements. Matrices are effective for organizing data and evaluating it later. After development of requirements, project team evaluates conceptual, preliminary, and detailed design. Before implementation, a detailed design is compared to benchmark products. All the weaknesses, minor problems, and unclearness can still be corrected. After formal approval of a detailed design, it can be implemented. Implementation includes production, distribution, marketing, and labeling.

However, there are several barriers that can limit life cycle design process:

- Lack of data for determining life cycle impacts.
- Lack of motivation within a company.
- Decrement in total impacts may increase local impacts (Keoleian and Menerey 1993).

In order to succeed the implementation of Life Cycle Design, involvement of different stakeholders is essential. According to Behrendt et al. (1997), stakeholders, who are involved in life cycle design, can be divided into three levels according to the importance of their involvement (see Fig. 2.3).

The most important stakeholders of life cycle design are in a Design Team. Design Team elaborates and facilitates Life Cycle Design process. Usually it consists of designers, constructors, product managers, sales and marketing managers, and environmental and safety experts. The successful introduction of LCD depends on the commitment of product managers (Keoleian and Menerey 1993). They should control that product projects are applied in the environmental requirements. Marketing managers define market and environmental characteristics of a product by analyzing environmental performance, price levels, and customer profiles.

In the second level, the stakeholders of the whole product chain (production, operation, distribution, supply, packaging, etc.) are involved. Especially good cooperation with suppliers in environmental programs is important because it enables a company to have environmentally certified materials, and components. In the third level, customers, government, stockholders, environmental organizations can play a role of a driving force toward implementation of life cycle design.

However, in my opinion, there can not be the same common picture of stakeholders for Life Cycle Design for all companies. Each company should identify, prioritize, and involve stakeholders in Life Cycle Design process individually and according to the company's Life Cycle Design goals. For example, I can argue that a group of big customers and suppliers should be named not as

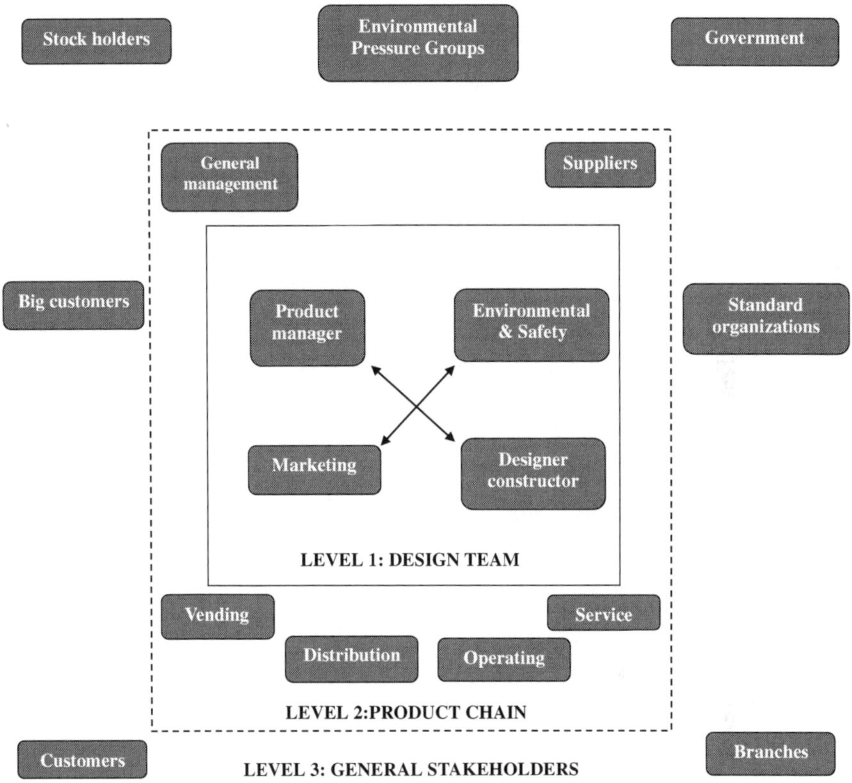

Fig. 2.3 Stakeholders for life cycle design, (modified by Behrendt et al. 1997)

external stakeholders but as a part of a Design Team. Big customers and suppliers should be involved in Life Cycle Design decision-making process (Johansson 2001). According to McAloone (1998), the supplier network can provide important information in choosing environmental alternatives of materials, components, and process. The customers are the one that can contribute in identification and determination of environmental criteria and profile of a product.

Existing studies have analyzed economic and environmental effects of selected policies, usually in partial equilibrium models, but comparison across policies is made difficult by differences in the design of those studies (Fullerton and Wu 1998).

The U.S. Office of Technology Assessment (1992) defines green design as a "process in which environmental attributes are treated as design objectives." The purpose is to reduce pollution at its source, that is, to "avoid the generation of waste in the first place". It also finds that "better product design offers new opportunities to address environmental problems, but that current governmental regulations and market practices are not sufficient to fully exploit these opportunities".

A variety of reforms have been proposed to deal with these perceived problems, both at state and federal levels. Packaging could be subjected to standards, taxes, deposit-refund systems, or recycling requirements. Other proposals would tax toxic substances, require a minimum percentage of recycled content in certain products such as newspapers, require manufacturers to "take back" certain products such as batteries, provide tax credits for machinery used in recycling, require local governments to collect household recycling at the curb, and require households to pay a price per unit of garbage. Table 2.1 lists 34 such policy interventions.

Table 2.1 shows how proposed policies target different stages of this life cycle, and our model shows how the stages are connected. Policies to affect product design will also affect product disposal, and vice versa. Another policy directed at consumers may similarly affect market prices and firm behavior.

2.1.1 LCA and Product Development

As environmental awareness increases, industries and businesses are assessing how their activities affect the environment. Society has become concerned about the issues of natural resource depletion and environmental degradation. Many businesses have responded to this awareness by providing "greener" products and using "greener" processes. The environmental performance of products and processes has become a key issue, which is why some companies are investigating ways to minimize their effects on the environment. Many companies have found it advantageous to explore ways of moving *beyond* compliance using pollution prevention strategies and environmental management systems to improve their environmental performance. One such tool is Life Cycle Assessment (LCA). This concept considers the entire life cycle of a product (Curran 1996).

According to Khan et al. (2002) LCA is one of the most important techniques for the successful implementation of a process or product development in the context of environmental sustainability. As Allen (1996) indicated, one of the most common uses of LCA is identifying critical areas in which the environmental performance of the product can be improved.

Life cycle assessment technique is unique because it encompasses all processes and environmental releases beginning with the extraction of raw materials and the production of energy used to create the product through the use and final disposition of the product. When deciding between two or more alternatives, LCA can help decision-makers compare all major environmental impacts caused by products, processes, or services.

Life cycle assessment is a "cradle-to-grave" approach for assessing industrial systems. "Cradle-to-grave" begins with the gathering of raw materials from the earth to create the product and ends at the point when all materials are returned to the earth. LCA evaluates all stages of a product's life from the perspective that they are interdependent, meaning that one operation leads to the next. LCA

Table 2.1 Policy options that could affect materials flows (source: U.S. Office of Technology Assessment 1992)

Life cycle stage	Regulatory instruments	Economic instruments
Raw material extraction and processing	(1) Regulate mining, oil and gas non-hazardous solid wastes under the Resource Conservation and Recovery Act (RCRA) (2) Establish depletion quotas on extraction and import of virgin materials	(1) Eliminate special tax treatment for extraction of virgin materials, and subsidies for agriculture (2) Tax the production of virgin materials
Manufacturing	(1) Tighten regulations under Clean Air Act, Clean Water Act, under RCRA (2) Regulate non-hazardous industrial waste under RCRA (3) Mandate disclosure of toxic materials use (4) Raise corporate average fuel Economy Standards for automobiles (5) Mandate recycled content in product (6) Mandate manufacturer take-back and recycling of products (7) Regulate product composition, e.g. volatile organic compounds or heavy metal (8) Establish requirements for product reuse, recyclability or biodegradability (9) Ban or phase out hazardous chemicals (10) Mandate toxic use reduction	(1) Tax industrials emissions, effluents, and hazardous wastes (2) Establish tractable emissions permits (3) Tax the carbon content of fuels (4) Establish tractable recycling credits (5) Tax the use of virgin toxic materials (6) Create tax credits for use of recycled materials (7) Establish a grant fund for clean technology research
Purchase, use, and disposal	(1) Mandate consumer separation of materials for recycling	(1) Establish weight/volume-based waste disposal fees (2) Tax hazardous or hard-to-dispose products (3) Establish deposit-refund system for packaging, hazardous products (4) Establish a fee/rebate system based on product energy efficiency (5) Tax gasoline
Waste management	(1) Tighten regulation of waste management facilities under RCRA (2) Ban disposal of hazardous products in landfills and incinerators (3) Mandate recycling diversion rates for various materials (4) Exempt recyclers of hazardous wastes from RCRA Subtitle C (5) Establish a moratorium on construction of new landfills and incinerators	(1) Tax emissions or effluents from waste management facilities (2) Establish surcharges on wastes delivered to landfills or incinerators

enables the estimation of the cumulative environmental impacts resulting from all stages in the product life cycle, often including impacts not considered in more traditional analyses (e.g., raw material extraction, material transportation, ultimate product disposal, etc.). By including the impacts throughout the product life cycle, LCA provides a comprehensive view of the environmental aspects of the product or process and a more accurate picture of the true environmental trade-offs in product and process selection.

An LCA allows a decision maker to study an entire product system hence avoiding the sub-optimization that could result if only a single process were the focus of the study. For example, when selecting between two rival products, it may appear that Option 1 is better for the environment because it generates less solid waste than Option 2. However, after performing an LCA it might be determined that the first option actually creates larger cradle-to-grave environmental impacts when measured across all three media (air, water, land) (e.g., it may cause more chemical emissions during the manufacturing stage). Therefore, the second product (that produces solid waste) may be viewed as producing less cradle-to-grave environmental harm or impact than the first technology because of its lower chemical emissions.

This ability to track and document shifts in environmental impacts can help decision-makers and managers fully characterize the environmental trade-offs associated with product or process alternatives. By performing an LCA, analysts can:

- Develop a systematic evaluation of the environmental consequences associated with a given product.
- Analyze the environmental trade-offs associated with one or more specific products/processes to help gain stakeholder (state, community, etc.) acceptance for a planned action.
- Quantify environmental releases to air, water, and land in relation to each life cycle stage and/or major contributing process.
- Assist in identifying significant shifts in environmental impacts between life cycle stages and environmental media.
- Assess the human and ecological effects of material consumption and environmental releases to the local community, region, and world.
- Compare the health and ecological impacts between two or more rival products/processes or identify the impacts of a specific product or process.
- Identify impacts to one or more specific environmental areas of concern.

An extensive explanation of LCA techniques has been carried out in Chap. 3.

The product development team should have an indication of the relative importance of the environmental impact in the product planning; otherwise it cannot balance environmental aspects against other aspects (Table 2.2). It is very important to understand the following mechanism in an eco-design process:

1. At the beginning of the process the most important decision are made; an important part of the environmental impact is defined at this stage. For example

Table 2.2 LCA use in different stages of the product development

	Activity	Information needed	Information generated
Product Planning	The company management sets the target for the design process. It is usually described as a market/technology combination	General information: policy consumer behavior	Primary goals and directions
Analysis phase	The design problem is analyzed, and the requirements are defined. LCAs of reference products, short what-if? Analysis	General data	Provisional rules of thumb that apply for this product definition of priorities
Idea-generation	Different creativity techniques are defined in order to generate as many new solution as possible short what-if? analysis	Eco-indicators	An indication of the type of product that seems to be desirable
Concept-development phase	The best ideas are elaborated further into a limited number of concepts short provisional LCAs	Eco-indicators	Information on the most important contributions to the environmental impact for the chosen concept
Detailed design phase	The best concept is reworked into detailed drawings. Answering very specific question on the contents of materials etc.	Very specific information	Verification of the goals

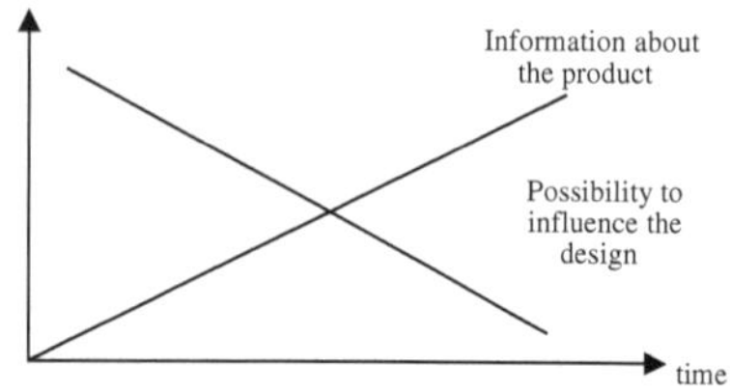

Fig. 2.4 Interaction between the possibilities to influence a design in relation to time

a manufacturer decides to make electric or diesel cars. At this stage there is no product definition. Therefore an LCA cannot be carried out.

2. At the end of the design process the product is defined to a large extent. It is possible to carry out an LCA, but it is hardly possible to influence the design. It is simply too late.

Figure 2.4 shows how during process more will become known about the product and better LCAs are possible. At the same time, however, the possibility of influencing its environmental impacts is reduced.

To overcome this dilemma the type of analysis that can be used should be carefully considered. There are several problems when applying LCAs in the design process:

1. There is not enough time to carry out an LCA for each decision.
2. Even if there was enough time it would be very difficult to produce LCAs. Product ideas are not yet products. In many case the ideas only refer to an operational principle. The idea has not been fleshed out: the material composition, weight, and dimensions are not fully known and, of course, even less is known about consumer behavior and disposal.

This does not imply that LCAs cannot be used. On the contrary, they are essential. It means the LCAs should be used differently at different product development stages. Once the product planning process has set the goals for the design process, it is important to understand the impacts of this type of product in general. LCAs of reference products are essential in this phase. These are existing products that perform a similar function to the desired product. Since many product development processes can be considered as redesign processes it is usually the current product that can serve as a reference product.

2.2 Design for Environmental Concepts

Design for environment (DFE) is defined as systematic consideration of design performance with respect to environmental, health, and safety objectives over the full product and process life cycle (Ray and Guzzo 1993).

This concept was created in 1992 by a number of electronic firms that were attempting to build environmental awareness in product development. The

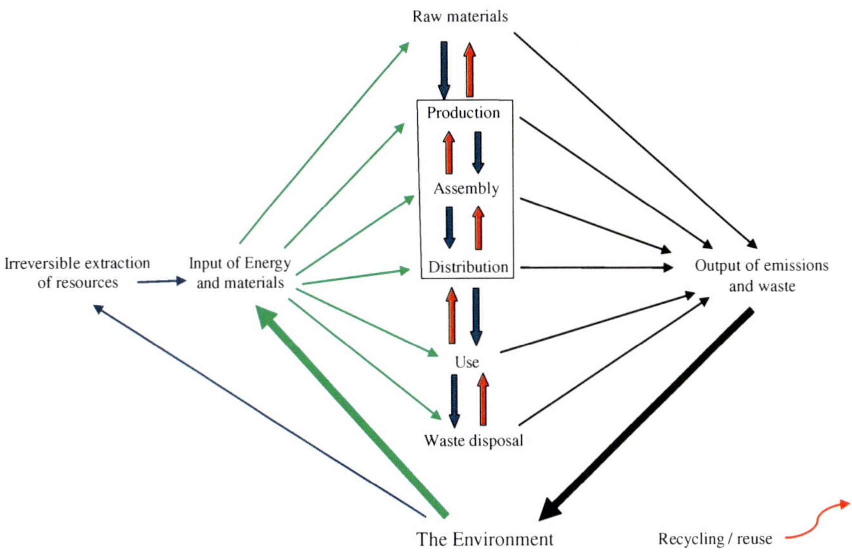

Fig. 2.5 Flow chart of materials: Adapted Life Cycle Concept (modified by Bakker 1995)

American Electronics Association was a first initiator of DFE (AEA 1992). After that, this concept started to be used widely by many industries that wanted to integrate environmental awareness into product design.

In the end, it should lead to sustainable production and consumption that can be achieved together with a number of other measures that are very important, for example, legislation. Life Cycle Design integrates environmental issues into product development by considering all product life cycle stages: raw material acquisition, manufacturing, use, distribution, and disposal (Keoleian et al. 1995).

There are many terms that relate to the life cycle design: eco-design, design for the environment, life cycle design, and environmentally conscious design and production (Brezet and Van Hemel 1997). The product life cycle is a model that contains and describes all the processes that are necessary for the extraction and processing of raw materials, production, distribution, consumption and disposal of the product (Fig. 2.5). Environmental impact is the material influence on the environment. Environmental criteria are product- and production process- oriented solution strategies that lead to less environment damaging products (Bakker 1995).

Environmental impact is any change to the environment which can be adverse or beneficial. This change comes from activities, products, or services of the company. It is important to notice that environmental impact can be positive. For example, production of heating (for district heating purposes) harms the environment, but the overall impact is less than if all households have their own boiler (Brorson and Larsson 1999).

Gertsakis et al. (1997) defined 15 strategies of "design for environment" (eco-design or Life cycle design as synonyms) concept: design for resource conservation, design for environmentally preferred materials, design for cleaner production, design for efficient distribution, design for energy efficiency, design for water conservation, design for minimal consumption, design for low-impact use, design for durability, design for re-manufacture, design for re-use, design for disassembly, design for recycling, design for degradability, design for safe disposal.

Moreover, Simon (2000) provides a concept of DFX, where X can be any environmental criteria. It depends on a company and designers to choose these criteria. It should be noted that the concept of design for environment covers whole life cycle of a product. Therefore, different "design for environment" strategies and tools apply to different life cycle phases.

Bakker (1995) describes a method of "design for environmental efficiency". In the Dutch manual for environment oriented product development it is called as "product analysis". The purpose of this method is to improve efficiency and effectiveness of product's functions, and reduce the environmental impact of the product. This method is based on existing design approaches, like for example value analysis (or value engineering), which objective is to increase the value of the product for the purchaser and reduce the costs for its producer. Recently Fiksel (2009) has developed a comprehensive guide that provides a powerful case for integration of environmental principles and into product development.

Some Design for Environmental guidelines with respect to life cycle strategies were followed (Table 2.3).

Design guidelines are among the most basic tools in design and the introduction of appropriate design guidelines is one of the easiest ways in introducing environmental issues in design. Guidelines can be taken into consideration with checklists in which a designer marks whether he or she has considered the corresponding guideline. Combined with a short course or dedicated education session, guidelines and checklists provided an effective means for exposing designers to environmental issues and ways to reduce the environmental impact of products.

The principles in product design shown in Table 2.3 can be analysed in depth using a methodological approach proposed by Tsoulfas and Pappis (2006).

- *Design and develop recoverable products, which are technically durable, repeatedly usable, harmlessly recoverable after use, and environmentally compatible in disposal*: materials with a high recycling rate and which have the least impact on the environment, both in use and origin, should be preferred. Where possible, environmentally safer substitutes should be used and the abuse of products should be actively prevented. Digital uses the 6R approach (Recycle, Reclaim, Refurbish, Remanufacture, Resell, and Reuse) on their used products (Dorgelo 1996). Xerox offers a 3-year total satisfaction guarantee on equipment containing reprocessed parts to demonstrate its confidence in the products, the same as that given on new equipment (Maslennikova and Foley 2000). In addition, Xerox designers choose a minimal number of materials from the Xerox material environmental index to simplify the eventual segregation of materials and to avoid hazardous materials. The index specifies the relative

Table 2.3 Design for environmental guidelines

Life cycle	Guidelines
Optimization of initial life-time	Reliability and durability
	Easy maintenance and repair
	Modular product structure
	User taking care of product
Selection of low-impact materials	Non-hazardous materials
	Non-exhaustable materials
	Low energy content materials
	Recycled materials
	Recyclable materials
Reduction of material	Reduction in weight
	Reduction in (transport) volume
Optimization of production techniques	Alternative production techniques
	Fewer production processes
	Low/clean energy consumption
	Low generation of waste
	Few/clean production consumables
Reduction of the environmental impact in the user stage	Low energy consumption
	Clean energy source
	Few consumables needed during use
	Clean consumables during use
	No energy/auxiliary material use
Efficient distribution system	Less/clean packaging
	Efficient transport mode
	Efficient logistics
Optimization of end-of-life system	Reuse of product
	Remanufacturing/refurbishing
	Recycling of materials
	Clean incineration

nature of various materials' impact on the environment and helps designers choose non-toxic materials that resist equipment to assert that products should wear during normal use and lend themselves well to reuse and recycling. IBM also develops design specifications for its new products to improve product's end-of-life material recovery (Germans 1996). Billatos and Nevrekar (1994) underline the Mercedes Benz design efforts, which include the selection of environmentally compatible and recyclable materials for components, the reduction of the variety and the volume of plastics used and the avoidance of using composite materials as much as possible. Hundal (1994) reports that another automotive company, BMW, has been trying to introduce more recyclable components in the original design so that it can produce cars out of 100% recycled parts. Finally, Rosenbach and Lindsay (2002) have reported many cases of the application of this principle in various companies.

- *Produce using minimum energy and materials*: the wasting of materials and energy either due to inappropriate design, or due to excessive number of defects should be avoided. Intel has worked in increasing the number of transistors in a single chip, which will result in fewer chips to build and fewer chips to dispose (Gungor and Gupta 1999).
- *Secondary raw materials should be given priority in usage*: primary raw materials should be used only in cases where there would be no stock of secondary ones. Furthermore, in many cases this policy is also moneysaving. Such a case is reported by Tsoulfas et al. (2002) and by Daniel et al. (2003), where lead is recovered and recycled from lead-acid batteries and then it is used for the production of new ones. In addition, Recopol Furniture report that they make furniture which incorporates up to 75% recycled resins and plastics that come from used appliances such as computers, vacuum cleaners, telephones, televisions, washing machines, and refrigerators, which would otherwise go to the landfill (http://www.wharington.com.au).
- *Use eco-friendly energy production, reduce water usage and keep control of pollution sources*: Using less energy is obviously good for the environment. It is also self-evidently good for business because it cuts companies' costs, and eventually avoids potential environmental liabilities. It is, therefore, a prerequisite to the long-term sustainability of business. To replace nonrenewable and polluting technologies, it is crucial to support the use of renewable energy resources, as well as to reduce energy consumption. The identification of where great amounts of energy are used could subsequently lead to redesign of the product or its use in order to make significant energy reductions. Major improvements in energy efficiency can often be achieved at little or no cost, even with net savings, through the use of targeted programs. Installed water-saving techniques and the use of closed re-circulating systems can lead to reduction of water use. In addition, the elimination of the stochastic factors, which affect pollution, may lead to greener production.
- *Use standardized parts*: such a policy ensures that these parts could be reused not only by the original producer, but also by a larger group of producers. For example, automotive companies use standardized screws, speedometers, etc. In most cases this policy is also money-saving. Pappis et al. (2005) report the case of containers that are standardized and can be used by different companies. Standardization is of major importance in Xerox and focus is paid standardizing components as much as possible between product families, thus simplifying and optimizing the opportunities for reuse (Maslennikova and Foley 2000).
- *Provide for easy disassembly of the product*: this would lead to cost, time, and energy savings. The opposite case would make the disaggregation of the product unpractical and costly. It has been reported that Chrysler, Ford, and GM researchers are trying to improve disassemblability features whilst improving the assembly ones (Gungor and Gupta 1999). BMW has been using a color coding scheme for differing plastic materials for the past 30 years, which allows the development of efficient dismantling and disassembly techniques (Hundal 1994).

- *Reduce by-products and get the best out of them:* during manufacture by-products are also produced. Some of them can and should be reduced and/or reused. The rest must be eliminated and disposed properly.
- *Packaging*: packaging design is important for attaining a company's environmental objectives. Though it serves certain needs related mainly to the distribution of the product (e.g. safe transportation), it is not part of the actual service offered by the product. In any case, it affects environment in many aspects. This is the reason why regulations concerning packaging constitute an essential part of governmental policies for environmental protection. The following principles may apply concerning packaging.
- *Limit packaging to the necessary size*: The opposite case not only is contrary to environment protection, but it also affects transportation negatively. Furthermore, environmentally safe packaging can be used as a marketing argument.
- *Design packaging for refilling or recycling and use standardized packaging when applicable*: there are examples of standardized bottles, crates, boxes, pallets, and containers, which may be used by different companies. In reorganizing the packaging policy, Xerox changed its packaging and established packaging-reuse centers in the UK, the Netherlands, and the US. In addition, it reduced the amount of internal packaging to minimize waste (Maslennikova and Foley 2000).
- *Collection and transportation*: despite the obvious environmental gain from used products' recovery, collection, and transportation of recovered products have an environmental cost. Minimizing such a cost is important in order to increase the total environmental gain from recovery. Principles applicable in this phase of the reverse supply chain are the following.
- *Formulate a policy for the recovery of used products*: such a policy favors the maximum utilization of used products. Companies may decide either to undertake the recovery of used products on their own, or to establish cooperation via local or more extended networks for the collection and recycling of similar products. United efforts may be more effective and provide higher recycling rates. Also, recovered products often suit more than one manufacturer. Leasing is a policy that has received much attention lately regarding its environmental dimension. Indeed, companies that lease their products instead of selling them have better chances in the management of their used products. Such cases are reported in detail in Fishbein et al. (2000).
- *Consider using existing forward supply chain facilities and transportation system as much as possible for the reverse supply chain*: transportation and the consequent environmental effects can be significantly limited if the recovery of used products can occur at the same time or in combination with the distribution of new products. The theoretical minimal average transportation distances can be determined using a tool for allocation and route planning. An application of this principle is reported by Vergitsi (2000), where the case of used beverage bottles of Hellenic bottling company (3E) is examined. The tracks that carry the beverages do not return empty to the warehouses of 3E,

since they carry the empty bottles from the consumer spots, in order to be reprocessed. Krikke et al. (2002) conclude that in the case of Printed Wiring Assemblies of Honeywell the forward supply chain is used in order to facilitate the returns of spent products.

- *Classify used products as early in the recovery chain as possible*: this eases the planning of storage of used products and redundant processes are avoided. Daniel et al. (2003) mention that using lead-acid batteries are classified in the electricians' shops, where they are bought by customers.
- *Treat hazardous materials safely*: it is necessary to ensure that the generation of hazardous wastes is reduced and also that adequate disposal facilities are available, for the environmentally sound management of hazardous wastes. In Ohio Manufacturer's Association Case Studies in Team Excellence (Ohio Manufacturer's Association 1992) the initiative of a Chrysler team from the Jeep plant in Toledo is reported, which was formed to respond to state legislation prohibiting the disposal of certain biodegradable and recyclable materials in landfills. The establishment of collection points and of a network of recyclers for such materials together with setting up returnable packaging systems with suppliers led to significant savings in an annual basis.
- *Recycling and disposal*: after its useful life, a used product may be either disposed or recycled (generally recovered). As in the phase of collection and transportation, recycling and disposal may significantly contribute to the total environmental gain and the attainment of the environmental goals of a company. Ideally, companies should borrow from natural cycles to design their systems as part of a larger natural cycle, where materials are borrowed from and returned to nature, without negatively affecting its overall balance. The following principles may be applied regarding recycling and disposal.
- *Close the supply loop by recycling effectively and efficiently*: the biological designs of nature provide a role model for sustainability. The goal is to work continuously toward closed-loop production systems and zero-waste factories, wherein every output is returned to natural systems as a nutrient or becomes an input for manufacturing another product (Executive Committee 2002). Designing for recyclability is essential but recycling becomes unproductive when the energy, materials and pollution used in collecting and processing used products exceed those used to produce the goods in the first place. Closing the loop by extending responsibility throughout the life cycle chain ensures total product and service stewardship. Mercedes Benz started taking scrap cars back in 1991 and has been performing the material recovery process as part of their environmentally friendly production program (Billatos and Nevrekar 1994).
- *Reduce the volume and amount of materials going to landfill and consider alternative uses of used products or wastes*: using appropriate techniques one can compact the scrap. In addition, smaller landfills can be used. Alternative uses of used products extend their life cycle. For example, used tires can be used as a protective in seaports, speedways, etc.

2.2.1 *Integration of LCA Technique and Design for Environmental Methods*

Many authors developed procedures and software in order to integrate LCA technique and DfE methods. Hideki Kobayashi (2005) presented a methodology and a software tool to establish an eco-design concept of a product and its life cycle by assigning appropriate life cycle options to the components of the product. He made a design support tool for efficiently planning product life cycles by using quality function deployment (QFD) and life cycle assessment (LCA) data. Bovea and Wang (2003) introduced a novel approach for identifying environmental improvement options by taking into account customer preferences. The LCA methodology is applied to evaluate the environmental profile of a product while a fuzzy approach based on the House of Quality in the QFD methodology provides a more quantitative method for evaluating the imprecision of the customer preferences.

In order to evaluate or select a solution idea or a design concept, other weighted rating methods are utilized (Pahl and Beitz 1988). In these methods, weighting of the evaluation criteria greatly affects the evaluation result. Total evaluations of eco-products have been reported in Williams et al. (1996) and in Zhang et al. (1998). In these evaluations, the weighting factors of the evaluation criteria were calculated using the analytic hierarchy process (AHP) (Satty 1980). Huang et al. (2004) combined three methods to evaluate the impact of packaging materials: (1) life cycle assessment (LCA), a quantitative method, to assess environmental loading, (2) analytic hierarchy process (AHP), a qualitative method, to obtain opinions from experts, and (3) cluster analysis to integrate the results of the former two methods. The authors developed this method to provide integrated information and avoid a bias toward either a qualitative or a quantitative approach.

The methodology proposed in this book, illustrated in Fig. 2.6, involves the integration of DfE and LCA techniques into those processes which are normally carried out when developing a new product: "Project definition", "Concept Development", "Prototype Assembly Testing" and "Field Test", and "Commercial Launch". Figure 2.6 shows the main phases of product development as outlined by several authors (Pahl and Beitz 1991; Olesen et al. 1996; Nielsen and Wenzel 2002). During each phase, aspects such as technical properties, ergonomic properties, economic properties, health properties, and environmental properties of the product are taken more or less into account and the final product usually comes out as a compromise between the different priorities.

The importance of eco-design in earlier phases has been emphasized, because decisions made in these phases greatly affect the environmental impact throughout the product life cycle (Frei and Züst 1997). The solution space, i.e. the degree of freedom to choose solutions (Jensen et al. 1998) and hence the potential for environmental improvements is large in the beginning of product development when ideas and conceptual solutions are open. However, it decreases gradually as the general product features are established and more and more details are

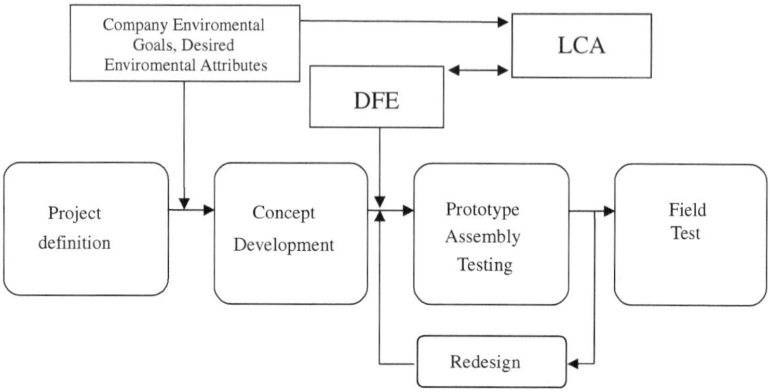

Fig. 2.6 Class product design

determined. The environmental improvement options are limited to production processes, logistics, recycling etc. when the production has been set up and the product is ready for the market.

The activities in the environmental part of the product development procedure shall be considered as a supplement to the traditional values (technical performance, economic performance etc.), and as a contribution to the overall competitiveness of the product.

First of all a manufacturer must examine the context in which the DfE is to be developed. He must identify the important and possible environmental goals within the process and fix the market objective that he wants to reach, whether it is local, national, or international. For example, prior to the introduction of legislation he may consider using this technique to arrive at the standards of eco-management contained in ISO 14000. He must assign responsibilities within the firm and must ensure that the process is controllable and traceable. He must involve all the firms, in particular the managerial staff, indicating DfE as a factor for improvement and market achievement. At the same time he must create awareness (for example by organizing seminars) in all the parties involved in the production process.

The methodology proposed in this study is a closed feedback cycle which is able to improve itself: this is based on a scheduled report system which indicates to the environmental management the need to change some characteristics of the product or the production processes or to address their study toward new products.

The means by which this concept can be integrated into the company is through its Design for Environment (DfE) program. The Design for Environment (DfE) tool considered in this book, is a methodological framework based on Life Cycle Assessment (LCA) thinking, which allows the integration of environmental parameters directly into the design of products and processes. DfE serves therefore as an environmental decision-making support for designers. In the design phase of a product the different problems and solutions should be assessed from a technical, economic, and ecological point of view at the same time.

LCA can be used in any phase of product development showed in Fig. 2.6, but the major potential exists in the project definition phase and the concept development phase and this paper.

- In the *project definition* step it is necessary to define a reference product, which can serve as a representative for the new product. Since new products are usually based on existing technologies in new compositions, it is in most cases possible to compose a useful reference product by putting existing units and technologies together. The environmental performance of a product or a service is determined as a sum of all impacts throughout the product's life cycle. Thus, from the beginning of the product development procedure the entire life cycle of the product system must be taken into account in order to achieve the best environmental performance of the new product (Nielsen and Wenzel 2002). Although much input can be obtained from the company's suppliers, LCA experts, public organisations and from public databases and literature, this procedure can be quite resource demanding for complex products. However, materials and processes, which from an initial judgement are found unimportant from an environmental point of view can be left out of considerations to keep the work in appropriate proportion. In this step it is necessary to make an LCA computer model of the product. The model must be constructed in a manner that allows the existing solutions in the reference product to be removed and replaced easily with others. Although the LCA is quite rough at this stage, special attention shall always be paid to possible emissions of toxic substances, which can occur in small but significant amounts in any stage of the product's life cycle. In this step the most important sources of environmental impact in the reference product's life cycle (environmental 'hot spots') are pointed out in order to identify potential focus areas for the further product development.
- In the *Concept development* step the designers have to determine whether some of the environmental 'hot spots' can be moderated or removed by modifying or replacing certain conceptual solutions in the reference product. In this step, all aspects such as economy, design, technical feasibility etc. must of course be taken into account to ensure that the new product becomes attractive in the market. At this stage existing and new conceptual solutions are compared with each other from an environmental point of view. Before any physical modifications of the product have been made, simulations in the LCA computer tool are used to test new conceptual ideas against each other. The purpose of this activity is to reveal how different conceptual solutions interact with the environment and hereby provide a basis for selection of optimal solutions. It is necessary to verify if environmental improvements are associated with increased production costs. Frequently decisions including both: environmental, technical, and economical aspects may very often lead to solutions with the same or even lower costs due to resource and tax saving (e.g. waste deposition duty). Costly environmental improvements can eventually be motivated by an anticipation of increased profits due to more goodwill toward the product and the company in general.

References

AEA (1992)–American Electronics Association, The hows and whys of design for environment, Washington, DC

Allen D (1996) Application of life-cycle assessment. In: Curran MA (ed) Environmental life cycle assessment, chapter 5. McGraw-Hill, New York, pp 5.1–5.18

Bakker C (1995) Environmental information for industrial designers, Technische Universiteit Delft

Behrendt S, Chr Jasch, Peneda MC, van Weenen H (1997) Life cycle design: a manual for small and medium-sized enterprises. Institut fur Zukunftsstudien und Technologiebewertung, Springer

Billatos SB, Nevrekar VV (1994) Challenges and practical solutions to designing for the environment. In: ASME design for manufacturability conference, Chicago, IL, 14–17 Mar 1994. pp 49–64

Bovea MD, Wang B (2003) Identifying environmental improvement options by combining life cycle assessment and fuzzy set theory. Int J Prod Res 41(3):593–609

Brezet H, Van Hemel C (1997) Eco-design: a promising approach to sustainable production and consumption. UN Environment Programme (UNEP)

Brorson T, Larsson G (1999) Environmental Management. EMS AB, Stockholm, p 34

Curran MA (ed) (1996) Environmental life cycle assessment. McGraw-Hill, ISBN 0-07-015063-X

Daniel SE, Pappis CP, Voutsinas TG (2003) Applying life cycle inventory analysis to reverse supply chains: a case study. Resour Conserv Recycl 37(4):251–340

Dorgelo R (1996) 6R-principle of Digital. In: Proceedings of the first international working seminar on reuse, Eindhoven, The Netherlands, pp 85–106

Executive Committee (2002) World Business Council for Sustainable Development. The business case for sustainable development: making a difference towards the Earth Summit 2002 and beyond. Corpo Environ Strategy 9(3):226–235

Fiksel J (2009) Design for environmental: a guide to sustainable product development. McGraw Hill

Fishbein BK, McGarry LS, Dillon PS (2000) Leasing: a step toward producer responsibility. INFORM, Inc, New York

Fraser of Allander Institute (2001) *An economic study of Scottish grouse moors*. Game Conservancy Scottish Research Trust & Game Conservancy Ltd, Hampshire, UK

Frei M, Züst R (1997) The eco-effective product design—the systematic inclusion of environmental aspects in defining requirements. In: Krause F-L, Seliger G (eds) Life cycle networks. Chapman & Hall, London, pp 163–173

Fuller D (1999) Sustainable marketing: managerial–ecological issues. Sage, Thousand Oaks

Fullerton D, Wu W (1998) Policies for Green Design. J Environ Econ Manag 36:131–148 Article no. ee981044

Germans RJ (1996) Reuse and IBM. In: Proceedings of the first international working seminar on reuse, Eindhoven, The Netherlands, p 119

Gertsakis J, Lewis H, Ryan C (1997) A Guide to EcoReDesig: Improving the environmental performance of manufactured products. Centre for Design at RMIT, Melbourne

Glazebrook B, Coulon R, Abrassart C (2000) Towards a product life cycle design tool. In: IEEE international symposium on electronics and the environment, pp 81–85

Gungor A, Gupta SM (1999) Issues in environmentally conscious manufacturing and product recovery: a survey. Comput Ind Eng 36:811–853

Huang CC, Ma HW (2004) A multidimensional environmental evaluation of packaging materials. Sci Total Environ 324:161–172

Hundal MS (1994) DFE: current status and challenges for the future. In: ASME design for manufacturability conference. 14–17 March 1994, Chicago, IL. pp 89–98

Jensen AA, Hoffmann L, Møller BT, Schmidt A, Christiansen K, Elkington J, van Dijk F (1998) Life cycle assessment—a guide to approaches, experiences and information sources. European Environment Agency, Copenhagen

Johansson G (2001) Success factors for integration of EcoDesign in product development–a review of state-of-the-art, International Graduate School of management and Industrial Engineering, Department of Mechanical Engineering, Linkoping University

Keoleian GA, Koch JE, Menerey D (1995) Life cycle design framework and demonstration projects, Environmental Protection Agency, EPA/600/R-95/107, July

Keoleian GA, Menerey D (1993) Life cycle design guidance manual: Environmental requirements and the product system, US Environmental Protection Agency, EPA600/R-92/226, January

Kobayashi H (2005) Strategic evolution of eco-products: a product life cycle planning methodology. Res Eng Des 16(1–2):1–16

Khan FI, Sadiq R, Husain T (2002) GreenPro-I: A methodology for risk-based process plant design considering life cycle assessment. J Environ Model Softw 17:669–692

Krikke HR, Pappis CP, Tsoulfas GT, Bloemhof-Ruwaard J (2002) Extended design principles for closed loop supply chains: optimising economic, logistic and environmental performance. In: Klose A, Speranza MG, Van Wassenhove LN (eds) Quantitative approaches to distribution logistics and supply chain management, Series: lectures notes in economics and mathematical systems. Springer

Maslennikova I, Foley D (2000) Xerox's approach to sustainability. Interfaces 30(3):226–233

McAloone T (1998) Industry experiences of environmentally conscious design integration: an exploratory study. PhD Thesis, School of Industrial and Manufacturing Science, The CIM Institute, Cranfield University, Cranfield

Nielsen PH, Wenzel H (2002) Integration of environmental aspects in product development: a stepwise procedure based on quantitative life cycle assessment. J Clean Prod 10:247–257

Ohio Manufacturer's Association Case Studies in Team Excellence, the award for team excellence in manufacturing sponsored by the Ohio Manufacturer's Association, 1991–1992

Olesen J, Wenzel H, Hein L (1996) Andreasen MM. Miljørigtig konstruktion. Danish Environmental Protection Agency, Copenhagen

Ottman J (1998) Green marketing: opportunity for innovation, 2nd edn. NTC/Contemporary Books, Lincolnwood

Pahl G (1991) Beitz W. Engineering design—a systematic approach. Springer, Berlin

Pahl G, Beitz W (1988) Engineering design. The Design Council, London, pp 118–139

Pappis CP, Rachaniotis NP, Tsoulfas GT (2005) Recovery and re-use of containers within Blue Container Line (BCL) S.A. In: Flapper SDP, van Nunen JAEE, van Wassenhove LN (eds) Managing closed-loop supply chains. Springer

Ray DL, Guzzo L (1993) Environmental overkill: Whatever happened to common sense? Harper Collins, New York

Rosenbach J, Lindsay C (2002) ''Greening'' electronics product design: a brief summary of government and private initiatives. Office of Solid Waste, US EPA

Satty T (1980) The analytic hierarchy process. McGraw-Hill, New York

Simon M, Poole S, Sweatman A, Evans S, Bhamra T, Mcaloone T (2000) Environmental priorities in strategic product development. Business Strategy Environ 9(6):367–377

Speer T (1997) Growing the green market. Am Demogr 19(8):45–50

Thornton E (1999) Enviro-cars: the race is on. BusWeek 8:74–75

Tsoulfas GT, Pappis CP (2006) Environmental principles applicable to supply chains design and operation. J Clean Prod 14:1593–1602

Tsoulfas GT, Pappis CP, Minner S (2002) An environmental analysis of the reverse supply chain of the SLI batteries. Resour Conserv Recycl 36(2):135–54

U.S. Congress, Office of technology assessment (1992) Green product by design: choices for a cleaner environmental U.S. Government Printing Office, OTA-E-541, Washington DC

Vergitsi C (2000) The reverse supply chain of glass bottles in the case of Hellenic Bottling Company (3E). Bachelor thesis [in Greek]. University of Piraeus, Department of Industrial Management

Williams R, Deshmukh A, Wang B (1996) Using analytical hierarchy process to select environmentally friendly produce shipping containers. Int J Environ Conscious Des Manuf 5(2):47–52

Zhang Y, Wang B, Zhang C (1998) Product concept evaluation using GQFD-II and AHP. Int J Environ Conscious Des Manuf 7(3):1–15

Chapter 3
LCA Process in the Eco-Design Process

3.1 Life Cycle Assessment

LCA has its roots in the 1960s, when scientists concerned about the rapid depletion of fossil fuels developed it as an approach to understanding the impacts of energy consumption. A few years later, global-modelling studies predicted the effects of the world's changing population on the demand for finite raw materials and energy resource supplies (Meadows et al. 1972). The predictions of rapid depletion of fossil fuels and resulting climatological changes sparked interest in performing more detailed energy calculations on industrial processes. In 1969, the Midwest Research Institute (and later, Franklin Associates) initiated a study of the Coca-Cola Company to determine which type of beverage container had the lowest releases to the environment and made the fewest demands for raw materials and energy (Franklin Associates 1991). In the 1970s, the U.S. Environmental Protection Agency (EPA) refined this methodology, creating an approach known as Resource and Environmental Profile Analysis (REPA). Approximately 15 REPAs were performed between 1970 and 1975, driven by the oil crisis of 1973. Through this period a protocol, or standard methodology, for conducting these studies was developed (Hunt et al. 1992). In the late 1970s and early 1980s, environmental concern shifted to issues of hazardous waste management. As a result, life cycle logic was incorporated into the emerging method of risk assessment, which was used with increasing frequency in the public policy community to develop environmental protection standards (Stilwell et al. 1991). Risk assessments remain controversial procedures: the public is often disinclined to trust them, especially when conducted after-the-fact to justify an activity or when performed by an organization with a vested interest in their conclusions. When solid waste became a worldwide issue in the late 1980s, the life cycle analysis method developed in the REPA studies again became a tool for analyzing the problem. In 1990, for example, a life cycle assessment was completed for the Council for Solid Waste Solutions, which compared the energy and environmental impacts of paper to that

M. Bevilacqua et al., *Design for Environment as a Tool for the Development of a Sustainable Supply Chain*, DOI: 10.1007/978-1-4471-2461-0_3,

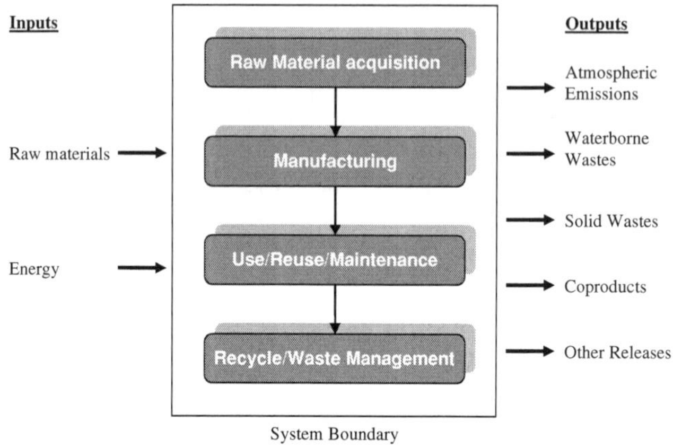

Fig. 3.1 Life cycle stages (*Source* EPA 1993)

of plastic grocery bags (Council for Solid Waste Solutions 1990). A similar study comparing disposable diapers to washable cloth diapers was also conducted. Environmental groups around the world have also adopted life cycle analysis; organizations such as Blue Angel, Green Cross, and Green Seal use and continue to improve LCA for the purpose of product labelling and evaluation. Thus, while initially limited to the public sector, LCA has been adopted by increasing numbers of corporations and no-profit organizations as an aid to understanding the environmental impacts of their actions. And as demand for "green" products and pressures for environmental quality continue to mount, it is quite likely that industrial life cycle analysis will become in the 1990s what risk assessment was in the 1980s.

Components of Life Cycle Analysis

The term "life cycle" refers to the major activities in the course of the product's life-span from its manufacture, use, and maintenance, to its final disposal, including the raw material acquisition required to manufacture the product. Figure 3.1 illustrates the possible life cycle stages that can be considered in an LCA and the typical inputs/outputs measured.

Specifically, LCA is a technique to assess the environmental aspects and potential impacts associated with a product, process, or service, by:

- Compiling an inventory of relevant energy and material inputs and environmental releases
- Evaluating the potential environmental impacts associated with identified inputs and releases
- Interpreting the results to help decision-makers make a more informed decision.

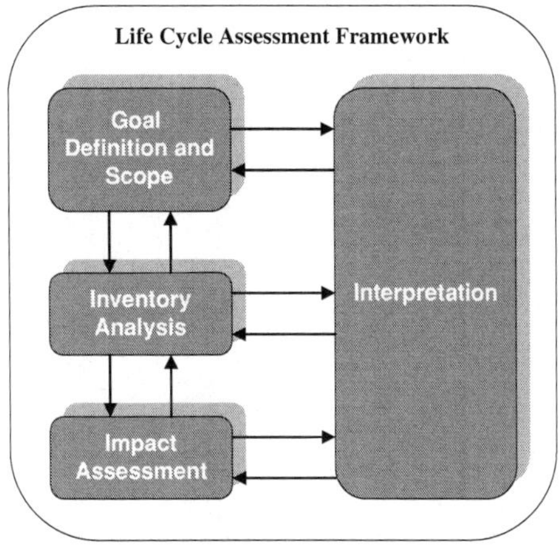

Fig. 3.2 Phases of an LCA (*Source* ISO 1997)

The LCA process is a systematic, phased approach and consists of four components: goal definition and scoping, inventory analysis, impact assessment, and interpretation as illustrated in Fig. 3.2:

1. *Goal definition and scoping*—define and describe the product, process or activity. Establish the context in which the assessment is to be made and identify the boundaries and environmental effects to be reviewed for the assessment.
2. *Inventory analysis*—identify and quantify energy, water and materials usage and environmental releases (e.g., air emissions, solid waste disposal, waste water discharges).
3. *Impact assessment*—assess the potential human and ecological effects of energy, water, and material usage and the environmental releases identified in the inventory analysis.
4. *Interpretation*—evaluate the results of the inventory analysis and impact assessment to select the preferred product, process or service with a clear understanding of the uncertainty and the assumptions used to generate the results.

3.1.1 Goal Definition and Scoping

Goal definition and scoping is the phase of the LCA process that defines the purpose and method of including life cycle environmental impacts into the decision-making process. In this phase, the following items must be determined: the type of information that is needed to add value to the decision-making process, how accurate the results must be to add value, and how the results should be interpreted and displayed in order to be meaningful and usable.

The following six basic decisions should be made at the beginning of the LCA process to make effective use of time and resources:

1. Define the goal(s) of the project
2. Determine what type of information is needed to inform the decision-makers
3. Determine the required specificity
4. Determine how the data should be organized and the results displayed
5. Define the scope of the study
6. Determine the ground rules for performing the work

3.1.2 Life Cycle Inventory

A life cycle inventory is a process of quantifying energy and raw material requirements, atmospheric emissions, waterborne emissions, solid wastes, and other releases for the entire life cycle of a product, process, or activity.

The level of accuracy and detail of the data collected is reflected throughout the remainder of the LCA process.

Life cycle inventory analyses can be used in various ways. They can assist an organization in comparing products or processes and considering environmental factors in material selection. In addition, inventory analyses can be used in policy-making, by helping the government develop regulations regarding resource use and environmental emissions.

The results can be segregated by life cycle stage, media (air, water, and land), specific processes, or any combination thereof.

EPA's 1993 document, "Life-Cycle Assessment: Inventory Guidelines and Principles," and 1995 document, "Guidelines for Assessing the Quality of Life Cycle Inventory Analysis," provide the framework for performing an inventory analysis and assessing the quality of the data used and the results. The two documents define the following four steps of a life cycle inventory:

1. *Develop a flow diagram of the processes being evaluated*

A flow diagram is a tool to map the inputs and outputs to a process or system. Figure 3.3 illustrates a detailed System Flow Diagram for Bar Soap. The more complex the flow diagram, the greater the accuracy and utility of the results. Unfortunately, increased complexity also means more time and resources must be devoted to this step, as well as the data collecting and analyzing steps.

2. *Develop a data collection plan*

The required accuracy of data was determined as part of the goal definition and scoping phase (discussed in Chap. 2). When selecting sources for data to complete the life cycle inventory, an LCI data collection plan ensures that the quality and accuracy of data meet the expectations of the decision-makers.

Key elements of a data collection plan include the following:

- Defining data quality goals
- Identifying data sources and types

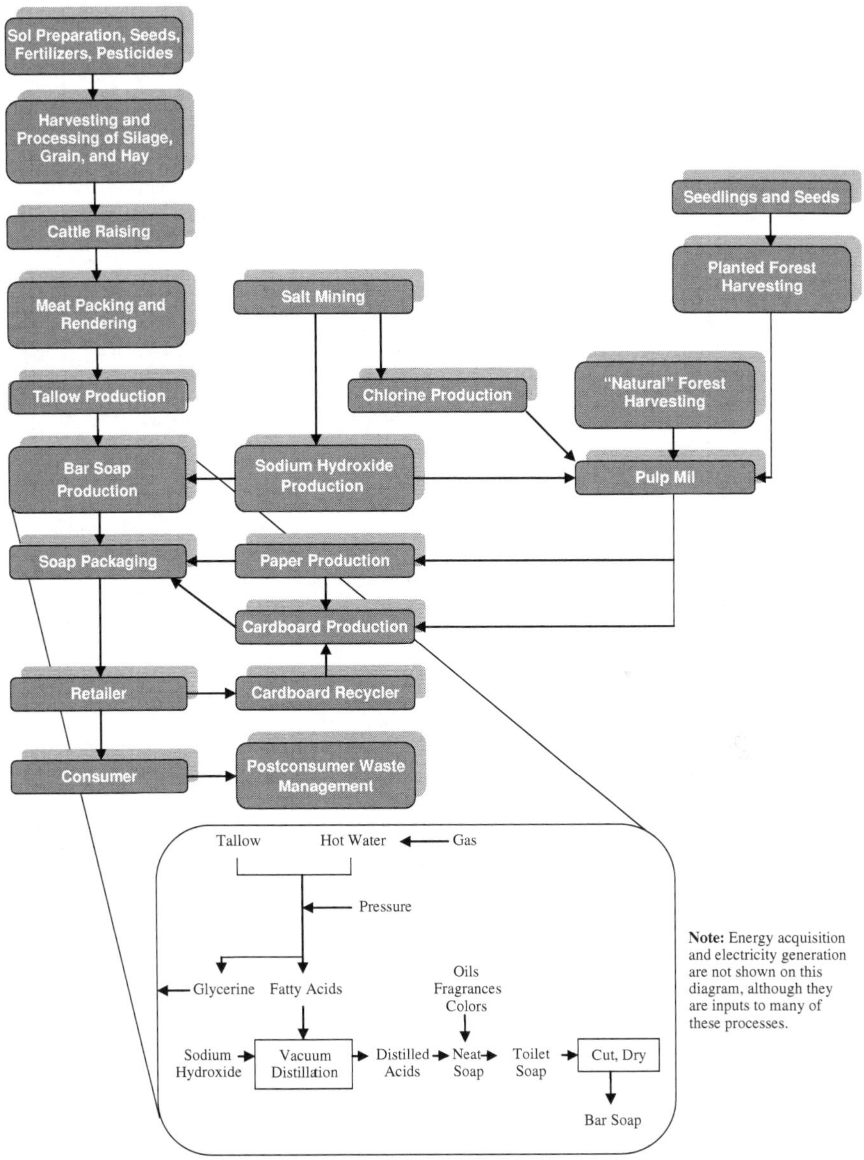

Fig. 3.3 Detailed system flow diagram for Bar Soap

- Identifying data quality indicators
- Developing a data collection worksheet and checklist.

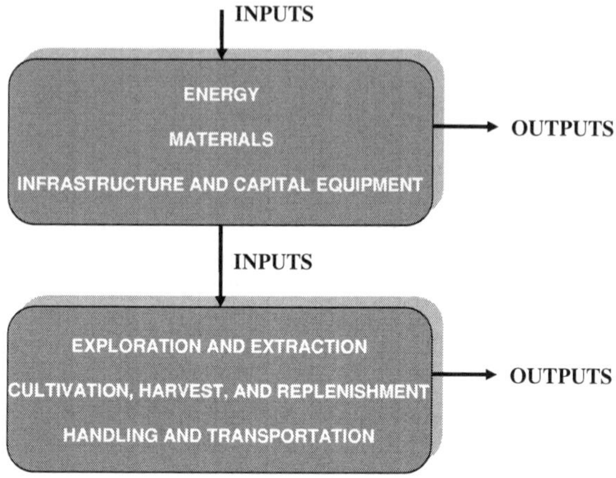

Fig. 3.4 Raw material acquisition subsystem

3. *Collect data*

Data collection efforts involve a combination of research, site-visits and direct contact with experts, which generates large quantities of data. As an alternative, it may be more cost effective to buy a commercially available LCA software package.

All industrial processes have multiple input streams and many generate multiple output streams. Usually only one of the outputs is of interest for the life cycle assessment study being conducted, so the analyst needs to determine how much of the energy and material requirements and the environmental releases associated with the process should be attributed, or allocated, to the production of each co-product. For example, steam turbine systems may sell both electricity and low-pressure steam as useful products. When co-products are present, the practitioner must determine how much of the burdens associated with operating and supplying the multi-output process should be allocated to each co-product. The practitioner must also decide how to allocate environmental burdens across co-products when one is a waste stream that can be sold for other uses. The guidance provided by the International Standards Organization (ISO) recognizes the variety of approaches that can be used to treat the allocation issue and, therefore, requires a step-wise approach (see text box on ISO 14041).

A thorough understanding of how an inventory analysis is conducted, and the limitations and assumptions inherent in the various stages is critical to effective use of LCA in decision making. The following is a synopsis of the various subsystems analyzed in an inventory analysis.

Raw Materials Acquisition

Data are collected for this subsystem on all activities required to obtain raw materials, including transportation of the materials to the point of manufacture (see Fig. 3.4).

Typically, raw materials are traced for the primary product and all primary, secondary and tertiary packaging. Managers should review the data to make sure equivalent comparisons are used. For example, a package containing recycled materials may need increased thickness to compensate for the decreased strength of recycled materials. In this case, managers must make a trade off between weight of materials that will someday become part of the waste stream and virgin material content. The inventory should also include all inputs of energy, materials, and equipment necessary for acquiring each raw material. Because this dramatically increases the complexity of the analysis, criteria must be determined to eliminate insignificant contributions. This may be done by establishing a threshold for inclusion. For example, any component contributing less than five percent of inputs might be ignored.

Ecosystems are impacted in many ways by the extraction or harvesting of raw materials, but only those effects that can be quantified, such as pesticide run-off from agriculture or soil loss from logging, should be included in the inventory. Effects that cannot be easily measured, such as loss of scenic or aesthetic value, may be covered in the more subjective impact assessment. At this point, attempts to quantify renewable or non renewable resources for inventory calculations are subjective, as quantifiable data is not publicly available. However, maintaining separate lists of renewable and non-renewable materials may be helpful if an impact assessment is later performed. Energy acquisition is actually part of the materials acquisition subsystem, but because of the complexity of the subject, it warrants its own analysis. Data collected should include all energy requirements and emissions attributed to the acquisition, transportation, and processing of fuels. This means that if gasoline is used as a transportation fuel, not only should emissions related to combustion be included, but also energy consumption and emissions due to extraction and refining. In the U.S., energy is derived from a number of sources including coal, natural gas, petroleum, hydropower, nuclear power, and wood. Utilities use many different types of energy sources to produce consumer purchase. Calculations follow the same procedure as in converting raw material to intermediate materials and include the same limitations. Data collected for fabrication of the final product includes the inputs and releases associated with filling and packaging operations. As this is a necessary step for virtually any product, this step focuses on differences between processes or materials being compared.

If the filling procedure is identical for the two products being compared, this step can be ignored. Both primary and secondary packaging must be included in the calculations, taking care to keep packaging per unit consistent between alternatives.

Transportation/Distribution

An inventory of the related transportation activities of the product to warehouses and end-users may be simplified by using standards for the average distance transported and the typical mode of transportation used (see Fig. 3.5).

Inventory of the distribution process includes warehousing, inventory control, and repackaging. Environmental controls such as refrigeration are components of

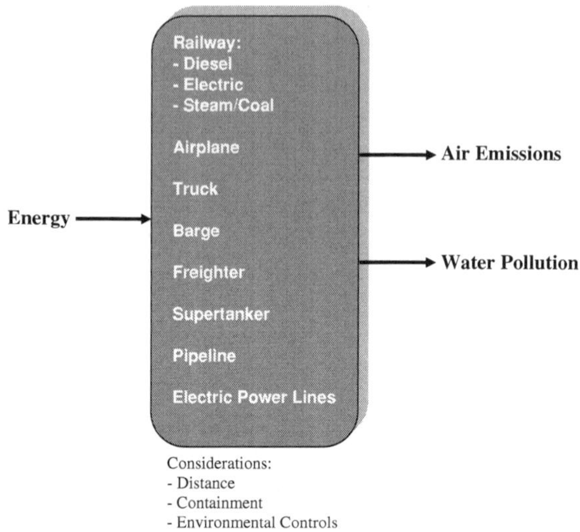

Fig. 3.5 Transportation/distribution system

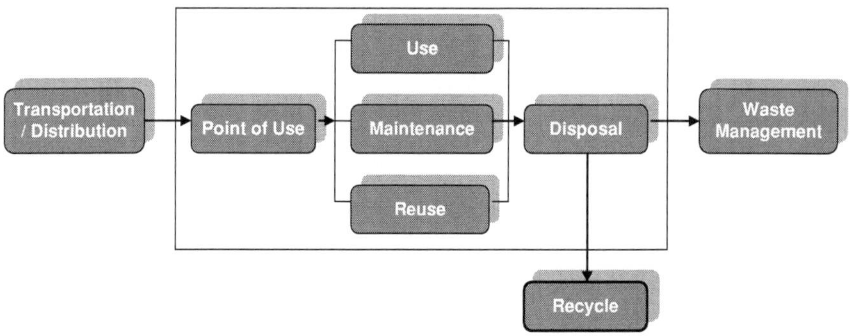

Fig. 3.6 Consumer use/disposal system

both transportation and distribution. As in previous stages, clear boundaries must be established to define the extent to which issues such as building and maintaining transportation and distribution equipment will be factored into the inventory results.

Consumer Use/Disposal

Data collected for this subsystem cover consumer activities including use (product consumption, storage, preparation, or operation), maintenance (repair), and reuse (see Fig. 3.6).

Issues to consider when defining the scope of the subsystem include:

- Time of product use before it is discarded
- Inputs used in the maintenance process
- The typical frequency of repair
- Potential product reuse options

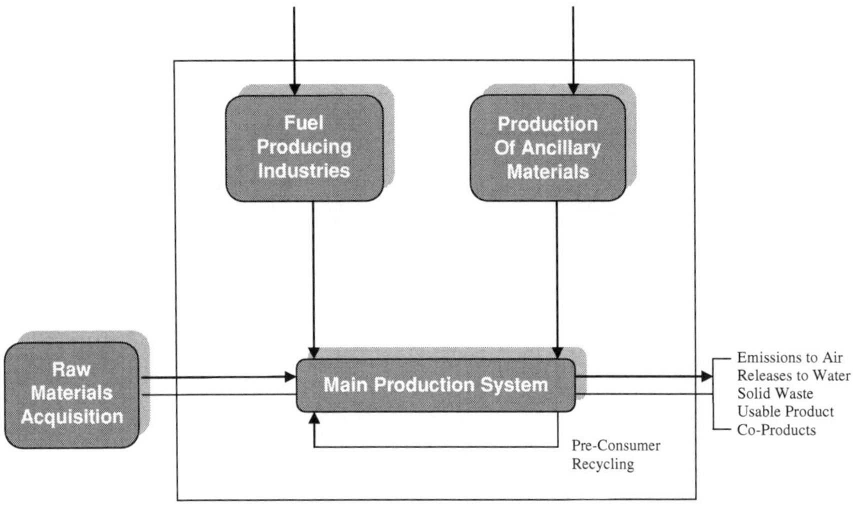

Fig. 3.7 Manufacturing and fabrication system

Managers should incorporate into the analysis any industry information on typical consumer usage patterns that may make the study's results more valid. For example, consumers may occasionally use two thinner paper cups to attain the strength of a single comparable polystyrene cup. Sources of data that may help this process include consumer surveys, published materials, electricity, so the energy analysis must include a determination of the fuel mix used to generate the electricity.

Generally, the national average fuel mix may be used, but industry-specific information is preferred. Some materials are made from energy resources and are therefore assigned an energy value. For example, plastics, made from petroleum and natural gas, release energy when burned. This energy value is credited against the system requirements for the primary product, resulting in a new energy requirement that is less than the total energy requirements for the system.

Manufacture and Fabrication

Data collected for this subsystem includes all energy, material, or water inputs and environmental releases that occur during the manufacturing processes required to convert each raw material input into intermediate materials ready for fabrication. This process may be repeated for several streams of resources as well as several intermediate cycles before final fabrication of the product (see Fig. 3.7).

Often co-products—outputs that are neither products nor inputs elsewhere in the system—are generated in the manufacturing process. Co-products are included in LCA until they are separated from the primary product being analyzed. Raw materials, energy, and emissions should be allocated between the primary product and the co-products by their proportionate weight or volume. If scrap within one subsystem is used as an input within the same subsystem, the raw material or

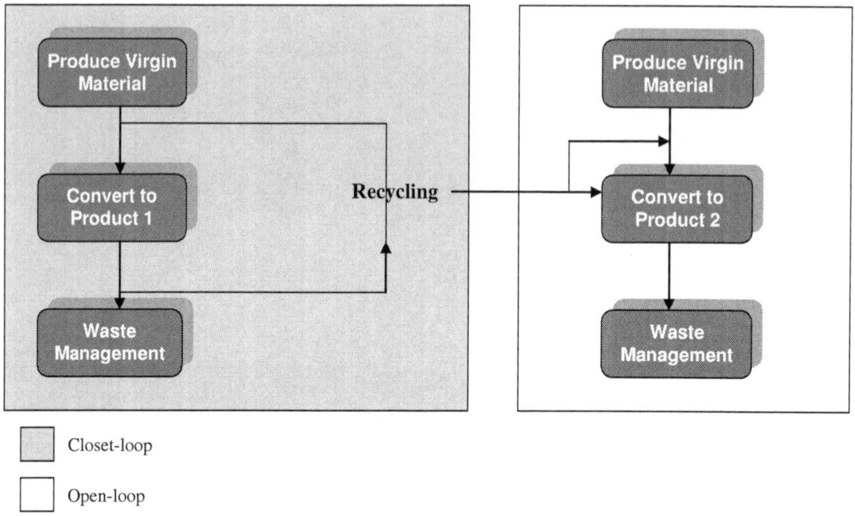

Fig. 3.8 Recycling subsystem

intermediate material required from the outside is reduced and should be factored into the analysis. If industrial scrap is used in another subsystem, it is considered to be a co-product and should be allocated to the same consumption and emission rates required to produce the primary material. Some scrap is simply discarded and should be counted as solid waste.

Differences in technology throughout the industry require certain assumptions to be made at this stage. Comparisons between different-size facilities, differing ages of equipment, different capacity-utilization rates, and differing energy consumption per unit of production must be made explicit. The data collected for final product fabrication assesses the consumption of inputs and the emissions required to convert all materials into the final product ready for and assumptions. Inventory reports must include documentation of assumptions including the timeliness of the data, potential biases, and other limitations.

Various disposal alternatives exist such as reuse, recycling, composting, incineration, and landfilling. Transportation and collection of post-consumer waste should also be included in the analysis. Inventories often use a national estimate of waste management methods, citing current averages for the percentage of waste disposed of by landfilling, recycling, and incineration methods. Recycling technology is expected to improve greatly in the future. Therefore, content levels and recycling rates should always be reported at current rates with documentation of study dates. Advances in technology will both increase rates and the number of products that are recyclable, altering both open- and closed-loop recycling options (see Fig. 3.8).

Open-loop recycling means that a product is recycled into a different product that is disposed of after use. In these cases, the resource requirements and

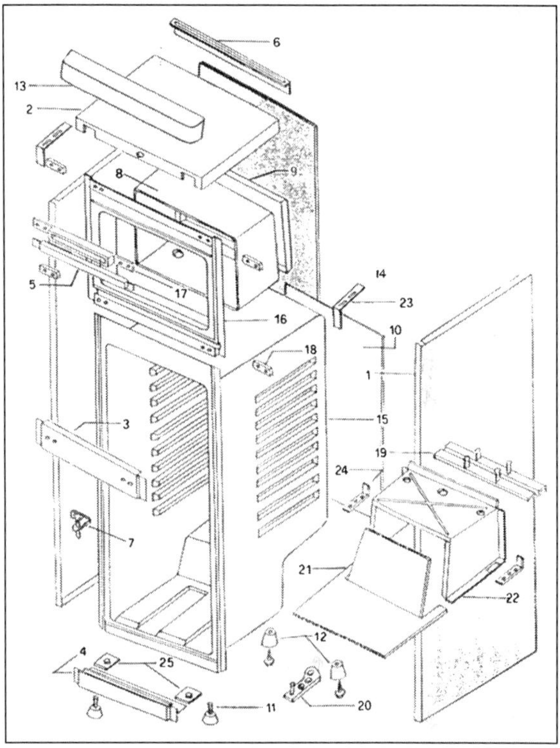

Fig. 3.9 Old model cold store

environmental emissions related to the recycling and final disposal of the recycled material is divided equally between the two products produced. Closed-loop recycling refers to materials that can be recycled into the same product repeatedly. This means that the more times the product is recycled, the less virgin material is required and the greater the number of cycles over which the resources and emissions can be allocated. The environmental effects of a closed-loop product will approach zero over the life of the product. For some products, a recycling infrastructure already exists, providing data on the collection, transportation, and processing of its materials. But for many products such information does not exist, leading to the use of data extrapolated from pilot programs or forecasts.

Wastes may be defined as materials that have no intrinsic or market value. Waste occurs in some form at every stage of the life cycle. Careful analysis of waste management issues is required as disposal options vary with the seasons, geography, and the technology used by a particular facility. Further complicating the inventory is the fact that many waste streams are combinations of materials derived from several subsystems, and that waste treatment facilities may produce a variety of releases including air, water, and solid wastes. For example, reported waterborne waste data should include an analysis of the water treatment system, the land associated with the treatment system, and atmospheric and solid wastes associated with the system. Information about emissions from solid waste is more difficult to find as there is no existing method to determine the emissions of a particular product once it has been mixed with municipal waste in a landfill or incinerator. If, however, a disposal process is being used for only one type of product (e.g., composting for yard waste or recycling for aluminum cans), accurate measures are available.

4. *Evaluate and report results.*

When writing a report to present the final results of the life-cycle inventory, it is important to thoroughly describe the methodology used in the analysis. The report should explicitly define the systems analyzed and the boundaries that were set. All assumptions made in performing the inventory should be clearly explained. The basis for comparison among systems should be given, and any equivalent usage ratios that were used should be explained.

3.1.3 Life Cycle Impact Assessment

The Life Cycle Impact Assessment (LCIA) phase of an LCA is the evaluation of potential human health and environmental impacts of the environmental resources and releases identified during the LCI. The results of an LCIA show the relative differences in potential environmental impacts for each option. For example, an LCIA could determine which product/process causes more global warming potential.

The following steps comprise a life cycle impact assessment.

1. *Selection and definition of impact categories*—identifying relevant environmental impact categories (e.g., global warming, acidification, terrestrial toxicity). Table 3.1 shows some of the mor e commonly used impact categories.

2. *Classification*—assigning LCI results to the impact categories (e.g., classifying carbon dioxide emissions to global warming).

The purpose of classification is to organize and possibly combine the LCI results into impact categories. For LCI items that contribute to only one impact category, the procedure is a straightforward assignment. For example, carbon dioxide emissions can be classified into the global warming category.

For LCI items that contribute to two or more different impact categories, a rule must be established for classification. There are two ways of assigning LCI results to multiple impact categories (ISO 1998):

- Partition a representative portion of the LCI results to the impact categories to which they contribute. This is typically allowed in cases when the effects are dependent on each other.
- Assign all LCI results to all impact categories to which they contribute. This is typically allowed when the effects are independent of each other.

Table 3.1 Common used life cycle impact categories

Impact category	Scale	Examples of LCI data (i.e. classification)	Common possible characterization factor	Description of characterization factor
Global warming	Global	Carbon dioxide (CO_2) Nitrogen dioxide (NO_2) Methane (CH_4) Chlorofluorocarbons (CFCs) Hydrochlorofluorocarbons (HCFCs) Methyl bromide (CH_3Br)	Global warming potential	Converts LCI data to Carbon dioxide (CO_2) equivalents Note: global warming potential can be 50, 100, or 500 year potentials
Stratospheric ozone depletion	Global	Chlorofluorocarbons (CFCs) Hydrochlorofluorocarbons (HCFCs) Halons Methyl Bromide (CH_3Br)	Ozone depleting potential	Converts LCI data to trichlorofluoromethane (CFC-11) equivalents
Acidification	Regional Local	Sulfur Oxides (SOx) Nitrogen Oxides (NOx) Hydrochloric Acid (HCL) Hydrofluoric Acid (HF) Ammonia (NH_4)	Acidification potential	Converts LCI data to hydrogen (H +) ion equivalents
Eutrophication	Local	Phosphate (PO_4) Nitrogen Oxide (NO) Nitrogen Dioxide (NO_2) Nitrates Ammonia (NH_4)	Eutrophication potential	Converts LCI data to phosphate (PO_4) equivalents
Photochemical smog	Local	Non-methane Hydrocarbon (NMHC)	Photochemical oxident creation potential	Converts LCI data to ethane (C_2H_6) equivalents
Terrestrial toxicity	Local	Toxic chemical with a reported lethal concentration to rodents	LC_{50}	Converts LC_{50} data to equivalents; uses multimedia modelling, exposure pathways
Aquatic toxicity	Local	Toxic chemical with a reported lethal concentration to fish	LC_{50}	Converts LC_{50} data to equivalents; uses multimedia modelling, exposure pathways

(continued)

Table 3.1 (continued)

Impact category	Scale	Examples of LCI data (i.e. classification)	Common possible characterization factor	Description of characterization factor
Human health	Global Regional Local	Total releases to air, water and soil	LC_{50}	Converts LC_{50} data to equivalents; uses multimedia modelling, exposure pathways
Resource depletion	Global Regional Local	Quantity of minerals used Quantity of fossil fuel used	Resource Depletion Potential	Converts LCI data to ratio of quantity of resource used versus quantity of resource left in reserve
Land use	Global Regional Local	Quantity disposed of in a landfill or other land modifications	Land Availability	Converts mass of solid waste into volume using an estimate density
Water use	Regional Local	Water used or consumed	Water Shortage Potential	Converts LCI data to ratio of quantity of water used versus quantity of resource left in reserve

For example, since nitrogen dioxide could potentially affect both ground level ozone formation and acidification (at the same time), the entire quantity of nitrogen dioxide would be assigned to both impact categories (e.g., 100% to ground level ozone and 100% to acidification). This procedure must be clearly documented.

3. *Characterization*—modeling LCI impacts within impact categories using science-based conversion factors (e.g., modeling the potential impact of carbon dioxide and methane on global warming).

The following is a list of several impact categories and endpoints that identify the impacts.

Global Impacts

Global warming—polar melt, soil moisture loss, longer seasons, forest loss/change, and change in wind and ocean patterns.

Ozone depletion—increased ultraviolet radiation.

Resource depletion—decreased resources for future generations.

Regional Impacts

Photochemical smog—"smog," decreased visibility, eye irritation, respiratory tract and lung irritation, and vegetation damage.

Acidification—building corrosion, water body acidification, vegetation effects, and soil effects.

Local Impacts

Human health—increased morbidity and mortality.

Terrestrial toxicity—decreased production and biodiversity and decreased wildlife for hunting or viewing.

Aquatic toxicity—decreased aquatic plant and insect production and biodiversity and decreased commercial or recreational fishing.

Eutrophication—nutrients (phosphorous and nitrogen) enter water bodies, such as lakes, estuaries and slow-moving streams, causing excessive plant growth and oxygen depletion.

Land use—loss of terrestrial habitat for wildlife and decreased landfill space.

Water use—loss of available water from groundwater and surface water sources.

Impact indicators are typically characterized using the following equation:

Inventory data $\times$ characterization factor $=$ impact indicators

For example, all greenhouse gases can be expressed in terms of CO_2 equivalents by multiplying the relevant LCI results by a CO_2 characterization factor and then combining the resulting impact indicators to provide an overall indicator of global warming potential. Characterization can put these different quantities of chemicals on an equal scale to determine the amount of impact each one has on global warming. The calculations show that ten pounds of methane have a larger impact on global warming than twenty pounds of chloroform.

Chloroform GWP factor value[1] $= 9$ Quantity

[1] Intergovernmental Panel on Climate Change (IPCC) Model.

Methane GWP factor value = 21 quantity
Chloroform GWP impact = 20 pounds × 9 = 180
Methane GWP impact = 10 pounds × 21 = 210

The key to impact characterization is using the appropriate characterization factor. For some impact categories, such as global warming and ozone depletion, there is a consensus on acceptable characterization factors. For other impact categories, such as resource depletion, a consensus is still being developed. Table 3.1 describes possible characterization factors for some of the commonly used life cycle impact categories.

4. *Normalization*—expressing potential impacts in ways that can be compared (e.g. comparing the global warming impact of carbon dioxide and methane for the two options).

Normalization is an LCIA tool used to express impact indicator data in a way that can be compared among impact categories. This procedure normalizes the indicator results by dividing by a selected reference value.

There are numerous methods of selecting a reference value, including:

- The total emissions or resource use for a given area that may be global, regional or local
- The total emissions or resource use for a given area on a per capita basis
- The ratio of one alternative to another (i.e. the baseline)
- The highest value among all options.

5. *Grouping*—sorting or ranking the indicators (e.g. sorting the indicators by location: local, regional, and global).

Grouping assigns impact categories into one or more sets to better facilitate the interpretation of the results into specific areas of concern. Typically, grouping involves sorting or ranking indicators. The following are two possible ways to group LCIA data (ISO 1998):

- Sort indicators by characteristics such as emissions (e.g. air and water emissions) or location (e.g. local, regional, or global).
- Sort indicators by a ranking system, such as high, low, or medium priority. Ranking is based on value choices.

6. *Weighting*—emphasizing the most important potential impacts.

The weighting step (also referred to as valuation) of an LCIA assigns weights or relative values to the different impact categories based on their perceived importance or relevance. Weighting is important because the impact categories should also reflect study goals and stakeholder values. As stated earlier, harmful air emissions could be of relatively higher concern in an air non-attainment zone than the same emission level in an area with better air quality. Because weighting is not a scientific process, it is vital that the weighting methodology is clearly explained and documented.

Although weighting is widely used in LCAs, the weighting stage is the least developed of the impact assessment steps and also is the one most likely to be challenged for integrity. In general, weighting includes the following activities:

- Identifying the underlying values of stakeholders
- Determining weights to place on impacts
- Applying weights to impact indicators.

Weighted data could possibly be combined across impact categories, but the weighting procedure must be explicitly documented. The un-weighted data should be shown together with the weighted results to ensure a clear understanding of the assigned weights.

Box A: The Most used Weighting Methods

- **Eco-indicator 99**
  The Eco-indicator 99 method comes in three versions, Egalitarian, Individualist and the Hierarchist (default) version. Normalisation and weighting are performed at damage category level (endpoint level in ISO terminology). There are three damage categories:

1. Human health (unit: DALY = Disability adjusted life years; this means different disability caused by diseases are weighted)
2. Ecosystem quality (unit: PDF*m2 yr; PDF = Potentially disappeared Fraction of plant species)
3. Resources (unit: MJ surplus energy Additional energy requirement to compensate lower future ore grade)
   Damage assessment step means that the impact category indicator results that are calculated in the Characterisation step, are added to form damage categories. Addition without weighting is justified here because all impact categories that refer to the same damage type (like human health) have the same unit (for instance DALY). This procedure can also be interpreted as grouping. The damage categories (and not the impact categories) are normalised on an European level (damage caused by one European per year), mostly based on 1993 as base year, with some updates for the most important emissions. Please note that the normalisation set is dependent on the perspective chosen. The normalised damage categories can also be used with the triangle tool that is build into SimaPro. This is useful if two products are to be compared without weighting, in case the damage indicators for Product A and B are conflicting (A is higher on Human health and B is higher on Ecosystem Quality). In such a case the answer is dependent on the weighting factors

for Ecosystem quality, Resources and Human health. The triangle must be understood as a way to show all possible combinations of weighting factors (represented as a percentage in such a way that they add up to 100%). If damage categories have conflicting values, the triangle will display two areas. One area represents all weighting sets for which product A has a lower environmental load, the other area will represent all weighting sets for which B has a lower load than A.

- **Eco-indicator 95** The Eco-indicator 95 method was developed under the Dutch NOH programme by PRé consultants in a joined project with Philips Consumer Electronics, NedCar (Volvo/Mitshubishi), Océ Copiers, Schuurink, CML Leiden, TU-Delft, IVAM-ER (Amsterdam) and CE Delft. The characterisation conforms with the CML 92 method; however the toxicity scores are specified into heavy metals, carcinogenic substances, pesticides and winter smog. Normalisation is based on 1990 levels for Europe excl. former USSR. Weighting is based on distance to target. Criteria for target levels are:

  - one excess death per million per year
  - 5% ecosystem degradation.
  - avoidance of smog periods

- **CML 2 baseline method (2000)** The CML 2 baseline method elaborates the problem-oriented (midpoint) approach. The CML Guide provides a list of impact assessment categories grouped into:

  1. Obligatory impact categories (Category indicators used in most LCAs).
  2. Additional impact categories (operational indicators exist, but are not often included in LCA studies).
  3. Other impact categories (no operational indicators available, therefore impossible to include quantitatively in LCA).
     In case several methods are available for obligatory impact categories, a baseline indicator is selected, based on the principle of best available practice. These baseline indicators are category indicators at "mid-point level" (problem oriented approach). Baseline indicators are recommended for simplified studies. The guide provides guidelines for inclusion of other methods and impact category indicators in case of detailed studies and extended studies.

- **EPS 2000** Environmental Priority Strategies in product design (default methodology). The EPS system is mainly aimed to be a tool for a company's internal product development process. The top-down development of the EPS system has led to an outspoken hierarchy among its principles and rules. The general principles of its development are:

1. The top-down principle (highest priority is given to the usefulness of the system)
2. The index principle (ready made indices represent weighted and aggregated impacts)
3. The default principle (an operative method as default is required)
4. The uncertainty principle (uncertainty of input data has to be estimated)
5. Choice of default data and models to determine them
   The EPS 2000 default method is an update of the 1996 version. The impact categories are identified from five safe guard subjects: human health, ecosystem production capacity, abiotic stock resource, biodiversity and cultural and recreational values.

- **EPD method**
  This method is to be used for the creation of Environmental Product Declarations or (EPD's) following the recommendations of the Swedish Environmental Management Council (SEMC). Note this is a preliminary version. In standard EPD's one only has to report on the following impact categories:

  Gross Calorific Values (GVC) (also referred to as the "Higher Heating Values")
  Greenhouse gases
  Ozone depleting gases
  Acidifying compounds
  Gases creating ground-level ozone (Photochemical Ozone creation)
  Eutrophicating compounds
  Except for the Gross Calorific Value (GVC) impact categories, all impact categories are taken directly from the CML 2 baseline 2000 method
- **TRACI**
  US Method, developed by the US EPA. Currently only contains characterization, normalization data are expected soon.
- **Impact 2002+**
  IMPACT 2002+ version is mainly a combination between IMPACT 2002 (Pennington et al. 2006), Eco-indicator 99, CML 2000 and IPCC.
- **Cumulative Energy Demand (CED)**
  Calculates the total (primary) energy use through a life cycle based (HHV) (Frischknecht et al. 2003).
- **IPCC Greenhouse gas emissions**
  IPCC characterization factors for direct global warming potential of airborne emissions. Three time perspectives are included: 20, 100 and 500 years. (Climate Change 2001).

7. *Evaluating and Reporting LCIA Results*—gaining a better understanding of the reliability of the LCIA results.

The selection of more complex or site-specific impact models can help reduce the limitations of the impact assessment's accuracy. It is important to document these limitations and to include a comprehensive description of the LCIA methodology, as well as a discussion of the underlying assumptions, value choices, and known uncertainties in the impact models with the numerical results of the LCIA to be used in interpreting the results of the LCA.

3.1.4 Life Cycle Interpretation

Life cycle interpretation is a systematic technique to identify, quantify, check, and evaluate information from the results of the LCI and the LCIA, and communicate them effectively. Life cycle interpretation is the last phase of the LCA process.

ISO has defined the following two objectives of life cycle interpretation:

1. Analyze results, reach conclusions, explain limitations, and provide recommendations based on the findings of the preceding phases of the LCA, and to report the results of the life cycle interpretation in a transparent manner.
2. Provide a readily understandable, complete, and consistent presentation of the results of an LCA study, in accordance with the goal and scope of the study.

3.2 Case Study: LCA as a Tool in "Design for Environmental": A Comparative Study Between Domestic Refrigerators

In this case study a comparative Life Cycle Assessment between two different models of domestic refrigerators has been carried out, with the eco-indicator methodology in the impact assessment phase. The study of life cycle enabled the analysis and the subsequent comparison of the results about environmental impact, focusing with special attention on those of energetic nature, coming from the comparison between a refrigerator (Old Model) and its successive corresponding model (New Model), which was structurally modified by means of a product's redesign. The study covers the entire life cycle of the two models: pre-production, production, use and disposal. It is important to underline the disposal problems of the domestic devices. The annual quantity to handle is about three million of pieces, only in Italy. These values are important to understand the impact of domestic appliances from environmental and logistic point of view.

Goal and Scope Definition

The goal of the work is to compare the environmental profiles of two domestic refrigerators (the old and the new model) in order to identify the one with the best environmental profile and the hot spots of the two systems.

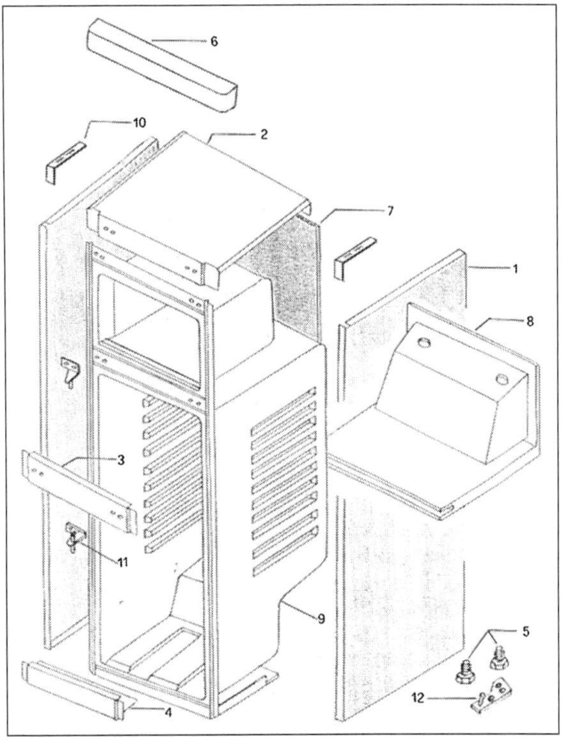

Fig. 3.10 New model cold store

The products considered in this work have a storage capacity of 240 l and a freezing capacity of 2.5 kg/24 h. Main functional groups of the two refrigerators are: internal accessories, Freezer and Fridge door, Cupboard, Cooling group and electrical components.

The cupboard is the refrigerator component that presents more changes from aesthetic and functional point of view. While in the New Model (Fig. 3.10) the fridge-freezer cell is produced with plastic material and is composed of a single body, in the Old Model (Fig. 3.9) the fridge-freezer cell is composed of different parts: the fridge space produced with plastic material, freezer space composed of an aluminum butt, of an aluminum foil, of an aluminum plate for evaporation, of a cupboard frame and of a several metal supports.

The study covers the entire life cycle of the two systems; only few operations, for which no good data were available, have been excluded. It was considered as a functional unit the refrigerator capacity to storage a liter of products.

For both the systems the following assumptions have been made:

1. the emissions of the electric energy production are those relative to the Italian energy system;
2. for the distribution phase of the two refrigerators we have considered an average space between the production site, near Ancona and the transit point located in Manchester, for a total distance of 1706 km.

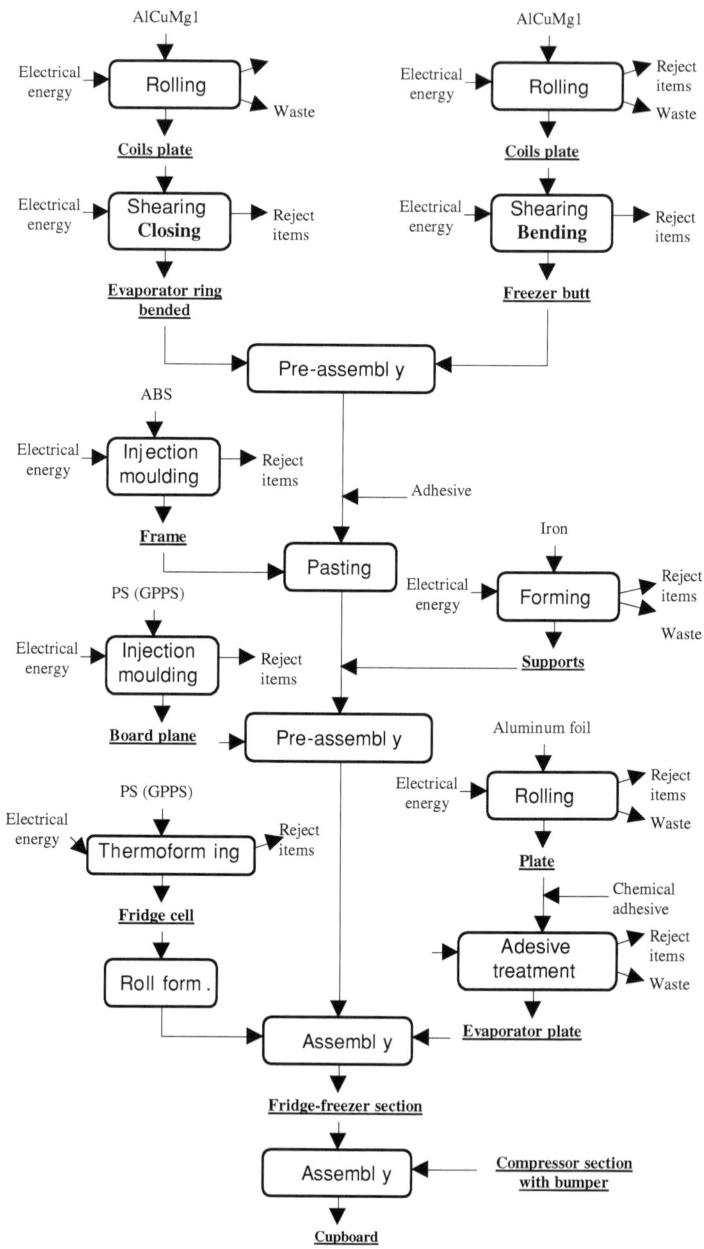

Fig. 3.11 Production process of the "cupboard" component (FRIDGE OM)

The average life of the refrigerators has been assumed to be 10 years on the basis of consumer preferences and of their technical characteristics. The impact assessment methodology used is the "problem-oriented" in which, as stated by the

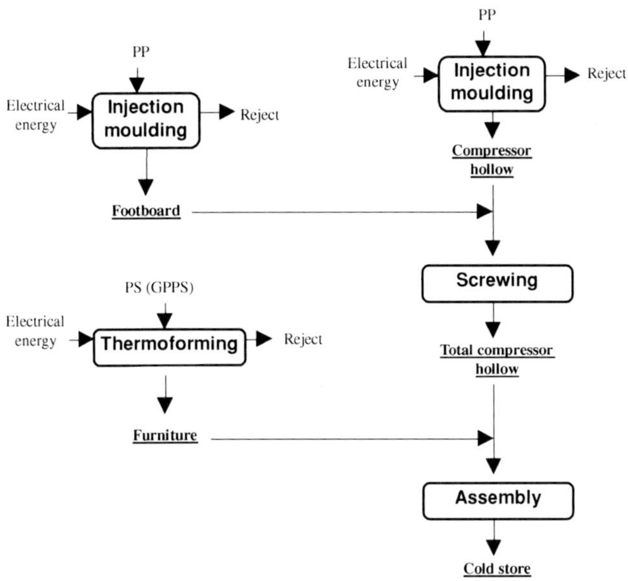

Fig. 3.12 Production process of cold store (NM)

Table 3.2 Summary of values

Model	Weight (g)	n° components	Trinkets weight (g)	Trinkets quantity	Tot weight (g)	Tot n° components + trinkets
OM	46,500	117	81.16	34	46581.16	151
NM	44,510	99	43	21	44553	120

ISO series 14040, the inventory data are associated with specific environmental impact categories in order to understand those impacts.

The impact categories which have been considered are: depletion of abiotic resources; global warming; ozone layer depletion; human toxicity; aquatic toxicity; acidification; nutrification; photochemical oxidant creation. The characterization factors used are those stated in Heijungs et al. (1992) with the exception of those cases in which factors have been updated (i.e. new global warming potential factors suggested by the IPCC). The normalization factors are those on the European scale published by the Directoraat-Generaal Rijkswaterstaat (1997). The weighting factors are those published by the NOGEPA (Netherlands Oil and Gas Exploration and Production Association) panel by Huppes et al. (1997).

Life Cycle Inventory

The study covers the entire life cycle of the two models: pre-production, production, use and disposal. The first operative phase of the work involved the analysis and weight calculation of raw materials and the study of the production processes of the two refrigerators. In Table 3.2 are shown the results obtained.

Fig. 3.13 Disposal phases

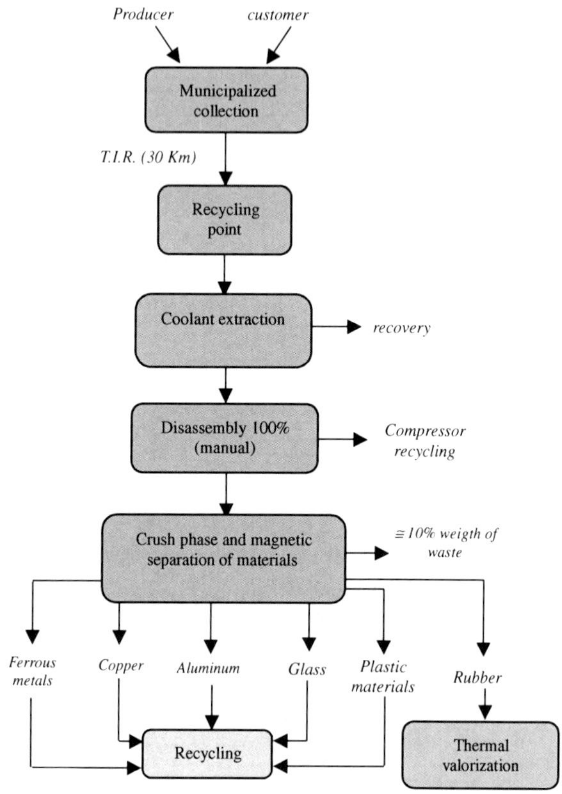

Every unit of the process associated to a component requires a certain amount of raw material, energy and water consumption like input and it is necessary to consider a certain amount of waste, rejects items and pollutant like output. On the basis of the values obtained in the firm these parameters were evaluated for all functional groups that characterize the two refrigerators.

In Figs. 3.11 and 3.12 are shown, as example, the diagrams of the production process of the components of the functional group called "cupboard" for the fridges Old Model and New Model.

A comparison between the two diagrams shows an important simplification of the production processes. Like most domestic appliance also for the refrigerators the main impact is due to the use phase. Laboratory test were performed to evaluate with great accuracy energy consumption and all performance of the two products. The disposal phases are shown in Fig. 3.13. We have considered for the two refrigerators a disassembly percentage of 90. This phase is composed of a logistic chain organized with several manual activities carried out through electrical device (compressed-air screwer, …). Disassembly times for the two refrigerators are about 30 min for the fridge "New Model" (composed of a moderate quantity of parts) and 60 min for fridge "Old Model" (composed of numerous parts).

Table 3.3 Contribution of impact categories

Impact categorie	Pre-prod. + prod. (OM)	Pre-prod. + prod. (NM)	Use (OM)	Use (NM)	Disposal (OM)	Disposal (NM)
Greenh. (GWP)	0.0227	0.02	0.415	0.383	0.0154	0.0146
Ozone (ODP)	0.00315	0.00227	0.001	0.00037	−0.00215	−0.0019
Acidif. (AP)	0.126	0.114	1.47	1.36	0.0175	0.0199
Eutroph. (NP)	0.0125	0.011	0.106	0.0977	0.000612	0.0003
Metal	0.0772	0.0602	0.0545	0.039	−0.0227	−0.0212
Toxins	0.0626	0.0394	0.0564	0.0342	−0.00621	−0.00521
W.smog	0.0363	0.0269	0.688	0.634	0.0164	0.0197
S.smog (POCP)	0.0393	0.0381	0.499	0.454	−0.0167	−0.0245
TOTAL	0.37975	0.31187	3.2899	3.00827	0.002152	0.00169

Life Cycle Impact Assessment

In Table 3.3 the contribution of impact categories to the eco-indicator of both the models are shown. This result can be better understood going through a closer examination of the contributions of the life cycle phases to the eco-indicators. The most important categories used are: Global warning potential (GWP), Nutrification potential (NP), Photochemical oxidant creation potential (POCP), Human toxicity (HT), Acidification potential (AP), Ozone depletion potential (ODP), Ecotoxicity aquatic/terrestrial and abiotic depletion potential.

In Fig. 3.14 the two systems have been compared. One can see that the Old Model scores worse for most of the impact categories.

In Fig. 3.14 can be easily noted that the use phase is the most responsible for the gap between the two systems. The production phase also has a relevant difference between the two systems.

The impact categories that are more involved in these three operations are:

- Acidification in the use phase (46%).
- Global warming in the production (38%) and use phases (31%), due to the relevant thermal energy consumption.
- Photochemical oxidant creation potential (POCP), Human toxicity (HT) in the use phase.

In Fig. 3.15 the environmental performance of life cycle phases are compared taking into account the contribution of the different functional units.

Improvements in production and disposal phases are essentially due to cupboard re-design.

Finally in this work the attention was focused on energy impacts of the two refrigerators. With the values summarized in Fig. 3.16 it is possible to underline the improvement in term of energy consumption obtained with the new model.

Fig. 3.14 Contribution of the life cycle phases to the eco-indicator

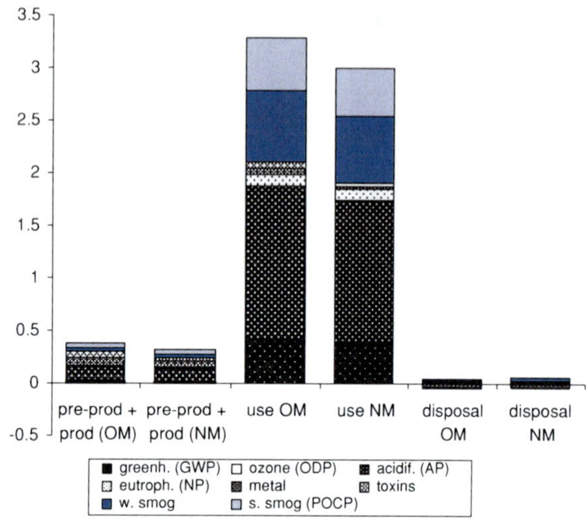

	RM+Prod+Distribution			Use			Disposal		
	Fridge OM (MJ)	Fridge NM (MJ)	%	Fridge OM (MJ)	Fridge NM (MJ)	%	Fridge OM (MJ)	Fridge NM (MJ)	%
Internal Accessories	248,48	243	2,20%	-	-	-	68,04	65,52	3,70%
Freezer door	101	101	0,00%	-	-	-	32,13	32,13	0,00%
Fridge door	409	409	0,00%	-	-	-	59,13	59,13	0,00%
Cupboard	1290	1140	11,60%	-	-	-	353,7	159,3	55,00%
Cooling group	782	791	-1,20%	-	-	-	690,3	684,9	0,80%
Electrical components	29	28	3,40%	-	-	-	8,51	4,8	43,60%
Transport	92,6	88,6	4,30%	-	-	-	1,31	1,25	4,60%
Municipal Waste	-	-	-	-	-	-	-57	-53	-7,00%
Disassembly	-	-	-	-	-	-	0,39	0,19	51,30%
Total	2958,57	2804,04	5,20%	42400	39200	7,50%	1156,51	954,22	17,50%
Total (MJ/litre)	**12,33**	**11,67**	**5,20%**	**176,67**	**163,33**	**7,50%**	**4,83**	**3,99**	**17,50%**

Fig. 3.15 Life cycle phases

The results of energetic impact are presented in MJ/liter and indicate the amount of energy, referred to a capacity of 1 L, necessary during the whole life cycle of refrigerators. The disposal phase allowed the company to obtain the most important reduction of energy in percentage term (17.5%).

From Fig. 3.16 one can see the relations between the phases of the two systems with the energy consumption. In the two systems the phase with the highest burden is use step.

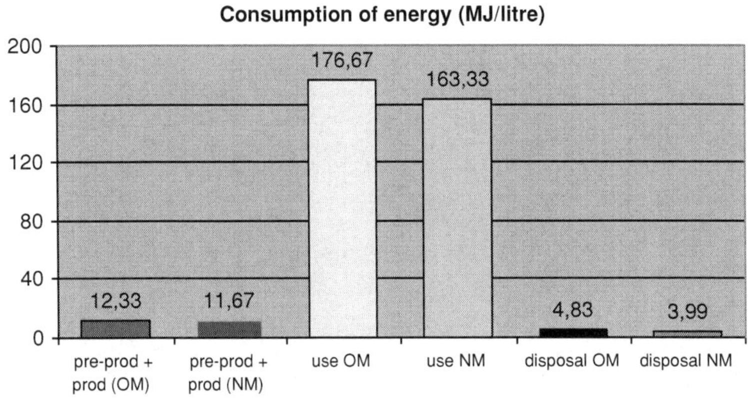

Fig. 3.16 Energy impact of two refrigerators

Discussion

The introduction of Design for Environment (DfE) methodologies in manufacturing firms allows attention to be paid to environmental aspects right from the start of the design stage leading to a reduction in the materials used and the waste products, avoiding any future weaknesses and inefficiencies.

DfE bears in mind the potential environmental impact throughout the life cycle of the product: emission of harmful substances, excessive use of energy or non-renewable energy sources. It also considers the life cycle of the materials from extraction to disposal. In this way the designers do not create just a product but a whole life cycle. In the practice of eco-design, life cycle assessment (LCA) provided the basic modeling framework for evaluating the environmental load and impact throughout the entire product life cycle from material acquisition to disposal.

It is important to understand that the early design phase can affect the environmental impact caused by products to a large extent. For example, in the case study, the new model of refrigerator the product was designed for disassembly and remanufacturing; parts in the product are more easy to identify, access, separate, handle and also have high wear resistance.

Through this re-planning it was possible to save raw materials, consequently to reduce the appliance weight and most of all to improve the disassembly stage, since the new model was primarily built through module assemblage rather than through integration of single components.

Moreover the reduction of electrical energy consumption of the new model in the use phase (409 KWh/year FRIDGE NM in comparison with 442 KWh/year FRIDGE OM), allowed the firm to improve environmental performance.

References

Climate Change (2001) IPCC third assessment report: the scientific basis. (www.grida.no/climate/ipcc_tar). Intergovernmental panel on climate change (IPCC)

Council for Solid Waste Solutions (1990) Resource and environmental profile analysis of polyethylene and unbleached paper grocery sacks. CSWS (800-243-5790), Washington DC

Directoraat-Generaal Rijkswaterstaat (1997) Drie referentieniveaus voor normalisatie in LCA, The Netherlands

EPA (1993) EPA's approach and progress in targeting indoor air pollution

Franklin Associates (1991) Product life-cycle assessment: guidelines and principles (EPA report no. 68-CO-0003)

Frischknecht R, Jungbluth N (2003). Implementation of life cycle impact assessment methods

Heijungs R, Guine′e JB, Huppes J, Lankreijer RM, Udo De Haes HA, Wegener Sleeswı′jk A, Ansems AM, Eggels PG, Van Duin R, Goede HP (1992) Environmental life cycle assessment of products: guide and backgrounds. CML, TNO, B&G, Leiden (The Netherlands)

Hunt R, Sellers J, Franklin W (1992) Resource and environmental profile analysis: a life cycle environmental assessment for products and procedures. Environmental impact assessment review

Huppes G, Sas H, de Haan E, Kuyper J (1997) Efficient environmental investments. In: SENSE international workshop session: environmental analysis and economics in industrial decision making 20 February, Amsterdam (The Netherlands)

ISO 14040 (1997) Environmental management—life cycle assessment—principles and frame work

ISO 14041 (1998) (E) Environmental management—life cycle assessment—goal and scope definition and inventory analysis

Jackson SL (1997) The ISO 14001 implementation guide: creating an integrated management system. Wiley, New York

Meadows D, Meadows D, Randers J (1972) Limits to growth. Universe Books, New York

Pennington DW, Margni M, Payet J, Jolliet O (2006) Risk and regulatory hazard based toxicological effect indicators in Life Cycle Assessment (LCA). Human and Ecotoxicological Risk Assessment 12(3): 450–475

Stilwell J, Canty R, Kopf P, Montrone A (1991) Packaging for the environment. American Management Association, New York

Chapter 4
Sustainable Product Assessment Tools

Sustainable product assessment tools are intended for the selection and prioritization of the potential environmental improvement of a product (Tischner 2000). They also support product designs and the development of ideas and target specifications.

Disparate impacts such as resource use, occupational and environmental health risks, and global environmental impacts have to be aggregated to a single score or at least lead to a single decision.

This need to cope with trade-offs between different kinds of environmental impacts has lead to often ad hoc decision-making rules.

The majority of the tools encompass all the product life cycle stages from the cradle to the grave. That is, the scope of the assessment includes the entire life cycle stages of a product including the use of raw materials, manufacturing, distribution, use, and end-of-life. According to Wong et al. (2010) the environmental product assessment tools can be classified into four types with respect to the presentation mode: matrices, checklists, spiderweb diagrams and parametric methods.

4.1 Matrix Assessment Tools

Matrix-type assessment tools include the material, energy and toxicity (MET) matrix (see box B), the AT&T matrix and target plot, and the Boeing process environmental matrix. Users assess products by filling out matrix elements which are mainly based on the ordinal scale (high–medium–low) (Lee et al. 2003; Sarkis 2001; Eagan and Weinberg 1997). The formats and evaluation criteria of the three matrix-type tools are quite similar. When compared with the life cycle assessment (LCA) method, the matrix-type assessment tools are relatively easy to use and are simple in terms of gathering data for the analysis and evaluation of a product

M. Bevilacqua et al., *Design for Environment as a Tool for the Development of a Sustainable Supply Chain*, DOI: 10.1007/978-1-4471-2461-0_4,

(Eagan and Weinberg 1997) as they only require users to have some background knowledge about the assessed product. The matrix-type assessment tools also allow users to give weight to specific life cycle stages or environmental parameters. This feature helps users assess existing products or product ideas in the product planning and development stage. The drawback of these tools, however, is that they are subjective and can be meaningless for standalone product assessment (Lee et al. 2003; Eagan and Weinberg 1997). The reason for subjectivity is that none of the matrix-type tools referred to here has guidelines to rate each element in the matrix. The ordinal scale assessment method implies that assessment results are far more influenced by human judgment than is the quantitative assessment method (Lee et al. 2003). Even though users have sufficient data on the products, it is difficult for them to rate the level of the environmental impacts. In addition, the total value summed up from the AT&T matrix and target plot or the Boeing process environmental matrix has no meaning for assessing the environmental performance if there is no reference product with which to compare it.

Box B: MET Matrix

The MET matrix is a result of the Eco Design demonstration program (Eindrapportage Eco Design programma, TNO Productcentrum, Delft 1994). The MET matrix's goal is to identify environmental "hot spots" within product life cycle and find the ways for environmental improvement of product and its processes. The MET matrix is based on the inventory analysis of the LCA methodology. It has three groups of environmental interventions: Material cycle (aspects that are related to a product's materials), Energy consumption (in production and use stages), Toxic emissions (during different life cycle stages of a product) (MET matrix). (Brezet et al. 1994), in his manual, provides an example of qualitative MET matrix for a copier (see Table 4.1.).
Once The MET matrix is ready, environmental effects should be prioritized and improvement options should be identified.
Limitations: It is still hard to decide environmental priorities from MET matrix. In this case, LCA study can be applied.

4.2 Checklists

Environmental product assessment checklists are qualitative and semi-quantitative in nature. Qualitative tools include the ABC analysis, the recycling checklist for the European Council's Directive on Waste Electrical and Electronic equipment (WEEE), the ecodesign checklist method (ECM), and the Philips's fast five checklist. Semi-quantitative tools include the Eco-Estimator and the Sony's green

Table 4.1 Qualitative MET matrix for a copier (Source: Brezet et al. 1994)

	Material cycle	Energy consumption	Toxic emissions
Production	-depletion of raw materials	-energy contents of the materials	-fire retardants
	-recycling of production wastes	-process energy	
Use	-consumption of paper	-energy consumption	-emissions of zone
	-toner waste	-energy during transport	
Disposal	-recycling of the machine		-selenium drum

product check sheet and product profile. Time required to complete the checklist-type assessment normally depends on the product complexity, data availability and users' knowledge of the product (Ernzer and Wimmer 2002; Wimmer 1999). Table 4.2 shows an example of checklist. Major weaknesses of the qualitative and semi-quantitative checklists lie in the questions formulated in the checklist. Although both qualitative and semi-quantitative checklists can be used as a reminder of the environmental issues during the product design and development, questions on most types of checklists are notholistic because most checklists do not allow assessment throughout the entire life cycle of a product (Wong et al. 2010).

4.3 Spiderweb Diagrams

Spiderweb diagrams for the environmental product assessment present the level of the environmental impact in graphical modes, including the eco-compass, the E-concept spiderweb diagrams, and the life cycle design strategy wheels (for instance see box C). They are mainly used for comparing major differences in the same environmental aspects between the product being assessed and the reference product. Of all types of environmental product assessment tools, visual presentation by the spiderweb diagram is the best-known method that has been used to interpret the environmental product assessment results. Because of the clear graphical view of the results, the environmental product assessment by spiderweb diagram is the best type of tool for choosing design alternatives in the product planning and development stage. In addition, the time needed to complete this method is relatively shorter than that of the matrix type or checklist-type assessment tools. The environmental impacts and parameters, however, are not assessed in this type of the tool. Most spiderweb diagrams are not suitable for the stand-alone product assessment, as a reference product is always needed. If the reference product is absent, the assessment result has no meaning.

Table 4.2 Checklists (Source: Wong et al. 2010)

Product name:	Hand-held vacuum cleaner
Date of assessment:	5th june 2008
Cost per unit: (for existing product)	……..
Expected product life time:	5 years
Assessment objectives:	To assess the environmental performance of raw material extraction
Market status of product:	–Existing product –New to the market –New to the business (has not been possessed by the business before) –Product with changed product parameter(s) –Product with improved function(s) or quality
Application of product, product idea, or end use of components/subassemblies: (product is not rated greater than 1000 V ac or 1500 V dc	–Household appliances
	• Cooling appliances(except fan) Fanning • Appliances for heating rooms • Cooking appliances with heat generation • Cooking appliances without heat generation • Appliances for caring or processing of clothes • Washing machines for clothes or dishes • Personal care products • Appliances for household cleaning -IT and communications equipment -AUDIO and visual devices -Lighting equipment and luminaries -Household electrical and electronic tools -Toys, leisure, and sports equipment -Medical devices for domestic applications -Monitoring and weighing instruments

(continued)

Table 4.2 (continued)

Product name:	Hand-held vacuum cleaner
Use of hazardous material(s)	-No -Yes: • Lead(except in CRTs) • Mercury • Hexavalent chromium • Cadmium • PCBs • Flame retardants (Pollybrominated biphenyls, PBB) • Flame retardants (Pollybrominated diphenylether, PBDE) • Radioactive substances • Asbestos • Beryllium
Presence of by-products during manufacturing or consumables during use: (Tick if confirmed)	-No -Yes • Assessment includes by-products and consumables • Assessment includes by-products only • Assessment includes consumable only • Assessment do not include by-products and consumables
Confirmation of parameters related to product: (please put a tick if it is confirmed)	-Number of components/subassemblies -Weight, size or volume of product/ components/subassemblies -Use of material (e.g. weight, volume, type) -Detailed design of product/components/ subassemblies -Consumption of energy, water, and other resources during use -Quantity and nature of consumables -Aesthetic design
Tick if the life cycle stage involves assumptions in this study:	Raw material extraction Usage Manufacturing End-of-life Packaging and distribution
Remarks:	

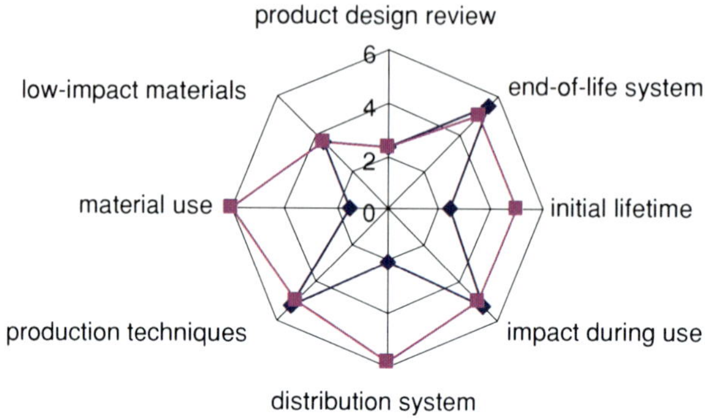

Fig. 4.1 Eco Design Strategy Wheel (*Source* http://www.io.tudelft.nl/research/dfs/ecoquest/welkom/Entry.html)

Box C: Eco Design Strategy Wheel

The Eco Design Strategy Wheel is an element of the Sustainable Value software which was developed by Delft University of Technology in the Netherlands. This tool is used in order to improve environmental performance of a product by prioritizing 7 environmental strategies: selection of low-impact materials, reduction of materials use, optimization of techniques, optimization of distribution system, reduction of impact during use, optimization of initial life time, optimization of end-of-life system (See Fig. 4.1). According to the statement at Delft University, the Eco Design Strategy Wheel is a "graphical representation of all possible Eco design strategies throughout the lifetime of a product" (TUDelft and EcoQuest 2004). According to Bakker (1995), the eco-wheel can be used in order to: (1) generate ideas, (2) select between strategies, (3) make environmental 'product portfolio'.

Limitations: It can be mostly used as a general framework or a starting point to identify the needs and generate the ideas of life cycle design strategy. The Eco Design Strategy Wheel usually does not describe the resulting separate environmental effects (Brezet and van Hemel 1997).

4.4 Parametric Assessment

The parametric-type assessment tool is another type of the quantitative method that assesses products from the environmental point of view. These tools are objective in nature. They have a common denominator where special factors or values are used for the estimation of the environmental impact of the product.

These factors are unique to each type of material and process. The factors are multiplied by the weight of each material and/or process type and yield the quantitative environmental impact value of the product (Ernzer and Wimmer 2002). In literature several different methods have been used:

- The health hazard scoring (HHS) system uses the analytical hierarchy process to weight workplace toxic effects and accident risks (Srinivasan et al. 1995). The HHS system focuses on various workplace hazards: inhalation and oral toxicity, skin and eye irritation, reactivity, and flammability. These impacts are roughly categorized based on information supplied by chemical vendors and weighted against each other using the analytical hierarchy process, a decision analysis method. HHS has been applied to the optimization of machining processes and the selection of process chemicals.
- The material input per service unit (MIPS) aggregates the mass of all the material input required to produce a product or service (Bringezu et al. 1994; Hinterberger et al. 1994). The MIPS is the simplest material-balance-based approach imaginable: the mass of material moved for the provision of a specific service is simply aggregated in kilograms. MIPS has been applied to services, products like catalytic converters and yogurt, and regional economies.
- The Swiss ecopoint (SEP) method scores pollutant loadings based on a source's contribution to an acceptable total pollution load and an environmental scarcity factor (Abhe et al. 1990). SEP sets every product-specific emissions stream in relation to the maximum that could be diluted to acceptable levels in a specific region and the current total emissions in that region. SEP has been applied in the LCA of standard consumer products.
- The sustainable process index (SPI) determines the area that would be required to operate a process sustainability, based on renewable resource generation and toxic degradation (an extension of the dilution volume approach) (Sage 1993; Narodoslawsky and Krotscheck 1995). SPI determines the area required to sustain a process based on harvest rates for inputs and the area required for the degradation of effluents given the condition that ambient environmental quality standards are not violated. SPI has been used to assess renewable resource technologies and to select process improvements in the electronics industry.
- The environmental priority strategies (EPS) characterize the environmental damage caused by equivalency potentials and expresse it in monetary terms, derived from environmental economics (Hanssen 1998; Rydh 1999). The EPS is based on SETAC life-cycle inventories of emissions, but is extended to use environmental economics to value product and process-specific impacts in monetary units. This requires the quantification of actual impacts or risks and the valuation of these impacts using market prices or willingness to pay.
- The Society of Environmental Toxicology and Chemistry's life cycle (SETAC LCA) IA method aggregates pollutants with similar impacts to equivalency potentials and uses decision analysis to assign weights to different adverse effects (ISO 14042: 2000 (E)). The method of critical volume gives the amount of clean water and air that would be needed to assimilate the emissions in order

to satisfy some quality standards for air and water (Lindfors et al. 1995). The Eco-indicator method weights the units of polluted air according to maximum accepted concentration (MAC values) health standards. The units of polluted water are based on standards for inlet water for drinking water companies (Goedkoop 1995). The Tellus method, which is based on controlling the costs of a number of pollutants, was used to establish prices for some criteria regarding air pollutants (Tellus Institute 1992).

Despite their use for the same purposes, the methods differ in what they try to achieve: in the effects they consider, in the depth of analysis, in the way values influence the final score, and in use of ordinal or cardinal measures of impact.

Two problem areas are identified: (1) to varying degrees, each of the methods has the potential to recommend an alternative that actually has a higher impact than other alternatives: (2) for some of the methods the data requirement is so extensive and the tolerance of imperfect data is so low that the application of the method for reasonably sophisticated products or processes would be too complicated.

Table 4.3 shows the main features of the proposed method. The goal is not only crucial for the decision of which method to choose for a specific purpose but also serves as a gauge by which to judge the performance of a method. The range of endpoints considered indicates the comprehensiveness of a method and the effort required for the analysis. A broad range of endpoints helps to avoid the inadvertent creation of new problems.

A method to get a more accurate interpretation of parameters, which influence the eco-balance, and probably has gained the widest acceptance, is the use of the so-called "impact categories and equivalency factors", which is based on the SETAC LCA method (Hertwich et al. 1997; ISO 14042: 2000 (E)). In this method, different approaches are used according to the type of the environmental problem. The IA for LCA as defined by the SETAC method, customized to the EDIP method, includes three main groups (resources consumption, ecological impacts, and impacts on the working environment) and may be global, regional or local (Wenzel and Hauschild 1997). Some of the above methods produce a single index trying to represent the total environmental effect of the examined system while other methods express the environmental effects with more than one index. For example, the methods of critical volume or MIPS provide their results in one end-point (ISO 14040: 1997 (E); ISO 14041: 1998 (E)).

4.5 Summing up the Engineering Perspective

1. Most references are conceptual; there are relatively few references that describe the diffusion of the tools, experience of how well they work in the product development process and how useful they are for actually reducing the environmental impacts of products.

Table 4.3 Main features of methods (source: Hertwich et al. 1997)

	HHS	MIPS	SEP	SPI	SETAC	EPS
What is the goal?	Reduction on occupational risks	Reduction of mass flow (by factor of 100)	Reduction of environmental local	Sustainability	Reduction of overall impacts	Increase of total welfare
What endpoints does the method include?	Oral and respiratory toxicity, carcinogenicity skin and eye irritation, flammability, reactivity	Mass flow	Resource use, global warming, toxicity, ozone depletion, ozone generation, etc.	Resource use, global warming, toxicity, ozone depletion, ozone generation, etc.	Biotic and abiotic resources, global warming, ozone depletion, human and ecosystem toxicity, ozone generation, acidification, eutrophication and deg	Human health, biological diversity production, resource, aesthetics
What is the basis of evaluation?	Potential effects	Mass flow	Relative pollution loads	Pollution loads and resource consumption	Potential effects	Actual effects or risks
Valuation through	Mired decision analysis	Implicit (or absent)	Implicit environmental quality standards	Implicit sustainability criteria	Explicit decision analysis	Explicit monetary valuation
How deep is the analysis?	Medium-shallow	Very shallow	Medium	Medium	Deep	Very deep
Measure units	Ordinal	Cardinal(kilograms)	Cardinal (ecopoints)[b]	Cardinal $(m)^2$	Ordinal[a]	Cardinal (ELU)
Is it site specific?	Yes	No	Yes	Yes	No (possible)[c]	No (possible)[e]
What is the information requirement?	Moderate	High	Moderate	Moderate	Very small	Small
Tolerance for imperfect information	Moderate	High	Moderate	Moderate	Very small	Small

(continued)

Table 4.3 (continued)

	HHS	MIPS	SEP	SPI	SETAC	EPS
Potential for undesirable outcomes[d]	Moderate	High[f]	High[f g]	Moderate[e]	Unknown (potentially low)	Unknown (potentially low)

a) The final measure will be ordinal, the measure that are compared in each impact category are equivalency potentials and have cardinal units, such as Kg CO_2 equivalent, Kg benzene equivalent, etc

b) It could be argued that through the weighting by a scarcity factor eco-points are ordinal units

c) The debate within SETAC as to the treatment of site-specific effects has not yet been concluded

d) A discovery of new effects is no: included here

e) A source of perverse outcomes (outside of the issue of uncertainty which inflicts all methods) is the implicit valuation both SEP and SPI treat a deviation from the acceptable load independent of the load would cause an impact of low concern (e.g. increase of one of many toxicants) or of high concern increase in CFC emissions-depletion of ozone layer

f) It is argued here that MIPS has a high potential to increase environmental impact,; it will be successful in reducing mass flows

g) The high potential is due to the variation of the geographical region in which different impacts are considered: otherwise the outcome is moderate

Most publications, with an empirical content, report on the testing of new tools. These are often developed in universities and tested by the researchers in a company case study. There are also reports of companies developing their own tools. However, these reports mostly describe the tools as such rather than how they work in practice, as a part of the business process.

2. The different tools are not equally 'green'. Some tools focus on the recycling step or on global warming impacts, while others take the whole life cycle into account. This variation partly represents different priorities on how environmental issues are best tackled.
   Partly, there is an intention that tools should complement each other, but it is somewhat surprising that no references were found that discuss how toolkits are best assembled in order to effectively reach environmental improvement. In contrast, toolkits seem to be put together with regard how different tools best fit with the work process (van Berkel et al. 1996).
3. The timing of the use of a tool during the product development process is important. It is generally recognized that the conceptual stage is the most influential one with regard to the product's environmental performance (although no empirical references have been found to support this). Consequently, many tool developers claim to address this particular stage. Some even try to lift certain environmental considerations of the later design stages to this stage (e.g. the recyclability map developed by Lee and Ishii (1997), and the Eco-Forecast tool, a Design for Disassembly for the end of conceptual design stage, developed by Luttrop (1997). And yet, designers feel that tools for the early design stages are lacking (Bhamra et al. 1999).
4. According to Ehrenfeld and Lenox (1997), it is DfE guidelines and checklists that are most commonly employed. Their explanation for this is that the diffusion of DfE is in the early stages. A number of firms have formalized DfE programs, but without formal sets of procedures, methodologies or routines. A more recent survey revealed that environmental aspects are mostly dealt with in the later design stages, although, most companies recognize the need for early integration of environmental aspects. Consequently, tools for the early design stages (pre-specification) were called for.
5. A few studies report on the effectiveness of tools, i.e. whether the tools have had a directly attributable influence to the environmental performance of the product. One document reports that in Danish companies, 30–50% environmental improvement of products have been implemented over a few years by companies working with the EDIP tools (Wenzel and Alting 1999).

References

Abhe S, Braunschweig A, Muller-Wenk R (1990) Methodik für Oekobilanzen auf der Basis okologischer optimierung. schriftenreihe Umwelt Nr. 133, Bundesamt fur Umwelt, Wald and Landschaft (BUWAL), Bern

Bakker C (1995) Environmental information for industrial designers, Technische Universiteit Delft

Bhamra TA, Evans S, McAloone TC, Simon M, Poole S, Sweatman A (1999) Integrating environmental decision making into the product development process, I. The earlier stages. In: Proceedings of EcoDesign'99 pp. 329–333 [Japan, Feb 1–3, IEEE]

Brezet JC, Horst Tvd, Riele HRM, Duijf GAP, Haffmans SP, Böttcher HE, Hoo SC, Zweers A, Verkooyen H (1994) PROMISE Handleiding voor Milieugerichte Produkt Ontwikkeling (PROMISE Manual for ecodesign), SdU Uitgeverij, Den Haag, The Netherlands

Den Haag Brezet H, Van Hemel C (1997) Eco-design: a promising approach to sustainable production and consumption. UN Environment Programme (UNEP)

Bringezu S, Hinterberger F, Schutz H (1994) Integrating sustainability into the system of national accounts: the case of interregional material flows. In: Proceedings of the International AFCET Symposium on Models of Sustainable Development: Exclusive or Complementary Approaches to Sustainability, Paris, France, March, pp. 669–680

Eagan P, Weinberg L (1997) Development of a streamlined, life-cycle, design for the environment (DFE) tool for manufacturing process modification: a Boeing Defense and Space Group case study. In: Proceedings of the 1997 IEEE international symposium on electronics and the environment, pp 188–891

Ehrenfeld JR, Lenox MJ (1997), The Development and Implementation of DfE Programmes. J Sustain Produc Des, 1(4):17–27

Ernzer M, Wimmer W (2002) From environmental assessment results to design for environment product changes: an evaluation of quantitative and qualitative methods. J Eng Des 13(3): 233–242

Goedkoop M (1995) The Eco-indicator 95. Final report, NOH report 9523. Pre Consultants, Amersfoort

Hanssen OJ (1998) Environmental impacts of product systems in a life cycle perspective: a survey of five product types based on life cycle assessment studies. J Clean Product 6:299–311

Hertwich EG, Pease SW, Koshland P (1997) Evaluating the environmental impact of products and production processes: a comparison of six methods. Sci Total Environ 196:13–19

Hinterberger F, Kranendonk S, Welfens MJ, Schimdt-Bleek F (1994) Increasing resource productivity through ecoefficient services. Wuppertal Papers Nr. 13. Wupprtal Institute, Duppersberg 19, 42103 Wuppertal, Germany

ISO 14040: 1997 Environmental management—Life cycle assessment—Principles and frame work

ISO 14041: 1998 (E). Environmental Management. Life Cycle Assessment—Goal and Scope Definition and Inventory Analysis

ISO 14042: 2000 (E). Environmental Management. Life Cycle Assessment—Life Cycle Impact Assessment

Lee BH, Ishii K (1997) Demanufacturing complexity metrics in design for recyclability. In: Proceedings of the IEEE international symposium on electronics and the environment pp 19–24

Lee J, Kim I, Kwon E, Hur T (2003) Comparison of simplified LCA and matrix methods in identifying the environmental aspects of products. In: Proceedings of EcoDesign 2003: 3rd international symposium on environmentally conscious design and inverse manufacturing pp 682–686

Lindfors L-G, Christiansen K, Hoffman L et al (1995) LCANORDIC, Technical Reports No. 1–9

Luttrop C (1997) Design for disassembly and ecoforecasting. A tool for interaction between management and design. In: Procedings of the 6th international conference on management of technology MOT'97. pp 348–349 [Göteborg, Sweden, June 25–28]

Narodoslawsky RB, Krotscheck C (1995) The sustainable process index (SPI): evaluating processes according to environmental compatibility. J Hazard Mater 41(2–3):383–397

Rydh CJ (1999) Environmental assessment of vanadium redox and lead–acid batteries for stationary energy storage. J Power Sources 80:21–29

Sage J (1993) Industrielle Abfallvermeidung und Deren Bewertung am Beispiel der Leiterplattenherstellung. Dbv-Verlag, Technische Universitet Graz

Sarkis J (2001) Greener manufacturing and operations: from design to delivery and back. Greenleaf Pub, Sheffield

Srinivasan M, Wu T, Sheng P (1995) Development of a scoring index for the evaluation of environmental factors in machining processes. Part 1 Health hazard score formulation. Trans NAMRI/SME 23:115–122

Institute Tellus (1992) The tellus packaging study. Tellus Institute, Boston

Tischner U, Schmincke E, Rubik F, Prosler M (2000) How to do EcoDesign?—A guide for environmentally and economically sound design. Verlag Form Germany

TUDelft (2004) Delft University of Technology, EcoQuest—website +software tool for electronic industries, http://www.io.tudelft.nl/research/dfs/ecoquest/welkom/Entry.html, [online, September, 2004]

van Berkel R, Kortman J, Lafleur M (1996) Issues in the development of improvement tools for environmental design of complex products. In: Proceedings of the IEEE international symposium on electronics and the environment p 230–235

Wenzel H, Alting L (1999) Danish experience with the EDIP tool for environmental design of industrial products. In: Proceedings of EcoDesign'99 (IEEE), pp 370–379 [Japan, Feb 1–3]

Wenzel H, Hauschild M (1997) Environmental assessment of products methodology Vol 1. Chapman & Hall, London

Wimmer W (1999) The ECODESIGN checklist method: a redesign tool for environmental product improvements. In: Proceedings of EcoDesign 1999: 1st international symposium on environmentally conscious design and inverse manufacturing, pp 685–688

Wong YL, Lee KM, Yung KC (2010) Model scenario for integrated environmental product assessment at the use of raw material stage of a product. Resour Conserv Recycl 54:841–850

Chapter 5
Case Study: The Domestic Cooker Hood "F77"

5.1 Structure of the Analysis

5.1.1 Functional Units

The main function of a cooker hood is to extract the fumes produced during cooking. The extraction system consists of a centrifugal blower composed of an impeller coupled to an electrical motor, both housed inside a scroll. The energy is provided to the exhausts to expel them. The system works in multiple conditions of air flow rate, velocity and pressure.

5.1.2 System Boundaries

The cradle-to-grave approach of an LCA involves the assessment of the environmental impact of each phase of manufacturing (from material extraction to product assembly); distribution (from production site to end user); working life; and End of Life (EoL) treatment (including recycling and disposal).

The cooker hood investigated in this work (Fig. 5.1) and its main components were manufactured and assembled in Italy. The product is prevalently sold in the German, French and Italian markets.

The analysis did not address the life cycle of the machinery used in manufacturing or distribution and excluded a number of processes whose share in the environmental impact was assumed to be negligible, i.e. manufacturing of the two lamps, the electronic board and the wires; and painting of the glass. The materials used in product assembly that contributed <1% to the weight of the cooker hood, such as screws, rivets, and packaging tape, were also excluded (Fig. 5.2).

M. Bevilacqua et al., *Design for Environment as a Tool for the Development of a Sustainable Supply Chain*, DOI: 10.1007/978-1-4471-2461-0_5,

Fig. 5.1 The cooker hood

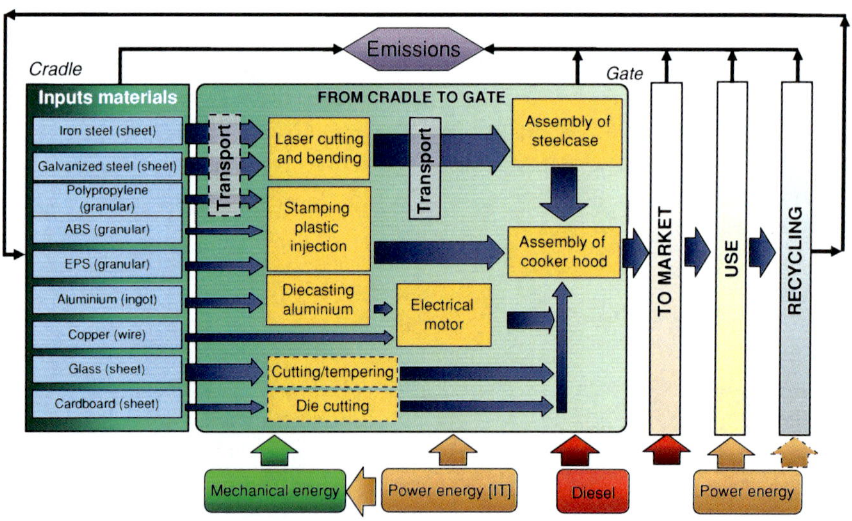

Fig. 5.2 System boundaries

5.2 Life Cycle Inventory

5.2.1 Production–Use–End of Life

The life cycle inventory was based on the production schedule and bills of materials of the cooker hood and on the study of the manufacturing process of each component. Data for the production of the steel components and the assembly phase were collected by the authors directly at the factory.

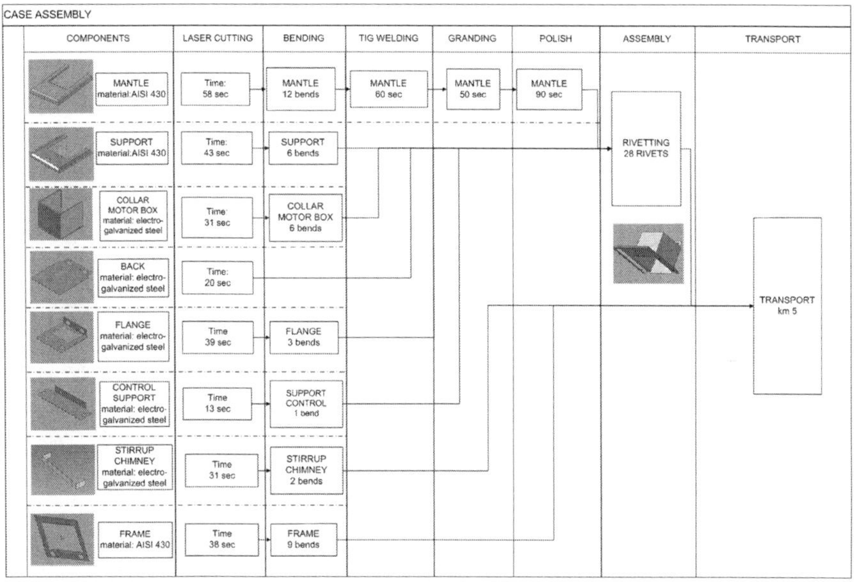

Fig. 5.3 Block diagram of the "case assembly" process

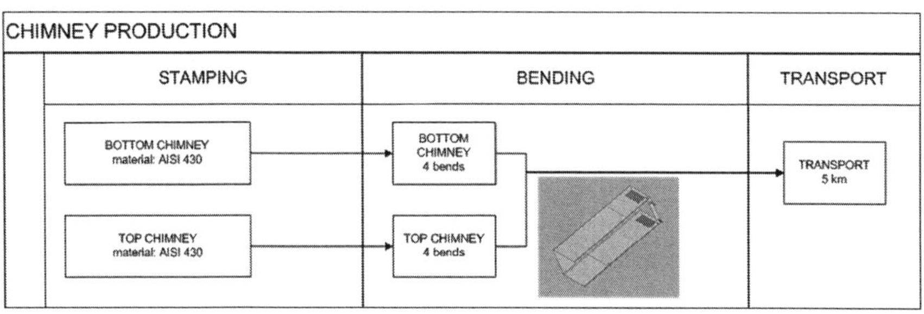

Fig 5.4 Block diagram of the "chimney production" process

The processes and components making up the assembled product are shown in the block diagrams in Figs. 5.3, 5.4, 5.5 and 5.6.

The main processes involved in case assembly are laser cutting, bending, TIG welding, grinding and polishing, assembly with rivets, and transport in medium-sized or small lorries, since all components are made by suppliers from the same industrial district (within 5 km, Figs. 5.3 and 5.4).

The plastic parts (polypropylene—PP, and acrylonitrile butadiene styrene—ABS) and the electrical motor are provided by suppliers based within a distance of 75 km (Fig. 5.5). The final assembly, packaging and distribution, and sale are performed by the cooker hood's manufacturer (Fig. 5.6).

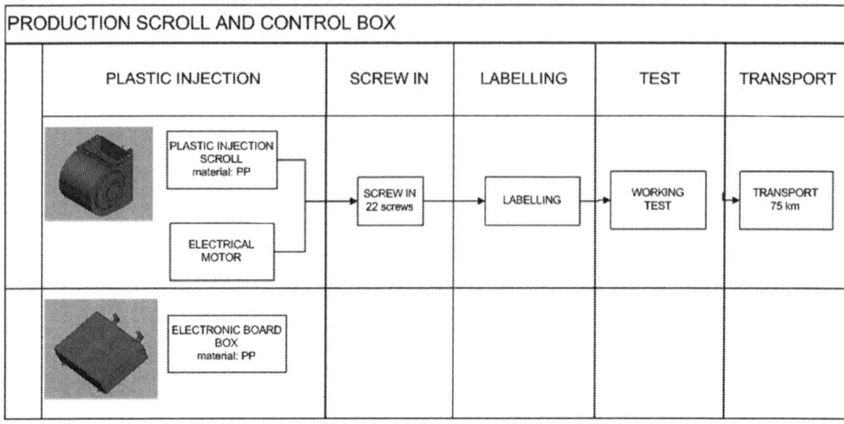

Fig. 5.5 Block diagram of the "scroll and control box" production process

5.3 Data: Sources and Assumptions

Data regarding the working of metal components and assembly operations were collected at the factory. The reference sources for all the other processes entered in the LCA are listed in Table 5.1.

The environmental impact of transport in the manufacturing phase was calculated based on the average distance of suppliers' production sites.

The cooker hood was assumed to be sold in north and central Europe (impact of distribution).

Use was assessed by comparing the cooker hood's operation in Italy and France, to account for the environmental effects of local power grid mixes. "Use" was assumed to be 2 h daily throughout the year over a lifespan of 10 years.

EoL analysis assumed that the product would be disposed of in Europe (landfill disposal/recycling) (Table 5.1; Fig. 5.7).

5.4 Life Cycle Impact Assessment (LCIA)

The inventory analysis provided the input data for LCIA—which is regulated by ISO 14040. The LCIA was conducted according to the EI99 methodology.

The EI99 methodology assesses life cycle impacts based on three categories of damage: ecosystem quality, human health, and resources. Since each category uses a specific measurement unit, the units must be normalized by weighting the various factors. The rigour of the scientific approach and subjectivity, connected to the evaluation of impacts, is balanced using Thompson's "Cultural theory" (Thompson 1997), which identifies three different perspectives of environmental impacts, individual, hierarchical, and egalitarian, based on the choice of the

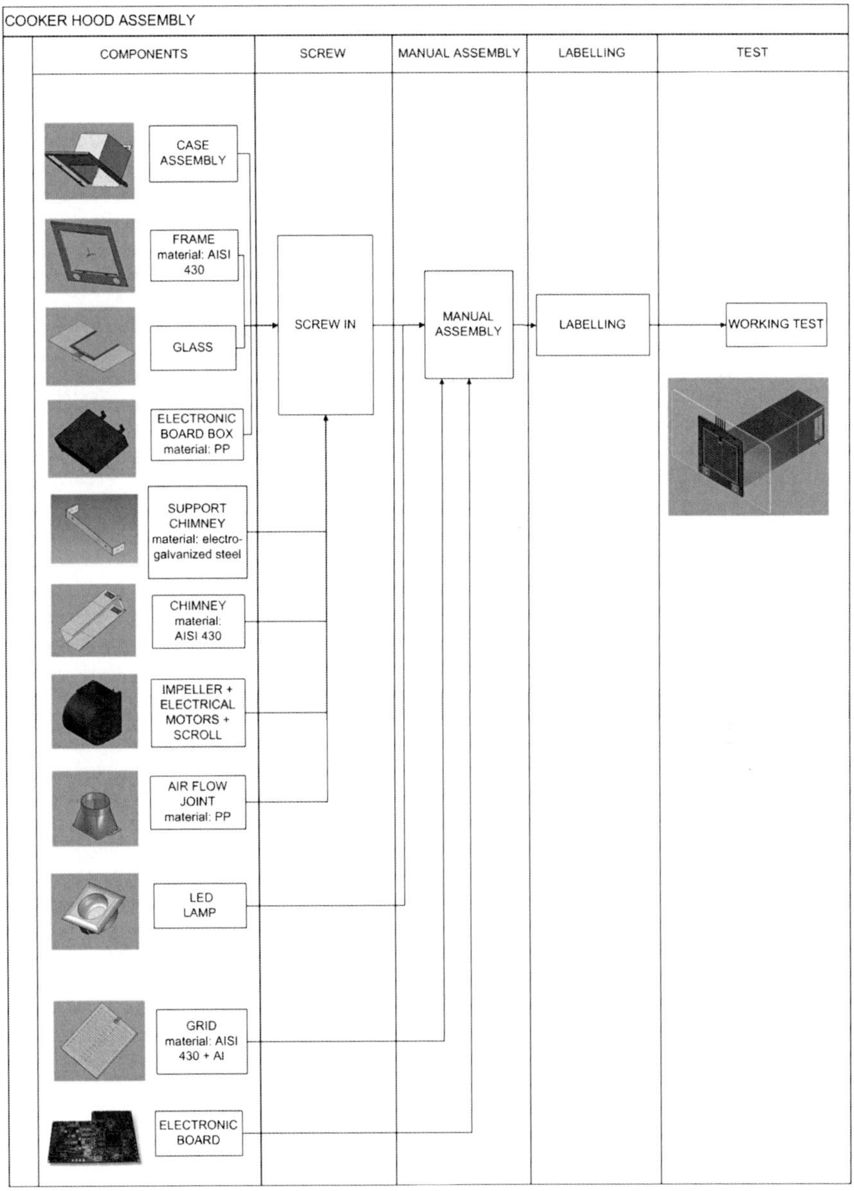

Fig. 5.6 Block diagram of the "cooker hood assembly" process

weighting factors assigned to each impact. The perspectives differ in time view and manageability. The individualist perspective is based on the short period and assumes that technology can avoid many problems; hierarchical views are based on a balance between short- and long-period perspective and of the thought that

Table 5.1 Data sources for the various life cycle phases

Life cycle stage		Data source	Reference area	Notes
Manufacturing	Extraction of resources and processing of inputs			
	Steel	PE international	Germany	
	ABS—PP	PE international	Germany	
	EPS	Buwal	Europe	
	Cardboard	Buwal	Switzerland	
	Glass		Europe	
	Aluminum	PE international	Europe	
	Assembly	Measured		
	Component transport	Measured		
	Power electricity mixer	PE international	Italy	
Component transport	Diesel production	ELCD/PE GaBi		
	Transport data	PE international		
	Distance of exports	Measured		
Distribution	Diesel production	ELCD/PE GaBi		
	Transport data	PE international	Global	
	Distance of exports	Statistics		
Use	Electricity consumption	Statistics		
	Electricity mixer	PE international	France	
EoL (End of Life) treatment	Landfill	PE international	Europe	
	Incinerator	PE international	Europe	
	Recycling	PE international		Included in the input materials processes

problems can be solved by appropriate policy. This study adopted the Egalitarian Approach (EI99-EA), which is based on the precautionary principle with a long period perspective. In this approach, the use and management of energy systems and energy policies are heavily affected by the exploitation of raw resources, which will be strongly impoverished in the long term.

5.5 Results and Discussion

According to the LCA, "use" of the cooker hood has the strongest environmental impact, followed by "manufacturing" (Figs. 5.8, 5.9). The national power grid mixes exert a strong influence through the environmental impact related to "use", both in terms of quantity (damage %) and quality (type of damage) (Figs. 5.10, 5.11).

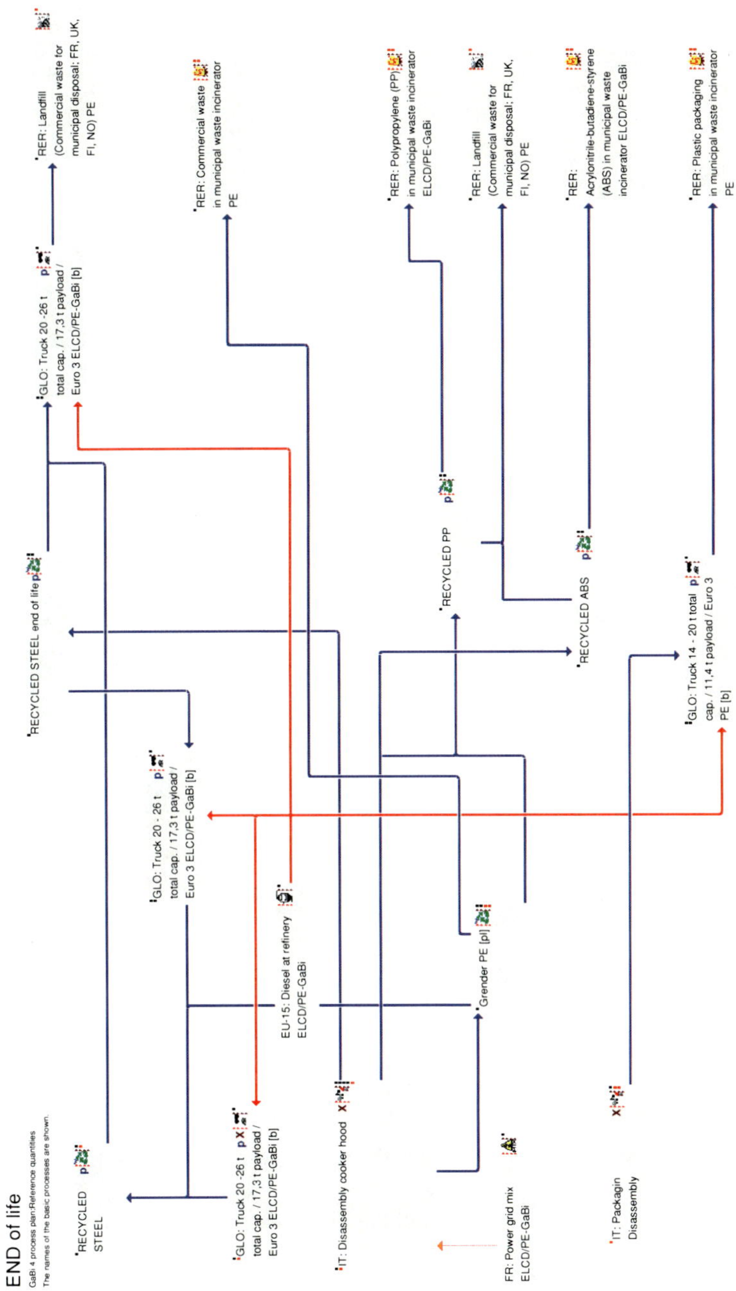

Fig. 5.7 End of life GaBi model

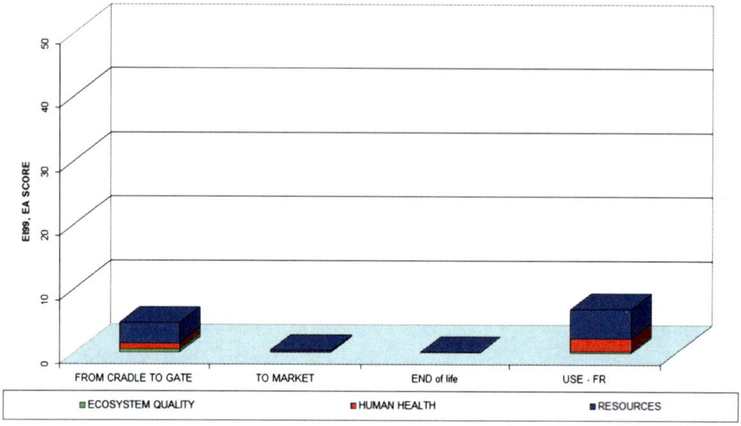

Fig. 5.8 LCA of a cooker hood sold in France

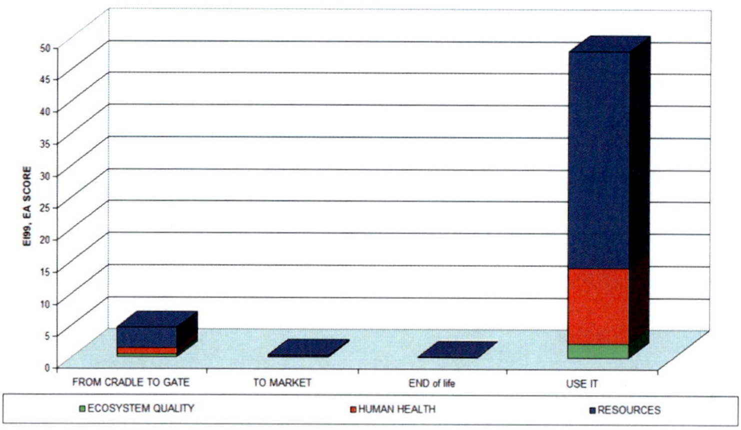

Fig. 5.9 LCA of cooker hood sold in Italy

5.6 Manufacturing

The manufacturing process as analyzed by cradle-to-gate LCA is modeled in Fig. 5.12.

Its main environmental impacts are due to the use of non-renewable resources, especially steel.

Figure 5.13 shows EI99-EA-calculated environmental impacts for each process shown in Fig. 5.12 and for the packaging process. Case assembly had the highest impact. For this reason this process was further investigated.

Figure 5.14 shows the impacts of each subprocess of "case assembly" on each eco-indicator. In particular, the use of steel is the main reason for the high EI99-EA

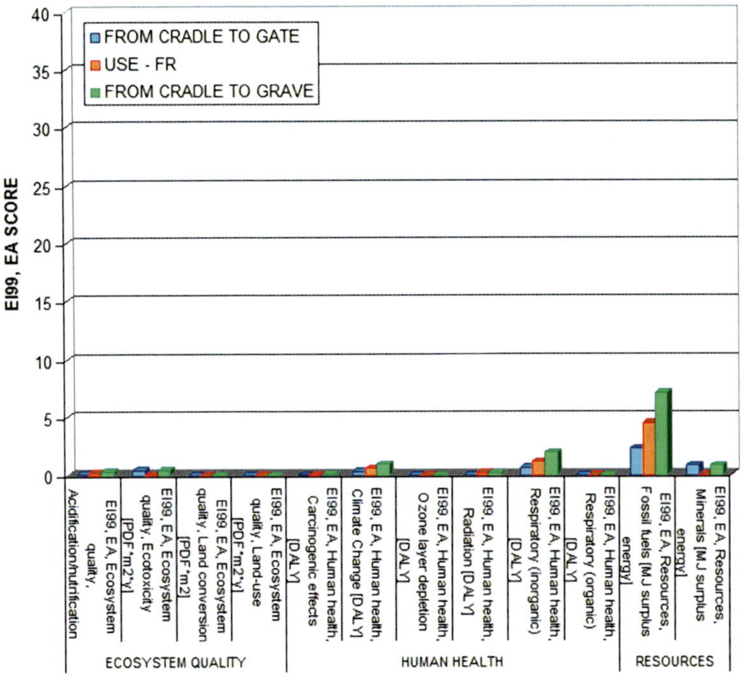

Fig. 5.10 EI99-EA of a cooker hood sold in France

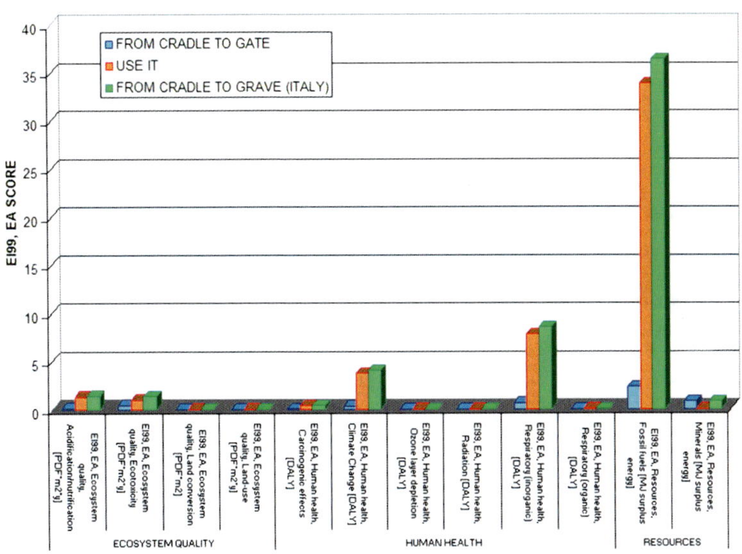

Fig. 5.11 EI99-EA of a cooker hood sold in Italy

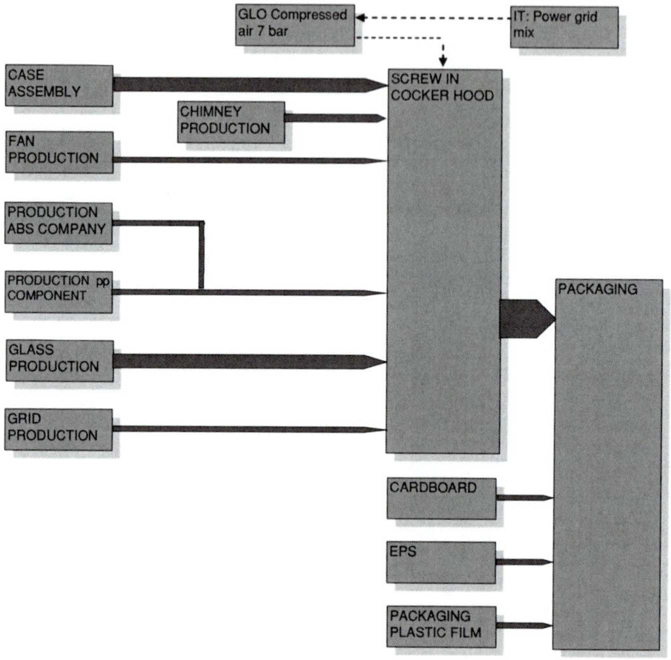

Fig. 5.12 Cradle to gate GaBi model

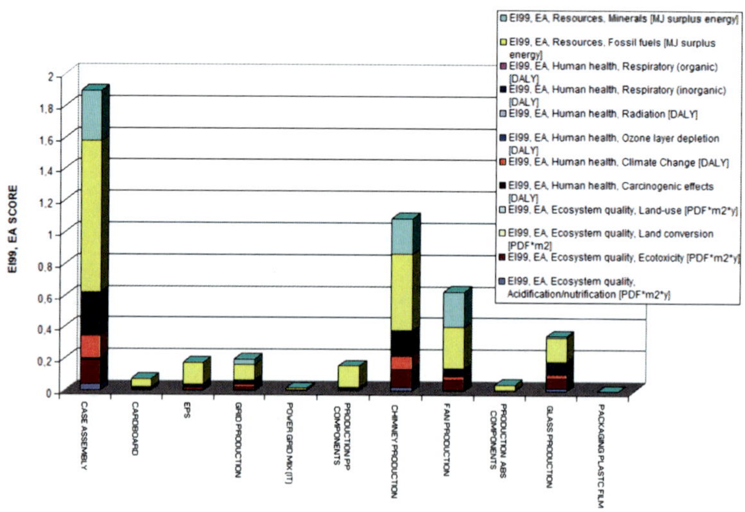

Fig. 5.13 EI99 EA, cradle to gate

score (Norgate et al. 2007). The subprocesses "Metal work—inox" and "Metal work—galvanized steel" strongly affect "Resources minerals" and "Resources fossil fuels".

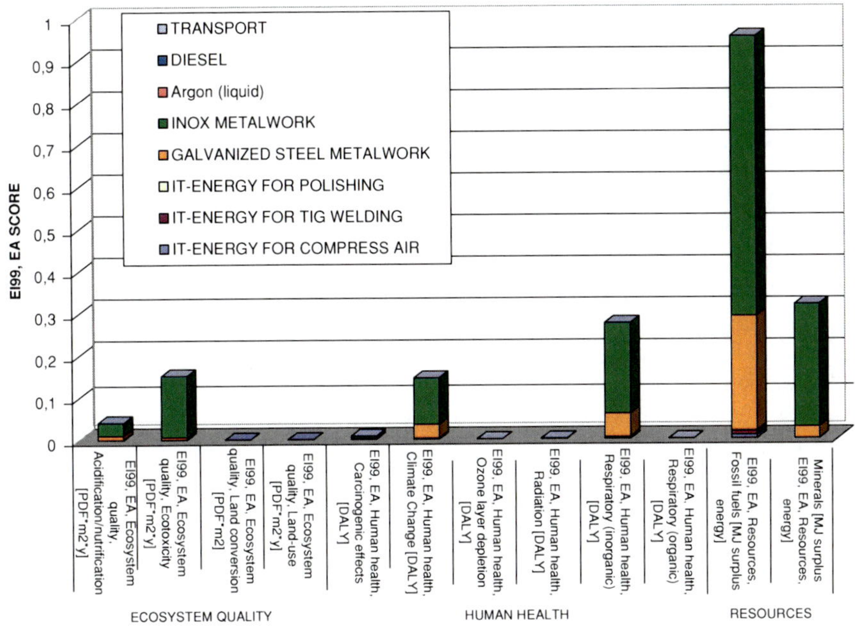

Fig. 5.14 EI99-EA of the "case assembly" process

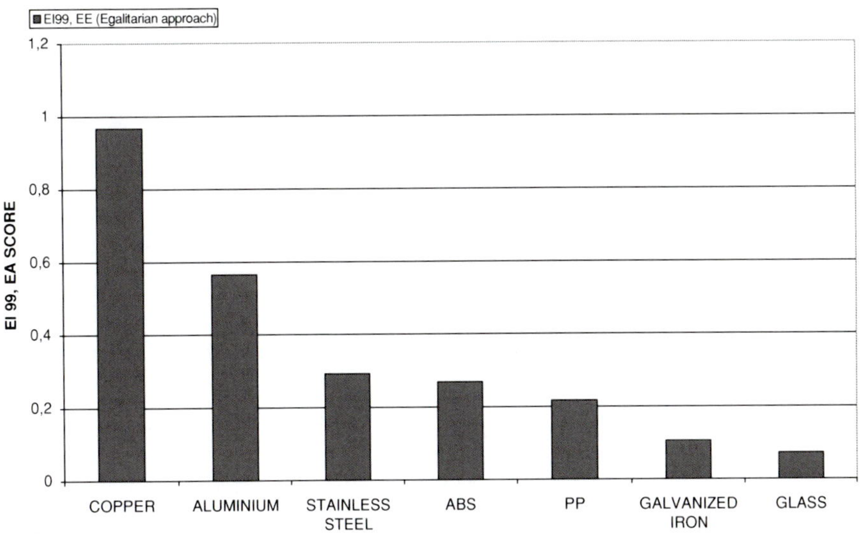

Fig. 5.15 EI99-EA: comparison of the materials used

Table 5.2 EI99-EA "use" scores in France and Italy

EI99-EA damage category	EI99-EA indicator	USE—FR	USE—IT	Difference
Total score		6.818200893	47.83678374	−41.01858284
Ecosystem quality	Ecosystem quality, acidification/eutrophication [PDF*m2*y]	0.211618653	1.265207867	−1.053589214
	Ecosystem quality, ecotoxicity [PDF*m2*y]	0.03880272	0.944501899	−0.905699179
	Ecosystem quality, Land conversion [PDF*m2*y]	0	0	0
	Ecosystem quality, land-use [PDF*m2*y]	0	0	0
Human health	Human health, carcinogenic effects [DALY]	0.069616599	0.33684286	−0.26722626
	Human health, climate change [DALY]	0.580784349	3.671672875	−3.090888526
	Human health, ozone layer depletion [DALY]	0.008001368	1.35628E−05	0.007987805
	Human health, radiation [DALY]	0.197824568	0.000352227	0.197472341
	Human health, respiratory (inorganic) [DALY]	1.195503316	7.775638894	−6.580135578
	Human health, respiratory (organic) [DALY]	0.000540384	0.004103448	−0.003563063
Resources	Resources, fossil fuels [MJ surplus energy]	4.514997192	33.83591289	−29.3209157
	Resources, minerals [MJ surplus energy]	0.000511744	0.002537215	−0.00202547

PDF*m2*y: potentially disappeared fraction of plant species per sq metre per year; DALY: disability-adjusted life years

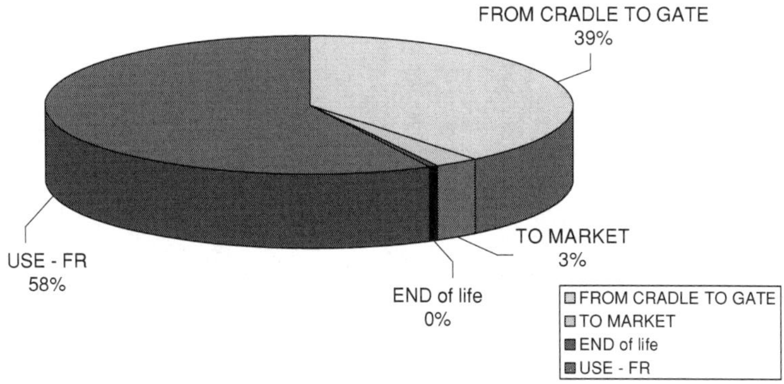

Fig. 5.16 EI99-EA—France

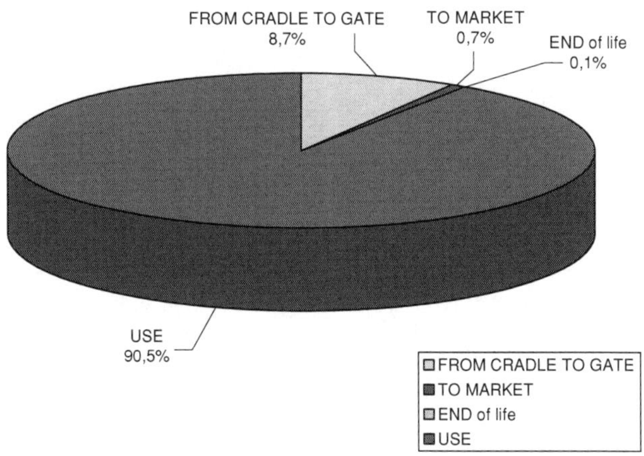

Fig. 5.17 EI99-EA—Italy

The environmental impact of "manufacturing" is affected more by the materials used than by their working processes. This provides an interesting comparison of the various materials. The different contributions to the EI99 score based on the materials used are compared in Fig. 5.15.

Product distribution is by lorry, from the factory in central Italy to northern Europe. Its environmental impact is lower compared to the other phases (Figs. 5.8, 5.9). Means of transport and cargo efficiency strongly affect the share of impact of each hood. We considered a lorry with a 22 ton payload and a cargo efficiency of 20%; the latter value was calculated by entering the weight of a packaged cooker hood and the volume occupied by each pallet.

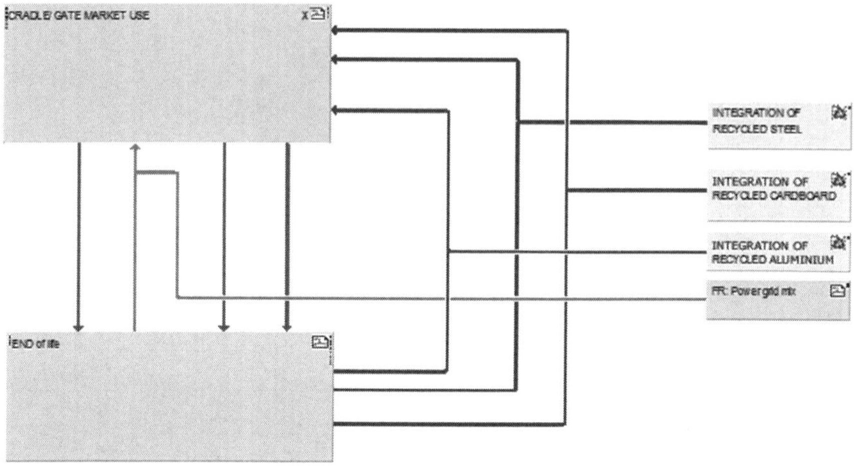

Fig. 5.18 Life cycle GaBi model

It is worth noting that packaging optimization is crucial for payload efficiency, also by reducing the environmental impact and cost of distribution. "Use" of the cooker hood is the phase that most affects the LCA, hence the EI99 score.

Total energy consumption over a 10-year lifespan was estimated to be about 4406 MJ.

The high EI99-EA score is related to the "Resources" category, particularly "Resources, Fossil Fuels" [MJ surplus energy], i.e. the amount of energy needed to extract and process the fossil (non-renewable) fuels used for electricity production (Table 5.2).

The environmental impact of this phase strongly depends on the country of use (Varun et al. 2009). Since the cooker hood is a final user of electric energy, the local power grid mix is the most influential parameter of both EI99-EA quality and quantity score.

Figures 5.16 and 5.17 show the different influence of "use" in France and Italy. The manufacturing process has the same absolute EI99-EA score (Figs. 5.10, 5.11), but its share of the environmental impact is greater in France (39%) than in Italy (8.7%) because of the greater weight of "use" in Italy (90.5% versus 58% in France). The same product, manufactured in Italy, thus has different EI99-EA scores in the two countries, leading to different environmental impacts of the life cycle phases.

As shown in Table 5.2, the EI99-EA "use" score is 6.82 in France and 47.84 in Italy. The indicator subscores demonstrate that the most influential EI99 item is "Resources, Fossil fuel [MJ surplus energy]", which is 4.51 in France and 33.84 in Italy. The difference (29.32 eco-indicator 99 points; EIP) is due to the power grid mix, which in Italy is based on non-renewable fuels. The nuclear power plants in the French power grid mix affects the indicator "Human health, Radiation [disability-adjusted life years DALY]".

Table 5.3 EI99-EA score improvement related to LED lamps in France

EI99-EA damage category	EI-99 EA indicator	From cradle to grave—(fr) halogen	From cradle to grave—(FR) LED	Improvement %
	Total score	13.04098742	11.83683683	9.23
Ecosystem quality	Ecosystem quality, acidification/ eutrophication [PDF*m2*y]	0.37461863	0.337245028	9.98
	Ecosystem quality, ecotoxicity [PDF*m2*y]	0.46097142	0.454118539	1.49
	Ecosystem quality, land conversion [PDF*m2y]	8.84211E−05	8.84211E−05	0.00
	Ecosystem quality, land-use [PDF*m2*y]	0.000508668	0.000508668	0.00
Human health	Human health, carcinogenic effects [DALY]	0.106626243	0.094331377	11.53
	Human health, climate change [DALY]	1.032968092	0.930396779	9.93
	Human health, ozone layer depletion [DALY]	0.009515618	0.00810251	14.85
	Human health, radiation [DALY]	0.234950513	0.200013061	14.87
	Human health, respiratory (inorganic) [DALY]	2.149326641	1.938190882	9.82
	Human health, respiratory (organic) [DALY]	0.001039174	0.000943738	9.18
Resources	Resources, fossil fuels [MJ surplus energy]	7.859709558	7.062323766	10.15
	Resources, minerals [MJ surplus energy]	0.810664437	0.810574058	0.01

PDF*m2*y: potentially disappeared fraction of plant species per sq metre per year; DALY: disability-adjusted life years

5.7 End of Life

The EoL phase, like "Market", contributes little to the environmental impact of the hood (Figs. 5.8, 5.9). The EoL phase was analyzed by hypothesizing product disposal in line with the reports of the consortiums that retrieve and recycle household appliances (Fig. 5.18) (Liu et al. 2009; EU Directive 2002/96/EC). The steel and glass in the case and chimney reach nearly 100% recycling efficiency, since they have been designed for disassembly. The ventilation system and the electronics are usually crushed and the materials separated later with a recycling efficiency that was estimated on the basis of literature data.

Table 5.4 EI99-EA score improvement related to LED lamps in Italy

EI99-EA damage category	EI-99 EA indicator	From cradle to grave—(it) halogen	From cradle to grave—(it) led	Improvement%
	Total score	61.30379075	52.85541967	14
Ecosystem quality	Ecosystem quality, acidification/ eutrophication [PDF*m2*y]	1.614280401	1.390834242	14
	Ecosystem quality, ecotoxicity [PDF*m2*y]	1.526624554	1.359817719	11
	Ecosystem quality, land conversion [PDF*m2*y]	8.84211E−05	8.84211E−05	0
	Ecosystem quality, land-use [PDF*m2*y]	0.000508668	0.000508668	0
Human health	Human health, carcinogenic effects [DALY]	0.421046868	0.361557637	14
	Human health, climate change [DALY]	4.669733058	4.021285305	14
	Human health, ozone layer depletion [DALY]	0.0001171	0.000114705	2
	Human health, radiation [DALY]	0.002602926	0.00254072	2
	Human health, respiratory (inorganic) [DALY]	9.891568505	8.51832646	14
	Human health, respiratory (organic) [DALY]	0.005231504	0.004506801	14
Resources	Resources, fossil fuels [MJ surplus energy]	42.35894112	36.38323946	14
	Resources, minerals [MJ surplus energy]	0.813047622	0.812599529	0

PDF*m2*y: potentially disappeared fraction of plant species per sq metre per year; DALY: disability-adjusted life years

Next steps of this work are focused on a sensitivity analysis in order to define design choices that allowed the company to decrease the environmental impact of the cooker hood F77. The analysis focused on

- Impact improvement with LED Light
- The use of electro-galvanized stainless steel
- Modification of distribution modality

5.8 Impact Improvement with LED Light

The cooker hood system is fitted with two halogen lamps (2 × 20 W). Since the LCA showed that its environmental impact is most affected by the electricity consumed during operation, one of the main eco-design changes proposed was their

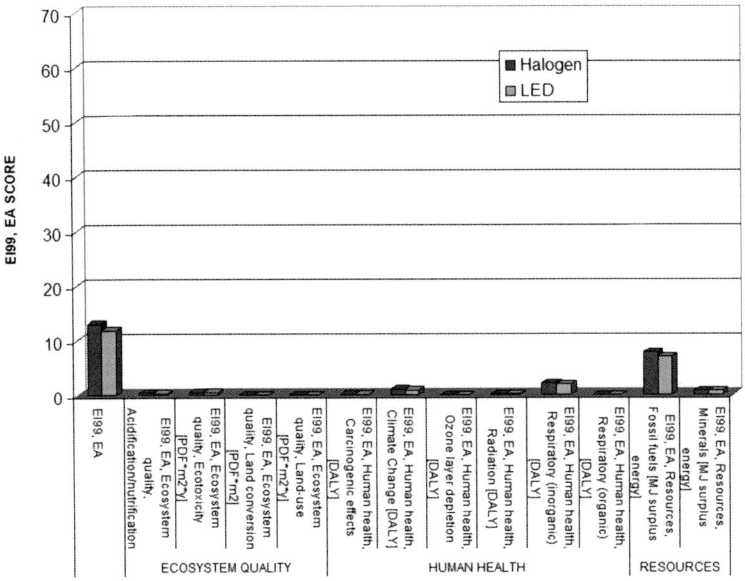

Fig. 5.19 Comparison between halogen and LED lamps in France

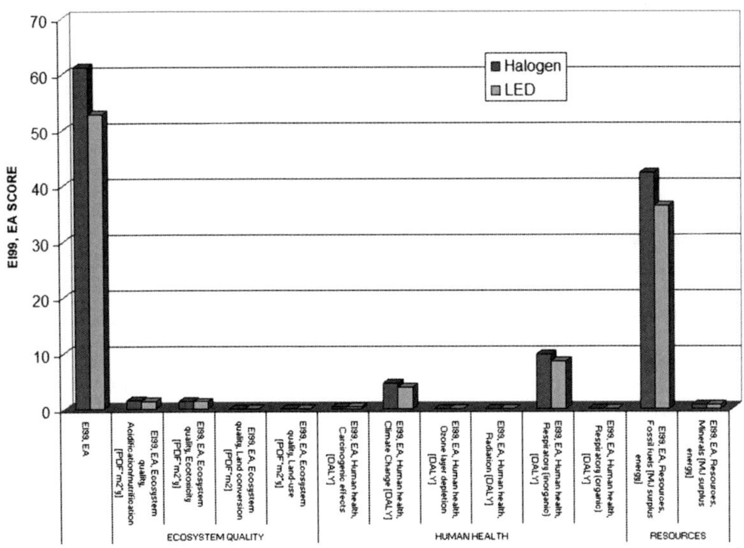

Fig. 5.20 Comparison between halogen and LED lamps in Italy

replacement with two LED lamps (2 × 5 W). This simple change can provide 75–80% reductions in energy consumption for lighting based on the LED technology used. Tables 5.3 and 5.4 show the improvement achieved for each EI99-EA

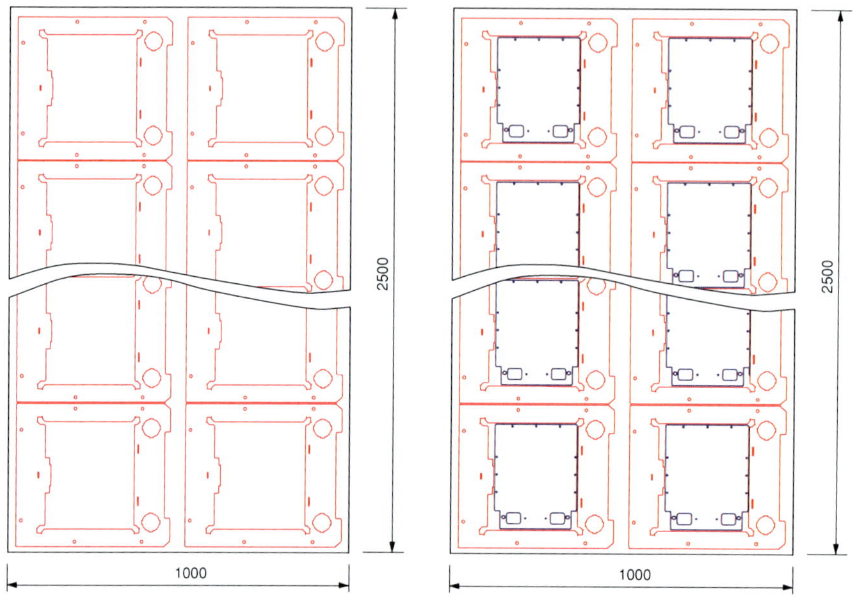

Fig. 5.21 Cutting foil old (**a**) and new (**b**)

in France and Italy. Figures 5.19 and 5.20 further demonstrate how the national power grid mixes influence eco-indicator quality and quantity, given that the LED lamps are associated with an improvement of about 9% in France and of 14% in Italy.

5.9 Use of Electro-Galvanized Stainless Steel

The company has a material waste of 41.4%. The redesign of cutting plan including the stainless steel components in the same foil of electro-galvanized one could allow the company to save a lot of material. The waste decreased to 27.5% and the need of steel decreased from 2.75 to 1.61 kg. Figure 5.21 shows the new and old cutting foils.

The new impact indicators are shown in Table 5.5.

5.10 Modification of Distribution Modality

By designing a custom-made package developed for cooker hood F77, it is possible to reduce the overall dimensions of the package (Fig. 5.22).

Table 5.5 New impact indicators

	New assembly solution	Old assembly solution	Improvement %
EI99, EE (Egalitarian approach)—OUTPUT	**0.587685759**	**0.613066288**	**4**
EI99, EA, Ecosystem quality, acidification/nutrification [PDF*m^2*a]	0.038537878	0.040821486	6
EI99, EA, Ecosystem quality, ecotoxicity [PDF*m^2*a]	0.150566782	0.151581983	1
EI99, EA, Ecosystem quality, land-use [PDF*m^2*a]	1.56E−05	1.56E−05	0
EI99, EA, Human health, carcinogenic effects [DALY]	0.007125152	0.007446561	4
EI99, EA, Human health, climate change [DALY]	0.128770251	0.138151148	7
EI99, EA, Human health, ozone layer depletion [DALY]	3.77E−05	3.91E−05	4
EI99, EA, Human health, radiation [DALY]	0.000925253	0.000959797	4
EI99, EA, Human health, respiratory (inorganic) [DALY]	0.26161755	0.273956765	5
EI99, EA, Human health, respiratory (organic) [DALY]	8.96E−05	9.39E−05	5
EI99, EE (Egalitarian approach)—INPUT	**1.19492784**	**1.269649973**	**6**
EI99, EA, Human health, climate change [DALY]	0.008929401	0.009140393	2
EI99, EA, Resources, fossil fuels [MJ surplus energy]	0.885281122	0.952782124	7
EI99, EA, Resources, minerals [MJ surplus energy]	0.300717316	0.307727456	2

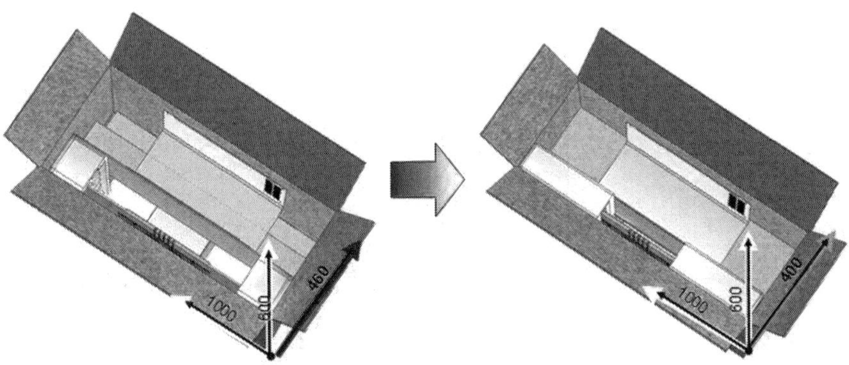

Fig. 5.22 Old and new package

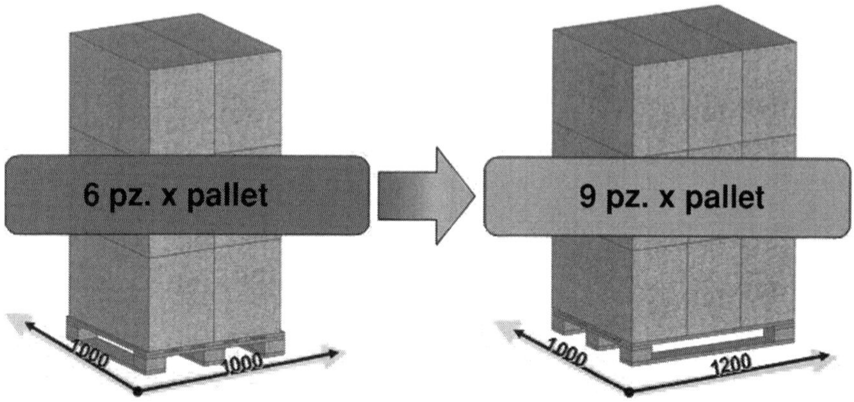

Fig. 5.23 Old and new pallet

Table 5.6 New impact indicators after modifying distribution modality

	To market new	To market	Improvement %
EI99, EE (egalitarian approach)—OUTPUT	**0.004943934**	**0.007128578**	**31**
EI99, EA, ecosystem quality, acidification/nutrification [PDF*m2*a]	0.000850382	0.001223145	30
EI99, EA, ecosystem quality, ecotoxicity [PDF*m2*a]	1.46E−05	2.09E−05	30
EI99, EA, human health, carcinogenic effects [DALY]	5.05E−05	7.24E−05	30
EI99, EA, human health, climate change [DALY]	0.000786967	0.001128511	30
EI99, EA, human health, ozone layer depletion [DALY]	7.52E−09	1.08E−08	30
EI99, EA, human health, radiation [DALY]	1.87E−07	2.69E−07	30
EI99, EA, human health, respiratory (inorganic) [DALY]	0.003239353	0.004680433	31
EI99, EA, human health, respiratory (organic) [DALY]	1.92E−06	2.84E−06	32
EI99, EE (egalitarian approach)—INPUT	**0.007508718**	**0.010766166**	**30**
EI99, EA, human health, climate change [DALY]	1.10E−07	1.58E−07	30
EI99, EA, resources, fossil fuels [MJ surplus energy]	0.007508325	0.010765602	30
EI99, EA, resources, minerals [MJ surplus energy]	2.83E−07	4.06E−07	30

The new solution allows to move 50% more products (Fig. 5.23) obtaining a decrease of environmental impacts of "TO MARKET" phase.

Using this solution the efficiency of transport phase increases from 20 to 30% obtaining an impact reduction of 31% (Table 5.6).

References

Liu Z, Pagani M, Zinniker D, DeConto R, Huber M, Brinkhuis H, Shah SR, Leckie RM, Pearson A (2009) Global cooling during the Eocene-Oligocene climate transition. Science. 323(5918):1187-1190

Norgate TE, Jahanshahi S, Rankin WJ (2007) Assessing the environmental impact of metal production processes. J Clean Prod 15:838–848

Thompson M (1997) Cultural Theory and integrated assessment. Environ Model Assess 2:139–150

Varun KB, Prakash R (2009) LCA of renewable energy for electricity generation systems—A review. Renew Sustain Energy Rev 13:933–1150

Chapter 6
Designers' Utilization of DfE and Requirements

Ritzen and Lindahl (2001) as well as Ernzer and Birkhofer (2002) state that despite the number of available DfE methods and tools, there are relatively few which are widely used by companies. One reason for this low utilization is that they are time-consuming; another is that many of them focus only on environmental issues (Ehrenfeld and Lenox 1997). For the enterprises and their engineering designers of course, the environmental issue is but one of many issues to be considered. When DfE methods and tools are used, these methods and tools are often not integrated in the product development process.

Even though more and more approaches focus on how to perform DIE, as well as on what is required for its successful integration, there seems to be a gap between the developers and the presumptive users. Bauman et al. (2002) have examined articles concerning methods and tools in the area of DfE. They found that there were relatively few references describing the diffusion of DfE methods and tools, the experience of how these tools and methods worked in product development, and how useful these methods were in actually reducing the environmental impacts of products. Bauman et al. (2002) conclude that most publications with an empirical content report on the testing of new DfE methods and tools, and that these are often developed at universities and tested by researchers in company case studies. This is supported by Tukker et al. (2000). They report that many of the DfE methods and tools are developed by researchers within universities or research institutes. In some cases, there is little or no testing of these methods and tools in industrial practice.

If the ambition is to integrate DfE into ordinary product development then there is also an essential need to involve and consider one of the main presumptive users of DfE methods and tools: the designer. In fact, it is the designer who is often the main practical executer of methods and tools used to develop a product. Even if designers do not always decide what method or tool to use, their use influences the outcome and the benefits from the use. When using a method or tool, it is

M. Bevilacqua et al., *Design for Environment as a Tool for the Development of a Sustainable Supply Chain*, DOI: 10.1007/978-1-4471-2461-0_6,

important to understand its various advantages and disadvantages. It is also important to know under what circumstances the method or tool's result is valid.

Lindahl (2005) carried out several interviews and questionnaires between designers from Swedish companies. The studies reveal that designers view themselves as users of methods and tools in quite low numbers, which the analysis of the results from the questionnaire emphasizes. The most utilized tools are various kinds of Computer-Aided Design (CAD) tools; if all CAD tools are considered as one tool, then there exist few additional methods and tools utilized independently. Among DfE-related methods and tools, the only ones mentioned during the studies have been EEA and LCA.

For DfE methods and tools, where are often problems in finding adequate data, the issue of reliable and relevant outcomes becomes even more difficult since much of the used data are instead based on validations, assumptions and limitations. "Freedom of action" appears, at least according to this research, to be important for the designers.[1]

Other important problems suggested by designers concern:

- Setup time. This is related to time efficiency is the conclusion based on the studies that a method or tool must not have an excessive setup time. For designer comprehension, complicated methods and tools must be daily or at least regularly utilized; otherwise, designers tend to forget how to utilize the method or tool's specific functions and the setup time increases. At the same time, this implies that the number of methods and tools a designer can utilize is limited.
- Not require excessive simultaneous collaboration. Collaboration takes time and coordination, influences the setup time and diminishes designers' freedom of action.
- Computer-based. When looking at which methods and tools are used more regularly, the conclusion based on the research and other existing research studies such as e.g. Norell (1992) is that the method or tool should be computer-based.
- Easy to adopt and implement.

6.1 Integration of Environmental Management System and Design for Environmental

Environmental management system (EMS) refers to the management of an organization's environmental programs in a comprehensive, systematic, planned, and documented manner. It includes the organizational structure, planning, and

[1] The paradox is that when the general design information is needed, it is not accessible, and when it is accessible, the information is usually not needed.

resources for developing, implementing, and maintaining policy for environmental protection.

The aim of the EMS is to identify and monitor the principal environmental aspects affecting our operations and coordinate all activities that have an impact on the environment. EMSs include the ISO (International Organization for Standardization) 14001 System, a widely adopted voluntary program to improve the quality of environmental management (see Box D). In addition to ISO 14001, a community eco-management and audit scheme ("EMAS") was introduced (see Box D). Compliance with EMAS is voluntary, and EMAS aims to encourage the improvement of environmental management and performance.

The integration of Environmental Management System and Design for Environmental is a topic developed by several authors. Today, more than 130,000 companies are using ISO 14000 and more than 5500 are using EMAS environmental management systems (EMS) in the world and this number is expected to continue to grow steadily. Therefore, it is interesting to study if and how standardized EMS's affect companies' environmental impact. It is the authors' experience that many companies, authorities and individuals regard a certification in accordance with ISO 14001 as a guarantee for good environmental performance. However, it appears to be too early to draw any general conclusions on the connection between standardized EMS and environmental performance because there are research findings pointing in both positive and negative directions (Ammenberg 2003).

Since environmental impacts are intimately connected to flows of materials and energy, and the most important flows, at least for manufacturing companies, are closely linked to products (Berkhout 1998) it seems urgent for management systems to encompass products and product development. Consequently, it is of great interest to illuminate how standardized EMS are related to Design for the Environment (DfE), i.e. to what extent they encompass the products and product development procedures.

There are many motives for integrating the two concepts. Firstly, DfE thinking might enrich EMS by contributing with a life cycle perspective. If EMS encompassed products' life cycles to a greater extent, they would be a better complement to the often facility-oriented legal requirements and authority control. Secondly, EMS might remove the pilot project character of DfE activities and lead to continuous improvement. Thirdly, integration could lead to successful co-operation, both internally and externally. Historically, DFE and EMS have, to a large extent, existed in separate spheres (Karlsson 2001). Reading through ISO 14001 it is clear that product development is not emphasized and that most product-related requirements leave substantial room for interpretation (Ammenberg and Sundin 2005).

Ries et al. (1999) state that, in spite of the inclusion of 'activities, products and services' in vital parts of the ISO 14001 standard, many companies have a very narrow perception of their environmental impacts, which is mostly limited to site-specific activities. All these facts show that there is an obvious risk that EMS are not directed at the most important environmental aspects. Environmental managers who Ries et al. (1999) have interviewed stated that companies generally have very limited knowledge regarding the environmental aspects of their products. Further studies by Ammenberg and Hjelm (2002) show that there are certified EMS without a strong link to products. From an environmental standpoint, it is important to regard the whole life cycle of a product. In this respect, 'DfE thinking' or 'life-cycle thinking' could function as an important complement to EMS by contributing a better understanding of which flows of materials and energy are most important, which would reduce the risk of sub-optimizations.

Every firm has its individual needs, capabilities, etc. Consequently, it is not possible to develop an integration model between DfE and EMS that expressly fits every company. On the contrary, a model for a wide application must address the integration of EMS and DfE on a general level, allowing flexibility.

Many of the integration models presented in the literature are based on the Deming cycle, i.e. the same management cycle as is the basis for ISO 14001 and ISO 9001 ('Plan-Do-Check-Act cycle; PDCA cycle'). The basic idea behind this approach is to facilitate the establishment of system that are compatible with many companies' existing or future management systems or practices, which seems wise. For example, Brezet et al. (2000) conclude that companies having a quality management system have major advantages in taking a structured approach to product-oriented environmental management.

In order to carry out a Product-Oriented Environmental Management Systems (POEMS), Ammenberg and Sundin (2005) proposed a general model shown in Fig. 6.1.

The described process is mainly focused on the first implementation of a POEMS, which could be carried out by companies with or without an existing EMS or other management systems.

Firstly, a review of the product portfolio from a life cycle perspective is carried out. This is supposed to contribute with knowledge about the products' environmental impact, which means that product-related environmental aspects are illuminated and it is determined which of these aspects are significant. Since this review to large extent should involve the products' life cycles it encompasses important actors in the supply chain, contradictory to many reviews within normal EMS (Ammenberg and Sundin 2005). Parallel to this process, a review of organizational aspects of DfE relevance should be conducted. This review brings knowledge together about the capabilities and weaknesses of the organization and includes an investigation and clarification of the product development process. It is advantageous if these two review phases yield information on the existing and future market (e.g. information on future customer needs).

Secondly, responsibilities and resources should be allocated and environmentally related procedures written for the product development process. It is recommended that environmental concerns should be incorporated from the beginning of the design process and handled as any other design parameter (Karlsson 2001). It is important to have support from top management and to ensure that corporate visions, strategies and policies are in line with the intentions of the environmentally adapted product development process. Based on the initial

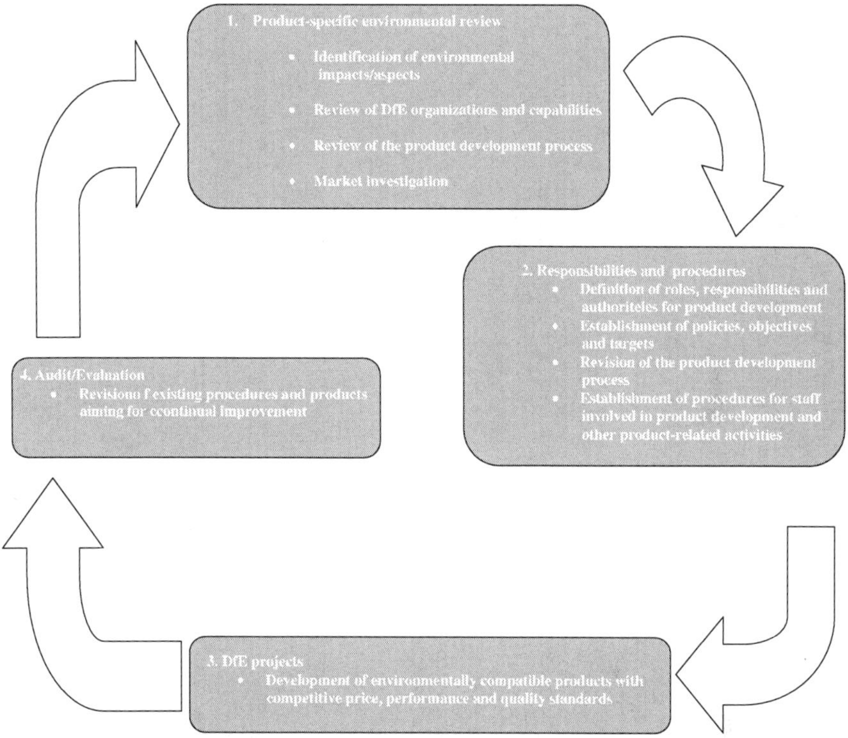

Fig. 6.1 POEMS model to integrate DfE and EMS. *(source:* Ammenberg and Sundin 2005*)*

reviews and relevant policies, product-related environmental objectives, and targets should be established.

Thirdly, DfE projects should then be performed at the operational level. These DfE activities should follow the procedures established in the second step and as much as possible use the information gained in the first step. If new products are designed that are clearly different from existing ones, new investigations will be needed. It should be observed that POEMS might include several parallel Deming cycles, as one concern the overall EMS and others are focused on single product projects. This implies that the third step may include several more or less separate PDCA cycles. Naturally, it is advantageous if the environmental concerns do not negatively affect the time for market introduction, price, and quality.

Fourthly, the POEMS activities must be evaluated/audited, which is supposed to lead to a good base for reaching continual improvement. Based on these audit procedures, measures should be taken to continually improve the product-related environmental performance.

Box D: Environmental Management Systems

ISO 14000

The ISO 14000 family addresses various aspects of environmental management. The very first two standards, ISO 14001:2004 and ISO 14004:2004 deal with environmental management systems (EMS). ISO 14001:2004 provides the requirements for an EMS and ISO 14004:2004 gives general EMS guidelines.

The other standards and guidelines in the family address specific environmental aspects, including: labeling, performance evaluation, life cycle analysis, communication, and auditing.

ISO 14001:2004 is a management tool enabling an organization of any size or type to:

- identify and control the environmental impact of its activities, products, or services
- improve its environmental performance continually
- implement a systematic approach to setting environmental objectives and targets, to achieving these and to demonstrating that they have been achieved.

ISO 14001 does not specify levels of environmental performance. The intention of ISO 14001 is to provide a framework for a holistic, strategic approach to the organization's environmental policy, plans, and actions. This standard gives the generic requirements for an environmental management system. The underlying philosophy is that whatever the organization's activity, the requirements of an effective EMS are the same. This has the effect of establishing a common reference for communicating about environmental management issues between organizations and their customers, regulators, the public, and other stakeholders. Because ISO 14001:2004 does not lay down levels of environmental performance, the standard can be implemented by a wide variety of organizations, whatever their current level of environmental maturity. However, a commitment to compliance with applicable environmental legislation and regulations is required, along with a commitment to continual improvement—for which the EMS provides the framework.

The EMS standard ISO 14004:2004 provides guidelines on the elements of an environmental management system and its implementation, and discusses principal issues involved.

ISO 14001 is a tool that can be used to meet internal objectives:

- provide assurance to management that it is in control of the organizational processes and activities having an impact on the environment

- assure employees that they are working for an environmentally responsible organization.

ISO 14001 can also be used to meet external objectives:

- provide assurance on environmental issues to external stakeholders—such as customers, the community and regulatory agencies
- comply with environmental regulations
- support the organization's claims and communication about its own environmental policies, plans and actions
- provide a framework for demonstrating conformity via suppliers' declarations of conformity, assessment of conformity by an external stakeholder—such as a business client—and for certification of conformity by an independent certification body (ISO Standards, 2009).

EMAS regulations

Companies that wish to adopt to comply with EMAS regulations, must:

1. carry out an initial environmental analysis to determine its initial position with regard to environmental conditions;
2. set its environmental policy, (i.e. the objectives and general principles for action with regard to the environment);
3. prepare an environmental program describing the measures used to achieve the specific objectives of the company's environmental policy;
4. implement an environmental management system, (i.e. that part of the management system, such as structure, planning, responsibility, practice, procedures, processes, and resources that permits the development, implementation, definition, and maintenance of the environmental policy);
5. carry out auditing, (i.e. systematic, periodic, documented and objective assessments of the performance of the company, the environmental management system, and of the environmental protection measures); and
6. prepare an environmental statement for the general public, which summarizes the environmental policy adopted, the environmental management system, the company and any significant environmental issues, the environmental objectives, and the company's environmental performance.

EMAS Regulation No. 761/2001 provides that a company's environmental statement must be certified by an independent accredited environmental auditor. Following review and certification, the company may request to be registered in the European EMAS list and use the EMAS logo.

Compliance with EMAS involves a series of benefits such as:

- internal reorganization and consequent growth in efficiency;

- reduction in costs due to rationalization in the use of resources and adoption of cleaner technologies;
- increase in employee motivation and participation, with consequent reduction in internal conflict;
- more trusting relationships with environment control and authorization bodies;
- less probable occurrence of events that may cause damage to the environment;
- greater certainty of compliance with environmental laws;
- good relations with the community, who perceives the company's commitment to environmental improvement;
- increased technical and scientific know-how, to be used for future improvement of environmental performance;
- increased ability to grow while defending the environment;
- greater guarantees of success in environmental actions because of the increased number of assessments;
- reduction of "red tape" to obtain or renew of authorizations/permits; and
- increase value of the company, because of the guarantees to correct environmental management.

6.2 Design for Environmental and Product Life Cycle Cost

Product Life Cycle Cost analysis involves evaluation of all future costs related to design, construction and/or production, distribution, operation, maintenance and support, retirement, and material disposal; that means all of the phases in the system life cycle (Fabrycky and Blanchard 1991). The process of systems engineering and the system life cycle are illustrated in Fig. 6.2.

Life Cycle Cost has been applied since the 1960s when the United States' Department of Defense stimulated the development and application of Life Cycle Cost to enhance its cost effectiveness. Defense systems, such as an aircraft or a special land vehicle, are ideal for Life Cycle Cost analyses since the Department of Defense mainly controls the entire life cycle (Sherif and Kolarik, 1981). Life Cycle Cost has moved from defense systems to industrial and consumer product areas, where each user controls only a portion of the actual life cycle of the system.

Life Cycle Cost may be defined as "the cost of acquisition, ownership, and disposal of a product over a defined period of its life cycle". Life Cycle Cost is a standard engineering economic approach to be used for choosing among alternative products or designs that provide approximately the same service to the customer (Lutz et al. 2006). Cost models may range from simple to complex, and are essentially predictive in nature. Parameters, such as the system's physical environment, usage demand, reliability, maintainability, labour, energy, taxes,

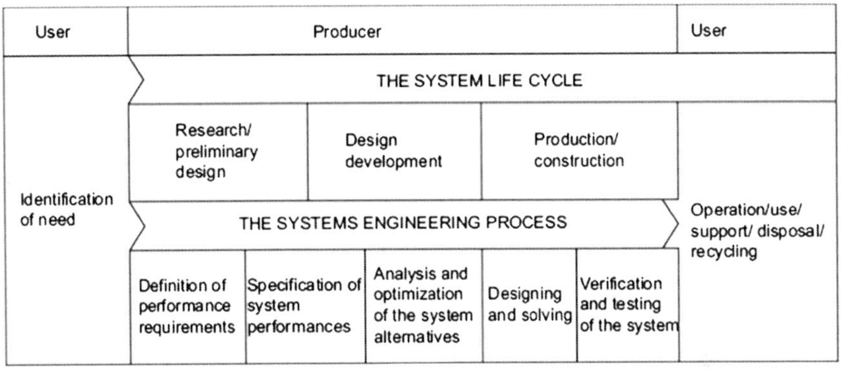

Fig. 6.2 The relationship between the systems engineering process and the system life cycle. (*Source:* Fabrycky and Blanchard 1991)

inflation, and the time value of money, may have a great impact on the life cycle costs. There are various ways of doing Life Cycle Cost, for example analogy models, parametric models, and engineering cost models. Choice of model is dependent on available resources, time, data, and the need for accuracy.

- Life Cycle Cost through an analogy model identifies a similar product and adjusts the differences in costs between the products.
- Parametric models utilize statistical methods, where the objective is to establish a functional relationship between changes in cost and the dependent factor(s) such as weight, lot size, etc. A parametric model is thus more advanced than an analogy model, which is only dependent on one single, dominant cost driver, with a linear relationship (Emblemsvåg 2003).
- The engineering cost model estimates capital and operational cost data when more detailed information is needed than found in an analogy or parametric model, however, complexity is increasing because costs are assigned to each system element at the lowest level of detail.

Following the publication of the World Commission on Environment and Development (WCED) report (1987), increased emphasis has been put on the use of economic instruments to modify the behavior of actors toward the environment. Environmental objectives are merged with economic policies, e.g., through economic valuation of the environment. Measuring environmental improvements in monetary may be done by estimating the costs of the improvements and the extra benefits. Benefit measurements may be interpreted in different ways, depending on stakeholders' preferences reflected in their "willingness to pay" (WTP). Sometimes WTP may excess the market prices, and this excess is known as consumer surplus. A complimentary concept is "willingness to accept" (WTA), which asks how much money should be paid to compensate for an environmental loss. Stakeholders may view gains and losses differently, and consequently WTP and WTA may differ. A significant difference may create problems difficult to resolve. Economic valuation is normally a part of cost-benefit analysis, multi-criteria analysis, and natural resource damage assessments (Schuijt 2003).

Environmental issues and consequences should be considered in Life Cycle Cost calculations. Life Cycle Cost has been suggested as a cost-accounting system to include environmental costs that normally appear outside the boundaries of traditional accounting systems, such as recycling and demolition costs. There are several variants of Life Cycle Cost tools that attempt to include environmental impacts as costs into the corporate accounting systems, for example (Gluch and Baumann 2004):

- Full Cost Accounting (FCA), which identifies and quantifies the full range of costs throughout the life cycle of the product, process, or activity.
- Total Cost Assessment (TCA), which is a decision-making tool that calculates the total cost of conducting business, environmental costs included, sometimes also used as a synonym for FCA.
- Life Cycle Cost Assessment (LCCA), which identifies environmental consequences and assigns monetary values to those consequences through evaluation of the life cycle costs.

One of the main problems with environmental accounting is that there is no standard definition of environmental costs and environmental cost savings. Environmental costs may include disposal costs, investment costs, and sometimes also external costs. Also, it may be difficult to determine the discount rates and the time horizon for discounting. In economic analyses it is often assumed that a given benefit or cost has a higher value now than in the future. This process of reduced importance is known as discounting, which creates problems when it is applied to environmental issues. An example may be that acidification is weighted less and less into the future, which means that if discounting occurs, the less important the losses due to acidification will be. Thus, discounting gives a bias against future generations and may seem inconsistent with sustainability.

6.2.1 Related Research Works

Several authors have proposed an approach of Design for Environment, including cost aspects and LCA. Warburg et al. (2001) used the method of Life Cycle Costing (LCC) to implement economic aspects in industrial decision making. Senthil et al. (2003) developed a Life Cycle Environmental Cost Analysis (LCECA) incorporating costing into LCA practice. This model prescribes a life cycle environmental cost model to estimate and correlate the effects of these costs in all the life cycle stages of the product. The newly developed categories of eco-costs are: costs of effluent treatment/control/disposal, environmental management systems, eco-taxes, rehabilitation, energy, and savings of recycling and reuse strategies. The mathematical model of LCECA determines quantitative expressions between the total cost of products and the various eco-costs. On the other hand, a method that evaluates an eco-product from the cost and environmental aspects independently has been reported (Biswas et al. 1997). In addition, a methodology based on the formulation of compromise decision support problems has been reported (Coulter and Bras 1999). Lye et al. (2001) proposed a methodology that scores the cost, quality, and environmental standing of four stages of the life cycle of a product. A self-learning algorithm is discussed that computes the best and worst values of the indices from a variety of similar products. However, these method requires a great amount of detailed information, making it inapplicable to the early phases of design.

Bovea and Vidal (2004) proposed a model that integrates the environmental, cost, and customer valuation during its design process.

This model is based on the combination of three methodologies: Life Cycle Assessment (LCA) methodology to evaluate the environmental requirements, Life Cycle Cost (LCC) to examine the internal and external costs of the product, and Contingent Valuation (CV) to quantify the customer's value in terms of his/her willingness-to-pay (WTP) for a product that incorporates certain environmental improvements.

The Life Cycle Costs methodology includes all internal and external costs incurred throughout the entire life cycle of a product. Internal costs (IC) are the costs for which the company is responsible over a particular period of time. They are also called total company costs and can be divided into conventional costs (CC), which include the direct costs borne by the company when manufacturing a product, hidden costs (HC), which are counted as general costs instead of being assigned them directly to the product or process, and less tangible costs (LTC), which are often not included in the company accounts due to their probabilistic nature.

External costs (EC) are costs for which the company is not responsible at a specific time, in the sense that there is no market or governmental regulation that assigns them to the company (i.e. the depletion of natural resources, impact on human health, ecological impacts, or infrastructures and buildings impacts). These costs are also called social costs since in the long term they fall back on society as a whole (Nasr and Varel 1997) and should be included in the company accounting mainly due to the following three reasons: international competitiveness, accountability beyond responsibility and avoid regulatory treadmill (Shapiro 2001). External costs are assessed by quantifying the negative effect of the damage caused by pollution (Craighill and Powell 1996).

Bovea and Vidal (2004) carried out an economic evaluation of users behavior toward products that include certain environmental advantages by applying Contingent Valuation techniques. CV is a survey-based method that is frequently used for placing monetary values on environmental goods not bought and sold in the marketplace. The authors used questionnaires to obtain the customer WTP for those environmental goods. The questionnaire is organised in three parts (Vidal et al. 2002). The introductory section includes general questions about the importance the respondent gives to environmental issues associated with the product being designed. The second part includes questions related to the WTP for a product that incorporates certain improvements to make it more environment-friendly, and the respondent is asked to choose between seven pairs of prices for each of the environmental advantages.

For each pair of prices the respondent must decide which item he or she would buy—the currently available version of the product (at its present market price) or an ecological alternative to the same product (at a price above the current market price). By progressively increasing the price of the ecological alternative in each choice, respondents are provided with a simple way of deciding the maximum price they would be willing to pay for the environmentally improved product. The third and last part of the questionnaire collects socio-economic information about the respondent concerning aspects such as age, sex, their relation with the object being studied, employment status, level of education, or family income. Fig. 6.3 shows the procedure of the proposed method.
where:

LCA_0 is the impact of currently version of the product in terms of mPt and using the EcoIndicator99 method;

LCC_0 is the Life Cycle Cost of currently version of the product;

$i = 1, \ldots, n$, are the environmentally preferable alternatives generated in the previous stage.

LCA_i is the impact of each alternative version of the product in terms of mPt (using the EcoIndicator99 method);

LCC_i is the Life Cycle Cost of each alternative version of the product

WTPi = consumer willingness-to-pay for each of the environmentally preferable alternatives by applying the CV methodology-based questionnaire

Two criteria are proposed for use in selecting environmentally preferable alternatives from the point of view of the society and the company:

- From society's point of view, any product cataloged as an ecological product must satisfy the condition that both impacts on the environment and external costs are lower than those of the initial product.
- From the company perspective, if the decision to manufacture an ecological product is not imposed by outside conditioning factors (e.g. by governmental regulation), then that product, besides fulfilling the above conditions, has to be profitable for the manufacturing company. The economical profit gained with the environmentally friendly design has to be, at least, equal to those obtained from the sale of the initial design. In order to select the ecologically re-designed alternatives of any product, the following condition must be fulfilled:

$$Ai = \max\{\text{WTP}i - \text{IC}i\} \leftrightarrow [(\text{LCA0} - \text{LCA}i) > 0] \wedge [(\text{EC0} - \text{EC}i) > 0]$$

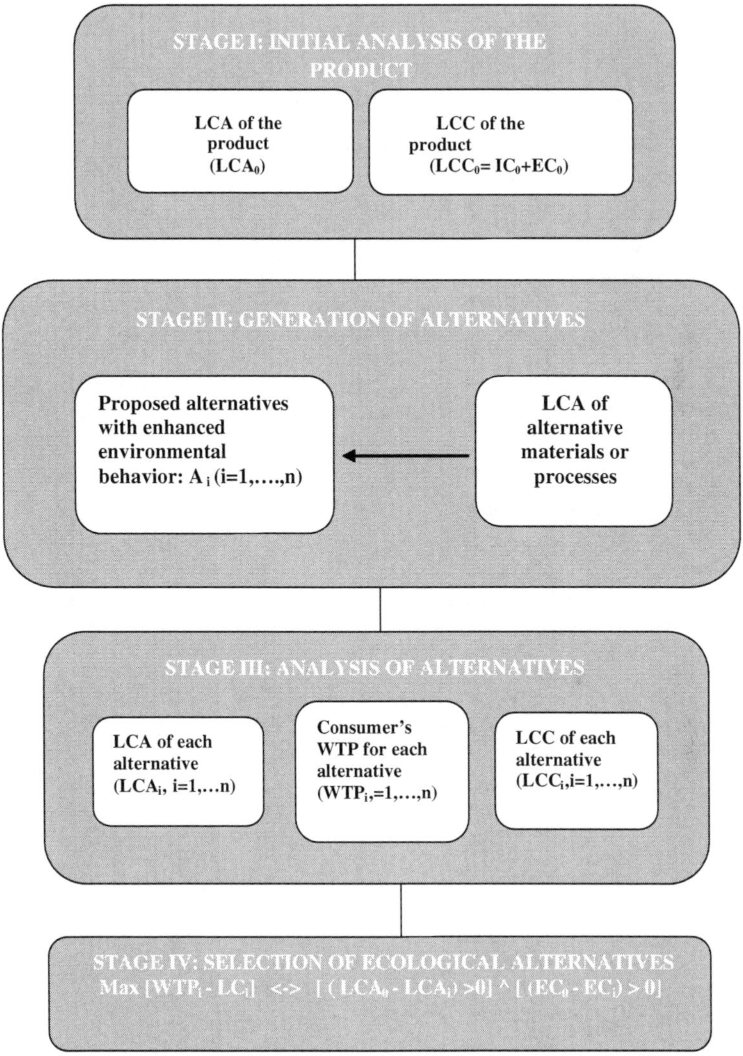

Fig. 6.3 Model for the eco-design of products

References

Ammenberg J, Hjelm O (2002) The connection between environmental management systems and continual environmental performance improvements. Corp Environ Strategy 9(2):183–192

Ammenberg J (2003) Do standardised environmental management systems lead to reduced environmental impacts? Dissertation, No. 851. ISBN 91-7373-124-2. Environmental Technology and management. Linkö ping University, Sweden

Ammenberg J, Sundin E (2005) Products in environmental management systems: drivers, barriers and experiences. J Clean Prod 13:405–415

Baumann H, Boons F, Bragd A (2002) Mapping the green product development field: engineering, policy and business perspectives. J Clean Prod 10(5):409–425

Berkhout F (1998) Aggregate resource efficiency. are radical improvements impossible? In: Vellinga P, Berkhout F, Gupta J (eds) Managing a material world Perspectives in industrial ecology. Kluwer Academic Publishers, Dordrecht, pp 165–189

Biswas G, Haftbaradaran H, Kawamura K, Hunkeler D, Lantz J, Shahinpoor M, Quinn T (1997) An environmentally conscious decision support system based on a streamlined LCA and a cost residual risk evaluation: fluorescent light bulb case study. Int J Environ Conscious Des Manuf 6(3):9–24

Bovea M, Vidal R (2004) Materials selection for sustainable product design: A case study of wood-based furniture eco-design. Mater Des 25:111–116

Brezet JC, Bijma A, Silvester S (2000) Innovative electronics as an opportunity for eco-efficient services, In: Reichl EHH, Griese H (eds), Electronics Goes Green 2000 + , Proceedings vol 1 pp 859–865

Coulter S, Bras B. (1999). Decision support for systematic product evolution. In: Proceedings of the 1999 ASME Design Engineering Technical Conferences, ASME, DTEC99/DTM-8747

Craighill AL, Powell JC (1996) Lifecycle assessment and economic evaluation of recycling: a case study. Resour Conserv Recycl 17:75–96

Ehrenfeld JR, Lenox MJ (1997) The development and implementation of DfE programmes. J Sustain Product Design (1):17–27

Emblemsvåg J (2003) Life-cycle costing Using activity-based costing and monte carlo methods to manage future costs and risks. Wiley, New Jersey

Emzer M, Birkhofer H (2002) Selecting methods for life cycle design based on the needs of a company intemational design conference—design. Dubrovnik, Croatia

Fabrycky WJ, Blanchard BS (1991) Life-cycle cost and economic analysis. Prentice Hall, New Jersey

Gluch P, Baumann H (2004) The life cycle costing (LCC) approach: a conceptual discussion of its usefulness for environmental decision-making. Build Environ 39:571–580

ISO Standards (2009) http://www.iso.org/iso/iso_catalogue/management_standards/iso_9000_iso_14000/iso_14000_essentials.htm (1/3/2010)

Karlsson M (2001) Green concurrent engineering. A model for DFE management programs. Doctoral Dissertation, The International Institute for Industrial Environmental Economics, Lund University, Sweden

Lindahl M (2005) Engineering designers' requirements on design for environment methods and tools. Industrial

Lutz J, Lekov A, Chan P, Whitehead CD, Meyers S, McMahon J (2006) Life-cycle cost analysis of energy efficiency design options for residential furnaces and boilers. Energy 31:311–329

Lye SW, Lee SG, Khoo MK (2001) A design methodology for the strategic assessment of a product's eco-efficiency. Int J Prod Res 39(11):2453–2474

Nasr N, Varel EA (1997) Total product life cycle analysis and costing. In: Proceedings of the 1997 Total Life Cycle Conference. Life Cycle Management and Assessment (Part 1). Society of Automotive Engineer (SAE), Inc., Warrendale PA, USA, p 9–15

Norell M (1992) St6dmetoder och samverkan i produktutveckling St6dmetoder och samverkan i produktutveckling. Stockholm, Sweden

Ries G, Winkler R, Zü st R (1999) Barriers for a successful integration of environmental aspects in product design. Proceedings of "EcoDesign '99". First international symposium on environmental conscious design and inverse manufacturing, Feb 1–3, Tokyo, Japan, 1999. pp 527–532

Ritzen S, Lindahl M (2001) Selection and implementation—key activities to successful use of EcoDesign tools. In: Proceedings EcoDesign 2001: 2nd international symposium on environmentally conscious design and inverse Union of ecodesigners (Association of EcoDesign Societies)

Schuijt K (2003) Economic valuation of the environment: an institutional perspective. In: Environmental management accounting. Purpose and progress. Kluwer Academic Publishers, Dordrecht, pp 387–403

Senthil KD, Ong SK, Nee AYC, Tan RBH (2003) A proposed tool to integrate environmental and economical assessments of products. Environ Impact Assess Rev 23:51–72

Shapiro KG (2001) Incorporating costs in LCA. Int J LCA 6(29):121–123

Sherif YS, Kolarik WJ (1981) Life cycle costing: concept and practice. Omega 9:287–296

Tukker A, Haag E, Eder P (2000) Eco-desig: European state of the art —Part I: Comparative analysis and conclusions—An ESTO project report. European Commission. joint research centre institute for prospective technological studies: Brussels, Luxembourg, p 60

Vidal R, Bovea MD, Georgantzis N, Camacho-Cuena E (2002) Es rentable diseñar productos ecológicos? El caso del mueble (Is profitable to design ecological products? The furniture case). Castellón, Spain: Ed. Universitat Jaume I [in Spanish]

Warburg N, Hemam C, Chiodo JD (2001) Accompanying the (re)design of products with environmental assessment (DfE) on the example of ADSM, IEEE international symposium on electronics and the environment, pp 202–207

World Commission on Environment and Development (WCED) (1987) Our common future. Oxford University Press, New York

Chapter 7
Case Study: Development of a Sustainable Product Life cycle in Manufacturing Firms

A major benefit of the DfE methodology proposed in this work is the possibility to use Life Cycle Assessment (LCA) data both during new product development and when modifying old products, with the aim of continuously reducing the overall environmental impact of products during their life cycle. This improvement cycle begins with the attempt to find new design solutions (for assembly and setup in the case of electrical distribution boards), continues with the calculation of the environmental Break Even Point (BEP) and with the assessment of the BEP for the expenses incurred by the client.

On the basis of these calculations and bearing in mind the technical specifications required by the clients and the work environment in which the product will be used, the designers will be able to make the most efficient choices from both the environmental and the economic points of view.

This work was developed thanks to the collaboration between the Sustainability Affairs Department of ABB Italia and the Department of Energy Studies of Marche Polytechnic University, Ancona and concerned the design of low tension electrical distribution boards carried out at the ABB SACE SpA factory in Frosinone, Italy. The aim of the firm was to create a procedure based on the integration of DfE and LCA methodologies to allow the assessment of improvement, in terms of environmental and economic impact, which could be attributed to a different assembly layout for the "electrical distribution boards" set of products. The procedure developed is general and can also be applied to other products and to other industrial realities.

This work is organized as follows: in the next section the procedure for development of a sustainable product life cycle is presented; in the "Procedure development" section the case study is discussed; in "Approach and parameters used in the LCA study" section are presented the approach and parameters used in the LCA study; in the "Application example" section the case study of a specific product is used to illustrate the application of the proposed method; finally discussion and conclusions are presented.

M. Bevilacqua et al., *Design for Environment as a Tool for the Development of a Sustainable Supply Chain*, DOI: 10.1007/978-1-4471-2461-0_7,

7.1 Research Approach

This study proposes a new approach to DfE and a new way of integrating this design methodology with both the LCA technique and the economic aspects. The methodology develops an improvement cycle identifying environmental and economic BEP between two design solutions. The combination of these two BEP curves defines four areas of economic and environmental advantage, therefore the designers will be able to make the most efficient choices from both the environmental and the economic points of view. This methodology can be used by the designers in the initial phases of development of a new product or in the redesign of an existing product.

The methodology can be developed through a procedure composed of six steps:

(1) **Definition of the order specifications and first trial layout**
The procedure begins with the analysis of product specifications requested by the client and the work environment conditions in which the product will be used. Therefore the designer can develop the first trial layout for the product. On the basis of the product specifications set it is possible to obtain as data output the basic parts lists (codes, assembly times, costs) for the product.

(2) **Calculation of input and output flows and reduction of the modular complexity of the product**
A Life Cycle Inventory (LCI) analysis is performed in order to evaluate input and output flows. Often the modular complexity of the product and the necessity to reduce design times as much as possible led to the need to select which items to include in the LCI. The traceability of all the suppliers and in particular of all the production processes of the components may prove to be an almost insurmountable limitation for the introduction of an integrated DfE system, both because of the material difficulty and the time needed to carry out the research and also because of some problems connected with the diffusion/spread of company know-how (it is necessary to have the executive designs and to know the production techniques adopted).

(3) **Validation of the model**
Subsequently, it is necessary to validate the choice to exclude some sets of codes even at the environmental level so as to be able to demonstrate that the flows excluded are not of particular importance for the possible impact categories. This check must be carried out by doing an LCIA study for the single category of product.

(4) **Definition of the alternatives and calculation of environmental BEP**
After determining the values of the flows involved, the time necessary for processing and assembling the components and the energy consumption of the firm, it is possible to calculate the environmental impact of the initial layout for the product. The improvement cycle continues with the attempt to find new design solutions and with the calculation of the environmental BEP between the old and new solutions. The calculation must be carried out for each impact category when there are variations in the manufacturing parameters.

(5) **Calculation of economic BEP**
Subsequently a BEP analysis of the costs incurred by the client must also be developed. The two BEP analyses are referred to the same parameter.

(6) **Choice of the best solution**
Finally, by comparing the two BEP values (environmental and economic) four possible alternatives are defined. The designer must choose the most efficient solution on the basis of the work conditions in which the product will be used.

7.2 Procedure Development

Figure 7.1 shows the steps of the procedure (1–6), which have already been referred to in the chapter "research approach", applied to "electrical distribution boards" set of products. The aim of this procedure is to improve the performance of the electrical distribution board, in terms of environmental and economic impact, not by intervening in the design of the single components, for example circuit breakers, but rather by improving the assembly (changing the electrical distribution boards module number, n) and trying to limit energy consumption due to internal dissipation during use.

In order to integrate this procedure with the normal design steps software has been used, and in some cases created, which is able to interact with the management software system of the firm used by the designers. In particular the newly created software programs, interacting with each other, perform the procedure steps 2–5.

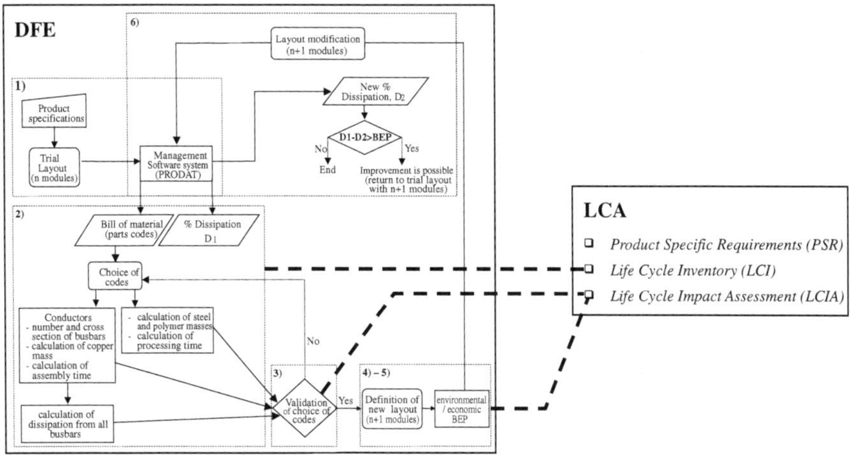

Fig. 7.1 Steps of the procedure

(1) The first step in the procedure is to integrate the management software system used by the designers (PRODAT) with the specifications of the product requested by the client, the work environment conditions in which the electrical distribution board will be used and therefore the minimum number of modules necessary to contain the components which the board needs (trial layout). On the basis of the product specifications set it is possible to obtain as data output the basic parts lists (codes, assembly times, costs) for the product and the percentage of maximum energy dissipation of the board (D1) calculated considering the trial layout.
(2) The calculation of the mass flow for the assembly stage was carried out by means of the basic parts lists. The modular complexity of the product and the necessity to reduce design times as much as possible led to the need to select which items to include in the LCI. The list of components is made up of over 30,000 codes to be managed in a modular way according to the specifications of the board required by the client. For this reason an error in calculation of the masses equivalent to 5% of the total weight of the product was considered acceptable. Bearing in mind these considerations, some specific elements for maneuver, such as circuit breakers, isolating switches and all other elements that are involved in their setup need not be analyzed from the point of view of calculation of the flow of mass.
(3) Validation of choice of codes. The high design complexity of the distribution board is compensated for at LCA level by the fact that the utilization stage is in fact the predominant one for environmental impact and therefore allows us to increase the error tolerance at the assembly stage. In the calculation of a single impact category referred to the whole life cycle an acceptable uncertainty equivalent to 5% has been set. This assumption originates from the awareness that there is a variance of between 5 and 10% in the LCI analysis data coming from different sources.
(4) Definition of the alternatives and calculation of environmental BEP. It is possible to calculate the environmental impact of the initial layout (n modules) and the environmental impact of the new layout obtained by distributing the components over several modules (n + 1 modules), and thereby reduce the dissipation during the distribution board utilization stage. The environmental impact due to the whole life cycle of the distribution board can be attributed to two different moments and causes:

- Impact due to the component building and board assembly stage;
- Impact due to the utilization stage of the board

For some types of boards the latter is responsible for more than 90% of the total impact per single category. This suggests that a possible solution for improvement could be brought about by a layout solution which increases the energy performance of the distribution board. At this point it is necessary to ensure that any improvements made at the level of environmental impact by these actions are not countered by an increase in the mass flows due to the increase in the raw materials used for the production of extra modules. The environmental BEP must therefore be calculated as a % dissipation of the maximum power of the module in a variety of working conditions.

(5) Calculation of economic BEP. The two BEP analyses are referred to the same parameter: the percentage of dissipation of the devices which must be reduced in order to ensure that the new layout is economically convenient.

(6) Modification of layout. The parameters of the new electrical distribution board layout are then integrated in the firm's management software system (PRO-DAT) which identifies the new dissipation (D2). Finally, by comparing the difference in dissipation between D1 and D2 and the two BEP values (environmental and economic) four possible alternatives are defined. The designer must choose the most efficient solution on the basis of the work conditions in which the electrical distribution board will be used.

7.3 Approach and Parameters Used in the LCA Study

The LCA module interacts at different stages of the DfE procedure: the first stage involves the transmission of the data necessary for carrying out the LCI stage, the second concerns the model validation stage, and the third concerns the comparison between the environmental impact of the various layouts of the product.

7.3.1 Product-Specific Requirements (PSR)

The first step for using the LCA technique is to create the product-specific requirements (PSR). The PSR defines the functional unit to be used, the system boundaries, and the results to be presented in the final report. PSR acts as a mutually agreed official document that all manufacturers of the same kind of product follow when developing their own Environmental Product Declaration. In this way the results presented in EPDs for similar products, irrespective of producer, will permit easy comparison of the relevant environmental parameters. Since no PSR existed for electrical distribution boards, the firm was obliged to develop guidelines for this device, in parallel with their own product-specific LCA analytical work.

In accordance with the guidelines for Goal and Scope Definition (ISO 14040 2006) the parameters defined were:

(1) *Object of analysis*. Low tension distribution boards (<1000 V;1250-6300 A) in accordance with the specifications in the CEI EN 60439 series regulations. The basic unit of an electrical distribution system, which we will identify as **module,** is made up of a box, a cover, and a series of internal partitions. The

modules can be installed as single units or in combinations of units, and take on particular layouts according to the equipment housed and the clients' specifications (level of IP[1]). According to the specifications required, the box must allow the subdivision of the module into areas: the fixed parts area the busbar area, and the cables area.

(2) *Function.* Distribution of electrical energy and control of mechanical devices in civil and industrial applications.

(3) *Functional Unit.* Due to differences in size, output, and efficiency between single machines it is not meaningful, for comparative purposes, to specify the resulting environmental impact per machine, but rather to relate this to a functional unit which is linked to the operational requirements. The functional unit chosen for the group of products "electrical distribution boards" is the capacity to distribute 1 MJ of electrical energy.

(4) *Physical limits of the system considered.* The parameters which are ecologically important must be set, in order to be able to decide whether to exclude mass or energy flows, or any emissions which may be present in the system being considered. For all the life cycle phases of the product the input and output must be specified as follows:

   (a) Materials and energy going from the environment to the system;
   (b) Emissions and waste products going from the system to the environment.

Our analysis is for internal use only, therefore it is possible to use an approach which does not include all the phases of the product life cycle. We will analyze only those aspects which we consider important. The production site investigated is involved mainly in assembly operations. The finished components arrive in the firm from other factories, except for some elements which are finished on-site because of their high level of design flexibility (for example the cutting of copper busbars). Bearing this in mind the life cycle investigation can be broken down as follows:

- *Distribution board component production stage.* This is considered as carried out by the firm being analyzed. The input and output flows are taken into consideration for the final environmental assessment. Production output represents the input for the distribution board assembly stage.

   (a) Extraction and production of raw materials: from an initial quantitative analysis of the product "electrical distribution boards" the number of components installed in the board proved to be, as previously mentioned, a problem to be addressed. For this purpose, for the materials used in the distribution board and assembled only in the firm, we

[1] The level of IP is a determinant of design because it defines the capacity of the board both to impede the access of foreign bodies such as dust or water, and to prevent any contact between the workers and the electrically charged parts.

adopted the simplification of assessing their contribution to the analysis only in terms of their weight, of the material used, and the technology required for producing them, without going into too much detail about the single elements. For example, for the materials in sheet metal we considered their weight and the fact that they were produced using cold rolling. This is one type of "cradle to gate" assumption. The transport from the various suppliers to the factory has also not been considered.

(b) Production stage emissions and waste; see point (a);

- *Distribution board assembly stage*. This is an integral part of the system.

  (a) The process of assembly of the components produced by the various suppliers. The predominant flows in this phase of the life cycle concern mass (input) coming from the previous stage and the energy consumption within the firm which are a satisfactory approximation of any finishing carried out (cutting, ...)

  (b) Emission and waste referred to the relative flows.

- *Distribution board utilization stage*. This is an integral part of the system.

  (a) The most important parameter for this stage of the life cycle is the energy which is dissipated during use. The expected life of the distribution board must be set so as to assess the environmental performance. The expected life has been set at 15 years. Transport to the final installation destination has been ignored.

- The *disposal phase* of the electrical distribution board has been ignored since the materials included in the analysis can all be recycled.

(5) *Time limits of the system considered*. limits to the system connected with the data collection reference time. All the data refer to production in 2004 in the factory studied.

(6) *Environmental limits of the system considered*. limits to the system connected with the data collection reference place. All the data concerning component production refer to European suppliers, as do the data which refer to utilization of the board. On the contrary the data which refer to energy consumption during the assembly stage are national.

(7) *Data quality*.
In this case direct data are:

- All the data which refer to the calculation of mass flows during the production and assembly stages;
- All the data which refer to the calculation of energy flows during the assembly stage;
- All the data which refer to the calculation of energy flows during the utilization stage.

  Indirect data are:

- All the data which refer to the component production stage.

(8) *Units of measurement*. SI units of measurement are used.

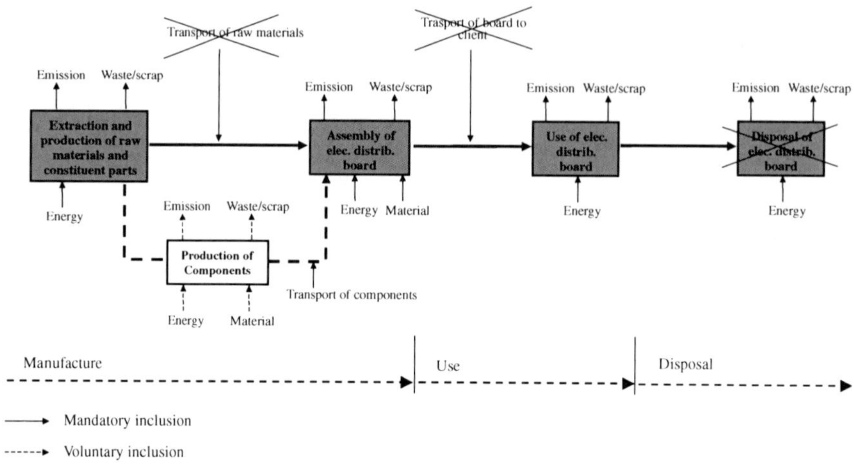

Fig. 7.2 Electrical distribution board life cycle flowchart

(9) *Allocations*. This sets the way in which the flows of mass and energy have been obtained. The calculations of mass flows have been obtained from the parts lists. The data for energy flows for the various production stages were obtained directly on-site by referring to consumption for the year 2004. The data concerning energy consumption during the utilization stage are partly obtained from the engineering guidelines used during design and partly calculated using studies of distribution board operations.

(10) *Level of analysis*. The analysis is a simplified LCA.

(11) *Acceptable error*. The maximum acceptable error for each single category of impact referred to the whole life cycle is 5%; this is in fact the variance which exists between the data for environmental impact when several databases are compared.

7.3.2 LCI: Life Cycle Inventory

The second stage of the LCA investigation is Life Cycle Inventory. In accordance with ISO 14041 this stage involves the collection of data concerning the processes and the various calculation procedures. The relationships between the system produced and the environment are defined; ISO 14041: "A system produced is a set of processes connected by flows or intermediate products which allow a specific function to be obtained". Information is collected from the product development office regarding material amounts, and machine performance such as power and efficiency. Data are added concerning emissions to air and water, generation of waste and energy consumption (Fig. 7.2).

7.3.3 LCIA: Life Cycle Impact Assessment

LCA calculation software called Eco Lab[2] has been used to process results. Central to the software is an extensive material database containing internationally agreed environmental characteristics. This software produces final reports showing quantified impact related to the global environmental threats (see Chap. 3):

- global warming potential (GWP);
- acidification (AP);
- depletion of the ozone layer (ODP);
- photochemical oxidant formation (POCP);
- eutrophication.

7.4 Application Example

In order to better explain the procedure proposed in the "Procedure development" chapter this case study illustrates its application to the design of a set of electrical distribution boards which will henceforth be referred to as "A". The simplest structure for this electrical distribution board consists in the use of one single module to allocate all the components. Analysis of the parts lists showed the mass flows which are involved in the assembly stage. Three different flows have been identified among the codes which are important for the analysis:

- Code by piece identifying components made up of cold rolled steel;
- Code by piece identifying components made up of electrolytic copper;
- Code by piece identifying components made up of polymers, in particular high impact polystyrene;

For the set of "A" products the mass and energy flows depend on the client's specific requirements and on the work environment in which the electrical distribution board will be used. For example, Table 7.1 shows the principal flows present during the various phases in the life cycle of a set "A" electrical distribution board, for which the requirements were a low protection index (IP), temperature of 35°C and range of current [0 A; 1250 A].

It is assumed that during working life the only flow is caused by the dissipation of electrical energy by means of heat, and that this makes the greatest contribution from the point of view of environmental impact. The heat developed is dissipated owing to the difference in temperature between the distribution board and the

[2] The computerized LCA-system Eco Lab issued by Nordic Port AB, Varbergsgatan 2C, S-412 65 Gothenburg, Sweden.

Table 7.1 Assembly stage flows

Steel mass (kg)	1.40E+02
Copper mass (kg)	5.10E+00
Polymer mass (kg)	1.12E+00
Electrical energy (MJ/s)	2.18E−01
Fuel oil (kg/s)	1.39E−02
Diesel fuel (kg/s)	3.62E−03
Assembly time (s)	8.35E+03

Table 7.2 Utilization stage flows

Dissipated energy main busbars (MJ)	6.48E+03
Dissipated energy module (MJ)	1.03E+05

external environment, thereby leading to heat exchange by radiation, convection and conduction.

Numerous factors influence this phenomenon, in particular:

- The size of the single modules of the board which help the exchange if their size increases;
- The level of IP: if this increases it obstacles flows toward the external environment;
- The low temperature of the environment in which the board is installed;
- The internal partitions;
- The arrangement of the sources of heat.

The environmental factors, such as temperature and level of IP depend on the needs of the client, but the size and the arrangement of the sources of dissipation inside the distribution board are the result of the layout chosen for the board itself, and must therefore be attributed to the designer.

As has already been explained in "Procedure development" chapter, in order to accelerate the LCI phase some groups of codes which were not considered to be important were excluded from the analysis, for example the functional elements for maneuver, such as switches, isolating switches etc. For the example already considered in Tables 7.1 and 7.2 the results are (Table 7.3):

The value of the total mass excluded remains under 5% (limit which was set in the definition of the PSR).

To quantify the various impacts of the flows the following method was applied:

(1) **Identification of any harmful substances emitted per impact category (classification)**. For example, for the Acidification impact (AP), expressed in kg of SO_2 equivalents, out of all the emissions caused by the cold rolling production process only those which concern this impact category are shown in Table 7.4 (the amounts emitted refer to one ton of product).

Table 7.3 Materials excluded

Materials	Mass excluded	Errors%
Steel (kg)	2.79	1.99
Copper (kg)	0	0
Polymer (kg)	0.2	17.85
Total (kg)	2.99	2.04

Table 7.4 Emission caused by cold rolling production

Output air + water	Quantity (g)
Ammonia NH_3	8,145
Nitrogen oxides No_x	4,540
Sulfur dioxides SO_2	6,180

Table 7.5 Harmful substances for AP

List of substances for AP	CF
Ammonia	1.60E+00
Nitrogen oxides (as NO_2)	5.00E−01
Sulfur dioxide	1.20E+00

The sources of the data used are mainly bibliographical and public and include the principal European LCI databases such as ETH-ESU, BUWAL, and APME (see references).

(2) **Identification of the CF (Characterization Factor)** for each single substance identified above (characterisation); according to ISO TR 14025 the harmful substances for this impact category are (Table 7.5):

(3) Calculation of the PI (potential impact) of the single raw material or the energy for each single impact (characterisation). In the abovementioned example the potential impact of the material "steel" on the AP category is 9.70E−03 (see notes[3]). Table 7.6 shows the potential impact (PI) of the flows for all the life cycle phases.

(4) For each category the total impacts, referred to the production and the component assembly stages, have been calculated (the result of the quantity of material or energy by the relative PI). The results are shown in Table 7.7 (see note[4]):

[3] The values shown in Tables 7.4 and 7.5 give the following results: $(1{,}60E{+}00 \cdot 8{,}145 + 5{,}00E{-}01 \cdot 4540 + 1{,}20E{+}00 \cdot 6180)/1000000 = 9{,}70E{-}03$ [kg of SO_2 eq./kg_{steel}].

[4] The values shown in Tables 7.1 and 7.6 give the following results: $9.70E{-}03 \times 140.05$ kg$+1.44E{-}01 \times 5.10$ kg$+1.93E{-}02 \times 1.12$ kg $= 2.12$ [kg of SO_2 eq.].

Table 7.6 Potential impact

	AP	NP	POCP	ODP	*GWP*
Steel	9.70E−03 (see footnote 3)	7.48E−04	8.64E−04	2.48E−07	3.18E+00
Copper	1.44E−01	0.00E+00	5.76E−03	0.00E+00	3.00E−01
Polymer	1.93E−02	1.57E−03	1.04E−03	2.97E-06	2.90E+00
Electical energy	3.60E−02	1.82E−03	8.84E−05	0.00E+00	1.60E-01
Diesel fuel	2.33E−01	3.67E−02	1.49E−03	0.00E+00	3.04E+00
Fuel oil	6.37E−01	6.25E−02	2.12E−03	0.00E+00	3.50E+00

Table 7.7 Total impact

Category	Total impact of the material during the manufacturing Stage
AP(Kg of SO2 eq.)	2.12E+00 (see footnote 4)
GWP100(Kg of CO2 eq.)	4.51E+02
NP (Kg of PO4 3- eq.)	1.07E−01
ODP (Kg of CFC11 eq.)	3.87E−05
POCP (Kg of C2H4 eq.)	1.52E−01

The values in Table 7.7 are the basis for checking the choice and for calculating the error made by excluding some codes. The exclusion of one code must also involve the assessment of its effects on a single impact category. In fact it is possible that, even if of little importance in terms of mass, the raw materials excluded may determine some important contributions for environmental impact. It is therefore necessary to calculate the CF_{limit} for each impact category. The necessary steps are as follows:

1. The overall contribution for each category of impact made during the distribution board production stage is defined. If a maximum percentage uncertainty (AU%) of 10% is accepted, then the negligible potential impact for that impact category will be (see note [5]):

$$AU = IMPACT_{TOT} \cdot AU\%/100 \text{ (for example AP)};$$

2. Calculation of the total mass excluded from the analysis = CE (cumulative excluded). For the module A being studied this is equal to 2.99 kg.
3. Calculation of the EF (emission factor): the emission factor is the sum of the quantities of all the important emissions for the impact category, given by Kg of substance excluded. Since in this case three substances were excluded (steel, copper, and ABS) the calculation of the EF was carried out as a weighted average of the EF of the three raw materials. From the abovementioned databases it is possible to extract all the substances emitted during the production cycles of the three raw materials. For each raw material and for each category

[5] AU (AP) = 2.12 × 0.1 = 2.12E−01.

Table 7.8 Emission factor values for the AP impact category

Materials	Emission factor (AP)
Steel	2.94E−03 (see footnote 7)
Copper	1.20E−01
ABS	5.80E−03

the harmful substances ("S") were identified and the quantities were multiplied by the relative CF. This number was divided by the number of the harmful substances ("n") and by the $CF_{average} = \Sigma_i^n CF_i/n$ (see note[6]), for those substances. In this way the EF for raw material and impact category was obtained:

$$EF_{(\text{raw material; impact *category})} = \sum_i^n (S_i \cdot CF_i) \Big/ (n \cdot CF_{average})$$

Table 7.8 shows the Emission Factor values for the AP impact category (see note[7]).

Three $EF_{(\text{raw material})}$ exist for each impact category. To determine the EF of the impact category these three values were averaged according to their importance in the module examined (see note[8]):

$$EF_{(\text{impact category})} = \left(\sum (EF_{(\text{raw material; impact category})} \cdot \left(Mass_{(\text{raw material})}\right)\right) / Mass_{tot}$$

4. Calculation of the CF_{limit} for each impact category (see note[9]):

$$CF_{limit} = AU/(CE \cdot EF_{(\text{impact category})})$$

Table 7.9 shows the values of the CF limit for all the impact categories.

In order to determine the errors caused by the exclusion of some codes it is necessary to verify for each impact category which substances emitted during the production cycles of the raw materials have a $CF > CF_{limit}$. The values found in the databases for the "A" product give the following results: NP, AP, and ODP do not have a CF greater than the limit. Therefore for these impact categories the error generated by the model is not meaningful. For the categories GWP 100 and POCP some substances have a greater CF and in these cases the error has been calculated:

[6] For example for the raw material steel and the AP impact category in Table 7.5 the result obtained is: $CF_{average} = (1.60E{+}00{+}5.00E{-}01{+}1.20E{+}00)/3 = 1.10E{+}00$.

[7] $EF_{(\text{steel; AP})} = (1.60E{+}00 \cdot 8.145{+}5{,}00E{-}01 \cdot 4540{+}1.20E00 \cdot 6180)/(3 \cdot 1000000 \cdot 1.10E{+}00) = 2.94E{-}03$.

[8] Tables 7.1 and 7.8 give the following results: $EF_{AP} = (2{,}94E{-}03 \cdot 1{,}40E{+}02{+}1{,}20E{-}01 \cdot 5{,}10E{+}00{+}5{,}80E{-}03 \cdot 1{,}35E{+}00)/146{,}5 = 7{,}04E{-}03$.

[9] For example the AP category gives: $CF_{limit} = 2{,}12E{-}01/(2{,}99 \cdot 7{,}04E{-}03) = 1{,}00E{+}01$.

Table 7.9 CF limit for all the impact categories

Category	CF_{limit}
AP	1.00E+01
GWP100	1.37E+04
NP	1.75E+01
ODP	4.00E+01
POCP	5.07E−03

Table 7.10 Percentage error for GWP100

Total neglected impact	8.86
% Error on production + assembly	1.96%
% Error on life cycle	0.015%

Table 7.11 Percentage error for POCP

Total neglected impact	0.012
% Error on production + assembly	8.02%
% Error on lifecycle	0.03%

$$\text{Neglected Impact}_{\text{raw material}} = \sum_{i}^{n} P_i \cdot Kg_{i(\text{raw material excluded})}$$

where P_i is the potential impact (PI) of the raw material excluded and Kg_i is the mass.

By adding the neglected impacts of the three raw materials it is possible to calculate the Neglected $\text{Impact}_{\text{Total}}$ and therefore the total % error.

$$\text{Total}\,\%\,\text{Error} = (\text{TOTAL IMPACT}/\text{Neglected Impact}_{\text{Total}}) \cdot 100$$

For the GWP 100 category the substances emitted by steel which are higher than the CF_{limit} are "HALON 1301" and "HALOGENATED HC" while several types of CFC emitted by ABS exceed the CF_{limit}. The results shown in Table 7.10 have been obtained by calculating the % error.

For the POCP category the substances emitted by steel and polymer which exceed the CF_{limit} are some "AROMATIC HC". The results shown in Table 7.11 have been obtained by calculating the % error.

In terms of environmental impact the error made by excluding the codes is much lower than the maximum (set at 5% in the PSR) for each impact category on the life cycle. For this reason the model can be considered valid for the subsequent analysis aimed at improving the environmental performance of the distribution board itself. This validation analysis carried out for the "A" set of electrical distribution boards was repeated with other types of distribution boards. The study was repeated for simpler and more complex distribution boards so as to cover all the possible arrangements, and in all cases the exclusion from the analysis of some groups of codes did not lead to meaningful errors. It can therefore be deduced that this approach can be considered valid for the design of all types of electrical distribution boards. The groups of codes which can be neglected in the assessment of the total impact have been defined so as to accelerate the DfE procedure.

7.4.1 Environmental Break Even Point

The electrical distribution board "A" has some fixed installation equipment without distribution bars and, in its simplest form, does not have any internal partitions (trial layout). The dissipations are calculated as the sum of those deriving from the equipment (module) and those from the copper conductors (main circuit breakers and distribution and off-set bars). Having defined the set of conductors, the electrical characteristics of the bars are set, and the dissipation becomes a function only of the supply current, which is a design feature. The density of the equipment and its layout within the module determine the % of the dissipative capacity of the module used, compared with the total. The environmental improvement attempt for product "A" consists in substituting single module set ups with solutions which have more than one module and verifying to what extent this dissipation % has been reduced when the equipment is arranged in a different way.

A possible solution to this problem is to find the BEP by defining to what extent the % dissipation of the module must be reduced in order to justify a solution which uses more than one module. The calculation must be carried out for each impact category when there are variations in the manufacturing parameters (level of protection IP) and in the supply currents. It is necessary to demonstrate that at the environmental level any improvements made by these actions (energy dissipated from the module) are not countered by an increase in the mass flows due to the greater amount of raw material used in order to manufacture more modules.

Any variation in layout which involves the arrangement of the equipment required by the client in two modules brings about the following increase in the mass and energy flows:

- The mass and energy flows during the manufacturing and assembly stages double: in fact even the copper busbars undergo this process because only the main ones are present and these must cover the whole width of the new layout. The cross section of the busbars and the number of conductors per stage do not vary because the supply current remains the same;
- The dissipations from the main busbars during use double for the abovementioned reason.

Table 7.12 Environmental BEP

Current (A)	BEP% (AP)	BEP% (NP)	BEP% (POCP)	BEP% (GWP100)
150	0.73	1.18	2.56	1.61
350	0.74	1.18	2.57	1.61
550	0.75	1.2	2.59	1.63
…				
…				
5400	2.88	3.24	9.29	3.73
5600	3.04	3.39	9.51	3.88
5800	3.2	3.55	9.87	4.04

To calculate the environmental BEP it is necessary to calculate for each impact category:

$$IMP_{mod2} = IMP_{tot1} - 2 \cdot IMP_{ass} - 2 \cdot IMP_{main\ busbars}$$

where:

- IMP_{tot1} is the impact of the whole life cycle in the standard layout (with one module in the case of product "A");
- IMP_{ass} is the impact of the whole life cycle before use in the standard layout (with one module in the case of product "A");
- $IMP_{main\ busbars}$ is the impact of the dissipation of the main busbars in the standard layout (with one module in the case of product "A");
- IMP_{mod2} is the value of the environmental impact of the equipment (module) which the re-designed distribution board must have at BEP.

The environmental BEP, expressed as a percentage of reduction in dissipation, is therefore calculated for each single impact category as: $1 - (IMP_{mod2}/IMP_{mod1})$; where IMP_{mod1} is the value of the environmental impact due to the equipment (module) in the standard layout. Analysis carried out with varying supply currents for the product in set "A", studied in the previous chapter, provided the results shown in Table 7.12.

The analysis of the ODP impact category was not carried out for any type of distribution board since the potential impact of electrical energy for this category is zero.

The most important category is POCP owing to the high CFC and halogen emissions in the mass flows. For this reason, when trying to obtain an improvement in all the categories by changing the layout, this category of impact was taken as a reference and the overall BEP line (Fig. 7.4) was calculated as (4 · BEP%(POCP)).

Fig. 7.3 BEPcosts, Module "A", IP 30/40

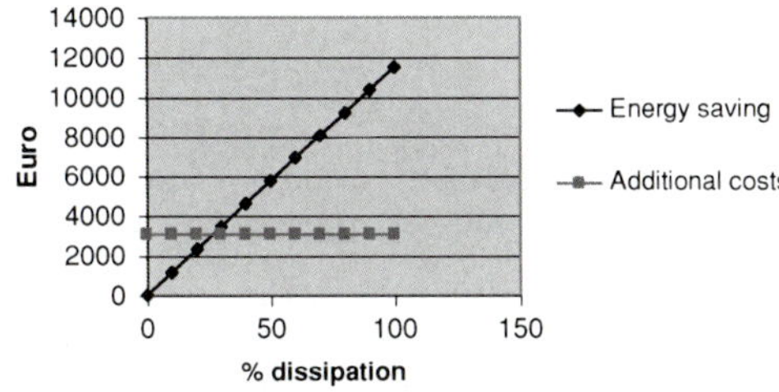

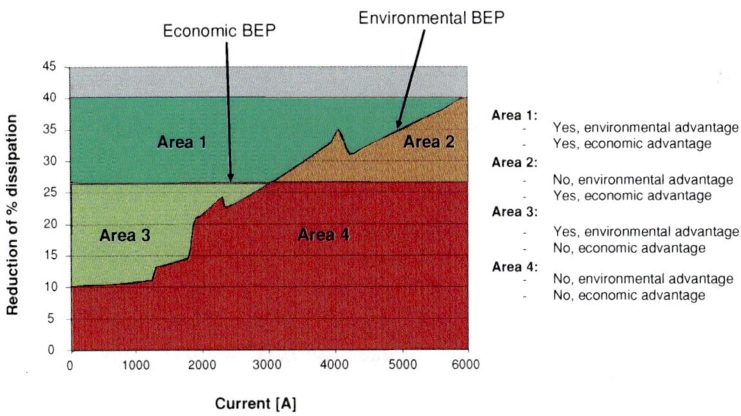

Fig. 7.4 Environmental BEP + Economic BEP analysis

7.4.2 Economic/Environmental Break Even Point

Any attempt to make manufacturing industries aware of the overall effects of their production processes must necessarily include an analysis of the expenses that the industries incur in order to reach their targets. It must be taken into consideration that modifications in layout, in order to obtain improvements in the various impact categories, generate an increase in the components of the electrical distribution board and an increase in the assembly time. On the other hand the percentage energy saving, considering a 15-year life cycle of the distribution board, reduces the economic burden on the client.

The parameters which have been taken into consideration are: costs of the components; costs of labor; costs of energy due to dissipations. Other possible sources of costs, such as the need for the purchaser to have a larger space available in order to be able to setup a distribution board with more than one module, have been ignored.

Figure 7.3 shows the savings and the additional costs involved in changing the layout of the electrical distribution board of the "A" group used as an example throughout this chapter, according to the varying % of reduction in dissipation.

The BEP for costs is obtained with a dissipation of 26.47%.

The overall results, shown in Fig. 7.4, indicate the presence of four areas of economic advantage resulting from the modification of the layout of the electrical distribution board. On the basis of the client's requirements and according to the conditions in which the distribution board will be used, the designer can easily choose the best solution.

The graph in Fig. 7.4 illustrates that with currents >3000 A, if it is possible to reach the environmental BEP there is also an overall cost benefit for the client. The extent of the economic benefit can be seen in Fig. 7.3. By repeating the analysis for other groups of electrical distribution boards we found that there is a better compromise between environmental improvement and economic competitiveness for modules which have a simple design and greater dissipation, while for the other modules it is not so easy to identify an arrangement which is sustainable from both points of view. The analysis provides different results according to the type of distribution board: the sets of simpler electrical distribution boards have relatively low environmental BEP values (for POCP the maximum reduction is 40%). In fact the total dissipation is almost entirely due to the module and the mass flows are low. The economic analysis highlights the same trend.

For other groups of electrical distribution boards with more complex layouts many conductors are present and the equipment is much less able to dissipate and therefore the % reductions in order to reach the BEP are very high, even with low supply currents. The costs analysis for these modules provides BEP values which can seldom be reached, varying from 80 to 90%.

7.5 Discussion and Conclusions

The aim of this work was to combine environmental and economic considerations with sustainable development. The study proposed a new way of addressing the issue of Sustainable Product Lifecycle by integrating DfE methodology and the LCA technique. In the procedure proposed the two modules of DfE and LCA interact at various times.

- The modular complexity of the product and the need reduce design times imposed a choice of items to be inserted in the LCI. This is a very serious problem which nearly all manufacturing firms have to resolve. In fact products with a great degree of modularity can frequently be found in manufacturing industries. The client can personalize these products to a great extent by choosing from a range of intermediate materials which are able to characterize the product purchased.
- The choices made in the previous step, when validated using LCA methodology, are used to create groups of codes which can be excluded from the study a priori, thereby further simplifying the tasks of the designer.

- The creation of an integrated system also allows for the rapid calculation of the impacts due to new solutions and for the definition of alternative environmental BEP solutions.

Following these steps the procedure proposed is general and can also be applied to other industrial realities.

The basic idea behind this DfE methodology is to bring environmental expertise directly to the designers, either by integrating the environmental expertise into the design process or by using a software tool that "speaks the language of the designers" and is integrated in their workflow. Thanks to the integration of the DfE tools in their daily workflow, the designers are able to assess the environmental performance of different design alternatives without being LCA experts and without additional effort. DfE tools calculate the environmental consequences of the design alternatives by using datasets and methodologies, which are integrated into the DfE database. The DfE tools provide algorithms and methodologies to calculate the environmental consequences of the current design alternative on the total life cycle of the product.

This work is part of a programe of activities which concern all the most important products of ABB Italia SpA aimed at obtaining Environmental Product Declarations (EPD). The EPD development process has created a great interest in environmental issues. It is important to point out that for a customer to be able to compare products from different suppliers, the product presentations must be standardized. The use of PSR provides a standardized method for creating an EPD to indicate environmental performance. It is also a long-term aim to use the EPDs to make the customers more aware of the environmental aspects connected with the products. These aspects are just as important as price, quality, delivery conditions, etc.

The environmental performance obtained thanks to the application of DfE (Design for Environment) methodology during the project design development stage are worthy of note.

The procedure setup may have future developments with the introduction of other DfE concepts:

- use of recyclable thermoplastic resins to partly replace the thermosetting resins;
- marking of the plastic components to help their identification and end of life recycling/recovery;

use of design solutions to simplify the dismantling of the distribution board at the end of its life, which, by allowing separation of the individual components, encourages its recycling and/or its correct waste disposal management.

Reference

ISO 14040 (2006) Environmental management-life cycle assessment-principles and framework, International Organization for Standardization, TC207

Chapter 8
Design a Sustainable Supply Chain

Every product generated, transported, used, and discarded within the supply chain (SC) causes a certain impact on the environment. This impact is a function of the material and energy consumed and of the wastes released in the product's whole life cycle, which in turn depend upon the type of the product and the technology used (Daniel et al. 1997).

Thus, it is important to examine all the procedures related to materials' flows, in order to opine regarding the environmental performance of SC. The first stage in the SC, including the selection of materials and methods for processing, may be regarded as product design. Product design and process technology typically determine the types of pollutants emitted, solid and hazardous wastes generated, resources harvested, and energy consumed (Sarkis 1995; Shrivastava 1995). During the next stage, the forward supply chain, first the materials are processed till the manufacture of the product. The manufactured items are usually packaged and then transported to distributors. Finally, the products end in final users e customers. There are cases though, mainly concerning job shop, where the products are directly transported from the producer to the end-user, without the intermediation of a distributor. The reverse supply chain starts when the product is no longer operable or when the end-user decides that he will not use it anymore. Given that products are not disposed uncontrollably, they are collected and transported to appropriate facilities, where a selection occurs: some of them are reprocessed and the rest are properly disposed.

The reprocessed items are finally redistributed and reused. In some cases the separation occurs away from the reprocessing facilities. A graphical representation of the materials flow is provided in Fig. 8.1.

Until lately, the main environmental emphasis has been on the manufacturing phase and to some degree on the disposal phase. This emphasis has given very good results, but at the same time, the numbers of products per household, energy consumption, and waste have increased greatly and have caused a larger environmental impact. In the past few years the environmental focus has shifted from

M. Bevilacqua et al., *Design for Environment as a Tool for the Development of a Sustainable Supply Chain*, DOI: 10.1007/978-1-4471-2461-0_8,

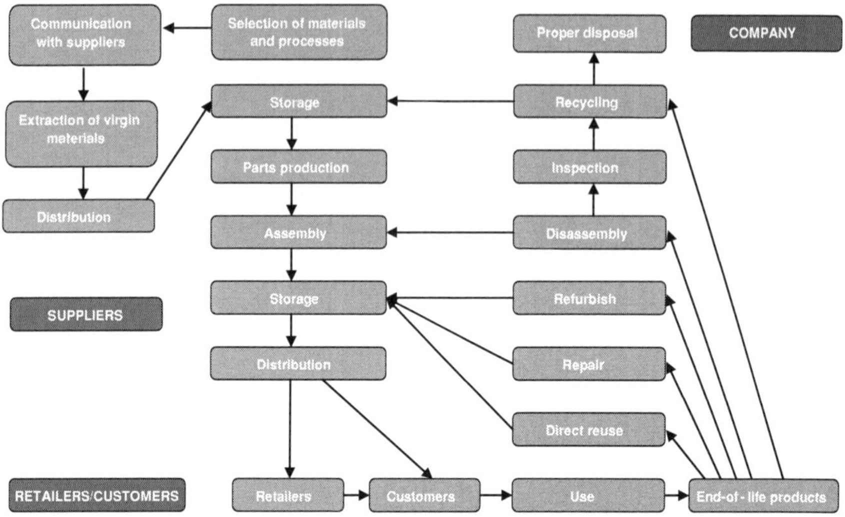

Fig. 8.1 The material flow (modified by Tsoulfas and Pappis 2006)

the manufacturing processes to the products. Therefore, the only alternative is to shift the focus to product design and minimize the environmental impact in the whole life cycle (Alting 1998). Indeed, the appropriate identification of all life-cycle stages of a product is necessary for the establishment or optimization of environmental policies. Any activity in the SC may have an undesired impact on the environmental chain. And vice versa, any disturbances in the ecological balance may affect production activities and social welfare in the long term. In this sense, the SC is connected at both ends with the environmental chain to establish a perpetual cyclical operation. The analysis of the cost-effectiveness of schemes for the recovery of products is difficult, however, it is useless to do so without looking at the whole process chain, combining the logistics aspects with the recovery ones (Nagel and Meyer 1999).

A set of approaches is applied in order to improve the environmental performance of SC. Cleaner production is the continuous application of an integrated preventive environmental strategy applied to processes, products, practices, and services to increase eco-efficiency and reduce risks for humans and the environment. Cleaner technologies extract and use natural resources more efficiently, generate products with fewer harmful components, minimize pollutant releases to air, water and soil during manufacturing and product use, and design durable goods that can be reused or recycled. Rather than, for example, capturing polluting substances after they have been produced (as with the end-of-process technologies), the goal is not to produce the harmful substance at all, or to produce less of it, or a less harmful one (Faucheux and Nicolai 1998).

Supply chain management (SCM) has had a substantial impact as a facilitator of globalization of the world economy.

It seems, however, that the society pays a high price for the economic advantages of globalization in terms of environmental shortcomings, which are today summarized by terms such as 'global warming, climate change, or carbon footprint'. Issues conceptualised under the umbrella of corporate social responsibility (CSR) have become common in many mission statements and annual reports from multinational corporations. CSR relates to transparency in financial reporting, sustainability reporting, and opportunities for stakeholder dialogue (Halldorsson et al. 2009).

The economic, social, and environmental dimensions are also recognised as the "three pillars of sustainability" (Hutchins and Sutherland 2008). The reactions to this development include the consideration of these issues in the design and operation of global supply chains.

A common denominator here is that environmental and social issues will not only affect the individual company but also the managed network of suppliers, producers, distributors, and customers; i.e., the supply chain will be influenced.

Some authors such as Dyllick and Hockerts (2002), Elkington (1998, 2001) highlighted the concept of Triple bottom line.

The triple bottom line consists of the following three parts:

1. Economy/profit: the economic dimension does not refer only to profitability. At any time, economically sustainable companies deliver cash flows that are sufficient to maintain liquidity and offer a constant, above-average return to the shareholders.
2. Ecology/planet: in the past, the environmental dimension has had the largest impact on sustainable development, as an eco-system represents the ultimate profit line. Dyllick and Hockerts (2002) define an ecologically sustainable company as a company that uses natural resources that are consumed at a rate below natural reproduction or at a rate below the developments of substitutes. Ecologically sustainable companies do not cause emissions that harm the environment, but are companies where managers limit the use of any type of resources as necessary and minimize any waste as much as possible.
3. Equity/people: the 'people' dimension could best be characterized as the company's social responsibility.

The social dimension refers to a growth strategy without decreased job quality and it reflects internal as well as external effects. According to Dyllick and Hockerts (2002), socially sustainable companies increase the human capital of individual partners as well as advancing the societal capital of their communities, in which they operate.

The consideration of the 3P (Profit, Planet, People) within the SCM concept will lead to sustainable SCM that can be defined as follows: Collaboration among supply chain members within all activities that concern the delivery of environmentally and socially responsible products and services to the end customer, as well as attaining acceptable profit and information in the supply chain Rabs and Bohn (2003). A sustainable supply chain as outlined in Table 8.1 includes the

Table 8.1 A sustainable supply chain (*Source* Halldorsson et al. 2009)

Supply chain stages	Triple bottom line dimensions		
	Environment/planet	Equity/people	Economy/profit
Supply of raw materials and components	Supplier evaluation and selection based on environmental profile, e.g., ISO 14000 Consolidation of shipments Sharing of information Use of eco-efficient transport modes Reuse of transport packaging materials Cooperation with suppliers to reduce environmental impacts	Supplier evaluation and selection based on social profile Training and education of logistics employees Ensuring codes of conduct at suppliers, e.g., safe working, no child labour, and no abuse of union rights	Transport savings Costs of supplier evaluation and monitoring Costs of internal and external audits of suppliers' compliance with codes of conduct Improved quality of products Reduced risk of damage to brand
Production	Elimination of waste and overuse of resources in the production process Environmentally friendly packaging Green design and manufacturing Eco-efficient production, e.g., waste from one company becomes input to another Replace hazardous materials and processes Recycling materials from used products	Automation of physical heavy work Minimization of specialized, repeating work Prevention of work Warehouse layout, that minimize picking distances In-service training of employees Improved staff recruitment and retention Job rotation and job enrichment	Improved working conditions may increase productivity Savings through resource minimization Economical gains through new product development Costs of certification, documentation and reporting
Distribution and reverse logistics	Choice of environmentally friendly distribution channels Choice of environmentally friendly types of transport Substitute information technology for physical transport Design effective return systems Reuse packaging materials	Reduced traffic congestion Education in energy saving driving Automation of loading and unloading Respecting driving and resting time rules	Savings due to consolidation of shipments to customers Savings due to increased capacity utilization of transport modes Higher prices for eco-friendly products Savings through increased reuse of materials and components

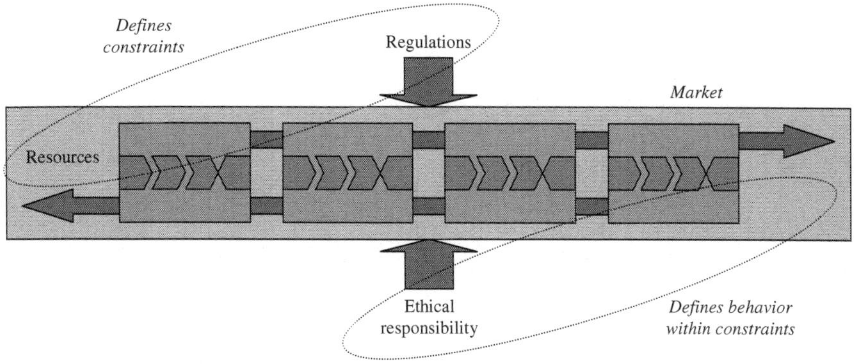

Fig. 8.2 Sources of environmental pressures affecting the supply chain (modified by Paquette 2005)

inter-organisational dimension as well as the value-added perspective, social and environmental issues.

8.1 Environmental Pressures and Supply Chain Response

Regulations represent just one source of environmentally motivated pressure, which affects supply chain decision-making. Although significant, this narrow frame of reference may be expanded to include three additional sources: resource availability, ethical responsibility of corporations, and consumer demands for environmentally advanced products and services. It is critical to understand the context and influence of pressures on the supply chain in order to respond effectively with technical and organizational innovation (See Fig. 8.2).

8.1.1 Regulations

Governments use a variety of regulatory instruments to address the environmental and health externalities associated with industrial production. These instruments include environmental directives, taxes and fees, and liability. All three affect the pricing and availability of products and services, and warrant consideration at the supply chain level. This section will describe the changing nature of environmental regulatory instruments as they may be applied to supply chain management.

Table 8.2 Summary of United States regulatory directives (*Source* Environmental Tax Policy Institute 2005)

Regulatory directives	Description
• Clean Air Act (1967, 1970, 1977, 1990, 1999)	Requiring development of National Ambient Air Quality Standards, Hazardous Air Pollution Standards, Motor Vehicle Emissions Standards, Fuel and Fuel Additive Standards, Aircraft Emission Standards, and authorizing provisions for ozone protection
• Clean Water Act (1972, 1988, 1981, 1987)	Authorizing regulation of wastewater facilities and non-point discharges and provisions for federal funding of municipal sewage treatment systems
• Resource Conservation and Recovery Act (1976, 1984, 1986)	Authorizing regulation and banning of the generation, storage, transport, treatment, and disposal of hazardous waste, as well as management of non-hazardous wastes
• Toxics Substance and Control Act (1976)	Authorizing regulation and banning of industrial chemicals that pose "unreasonable risk" to human health or the environment
• Comprehensive Environmental Response, Compensation and Liability Act (1980)	Allowing federal funding to remediate sites contaminated from prior unregulated disposal
• Superfund Amendments and Reauthorization Act (1986)	Authorizing the development of clean-up standards and provisions for increased public participation
• Emergency Planning & Community Right-To-Know Act (1986)	Authorizing the EPA to publicly report the release and storage of specified chemicals, and requiring emergency planning at the state level
• Pollution Prevention Act (1990)	Allowing provisions for agencies to support "cost effective" changes in production, operation, and raw material use through technical assistance and voluntary partnerships

8.1.1.1 Directives

The most commonly recognized examples of environmental regulation come in the form of directives, such as pollution limits, material bans, and fuel-economy standards. Regulatory directives set requirements for industry practices and performance. In the United States more than a dozen statutes form the primary legal basis for federal environmental regulations (for instance see Table 8.2):

Although this list comprises only a few of the more influential statutes from the supply chain perspective, it represents a discernible shift in the federal government's regulatory approach. Stringent "command and control" regulation of industrial point-source releases has given way to agency support for continuous environmental improvement and community risk management (Paquette 2005). While this shift has moved targets from "end-of-pipe" pollution control to process pollution prevention, current environmental regulations within the United States focus primarily on the *facility*. Facility personnel are responsible for implementing environmental health and safety activities, efficiency measures, and emergency planning. No formal mandate requires that environmental management processes and improvements extend beyond this domain. Further, while facility-focused regulations impact the cost of operations which very well may change the decisions of supply chain managers, they do not require that any factor beyond cost be explicitly considered.

Environmental regulations are increasingly focused on *consumer products*. Products embody the cumulative environmental impacts from production, use, and disposal. Therefore, regulatory directives aimed at improving the environmental attributes of individual products effectively impact the industry as a whole. In fact, product-focused regulation is ostensibly *supply chain* regulation, because changes to products drive changes to the design and operation of supply chains. Whereas regulations targeting manufacturing and transport activities at the facility level largely encourage either compliance or relocation of facilities (both of which are reflected in operation costs), regulations at the product level require new business processes both within the facilities that make up the supply chain and between them.

Today, there are at least three categories of regulatory directives that are focused on consumer products Table 8.3 shows the main aspects of these directives.

This discussion of product-focused directives is in no way exhaustive, rather providing a broad overview of present and future regulatory directions. Altogether, several broad conclusions may be drawn:

First, the global nature of today's markets and supply chains complicates regulatory compliance efforts. The broad and sometimes conflicting requirements of various regulatory bodies must be managed effectively, presenting an additional element of complexity to supply chain management. As such, there is considerable incentive to standardize environmental processes across the supply chain when possible. In the past decade, the United States has taken a much different approach to regulating industry than other nations—favoring environmental improvement through voluntary partnerships with corporations over more adversarial and legislative measures. While this shift may be preferable for supporting a market-oriented environmental response, it is likely that the more stringent regulations coming out of Europe and East Asia will set the standard for performance in all countries for better or worse.

Second, product-focused regulatory directives raise the stakes for the industry because they assign chief responsibility for environmental improvement to the most visible players in the production chain—the final manufacturers. A requirement that the product embody certain environmental attributes ensures that some level of improved environmental coordination occurred along the supply chain, regardless of whether or not the product was imported from a country with little to no environmental regulations. While regulations that required facility improvements affect operation costs along the supply chain, product-focused directives change the entire decision framework of the supply chain, influencing cost and adding environmental criteria to fundamental processes in sourcing, manufacturing, operations, distribution, and data management.

Table 8.3 Three categories of regulatory directives

Category	Description	Application
Performance requirements	Standards that address the environmental impact of products during their "use" phase are relatively established regulatory instruments, including product fuel economy, energy efficiency, and emissions standards	In the United States, sequential acts for National Energy Policy (1975, 1978, 1992) authorize the Department of Energy to regulate energy (and to a lesser extent) water efficiency in end-use equipment, appliances, and building systems, notably including Corporate Average Fuel Economy (CAFÉ) standards for passenger cars and light trucks. Use of such standards is increasing across the globe. The European Union passed the Directive on the "eco-design" of Energy Using Products (EU, Directive on the Eco-Design of Energy Using Products, 2005) which will harmonize and advance the already strict energy and water efficiency standards across the EU. It is likely that performance targets, as well as labeling and reporting requirements, will grow more stringent with time. These requirements place significant demands on product designers and also affect architectural, material, and process choices. Although it may appear that a change in product attributes has limited impact on the design and operation of the supply chain, a large body of research suggests that end-product design alterations affect the entire production system. Therefore, product innovation to meet mounting performance standards will affect fundamental supply chain functions—planning, sourcing, manufacturing, and marketing
Material mandates	Supply chain managers will be called upon to manage data, monitor supplier activity, and provide quality control while coordinating material transitions in existing	Mandates in the United States have moved beyond manufacturing emissions controls to regulate the use of select materials in consumer products. In concept, material

(continued)

Table 8.3 (continued)

Category	Description	Application
	product lines. Research increasingly correlates damage to the environment and human health to the use of toxic and hazardous materials	mandates are nothing new. The Food and Drug Administration has been regulating the materials of food, drugs, cosmetics, medical devices, and radiation-emitting electronics for over a century, representing a large portion of products that consumers purchase (US Food and Drug Administration, 2005). The Consumer Product Safety Commission sets guidelines for material use in consumer goods such as appliances, toys, clothing, and paint. Past mandates have focused on materials that may directly harm human health due to direct exposure, and include a variety of state and federal-level restrictions on products containing asbestos, lead, and mercury (US Consumer Product Safety Commission, 2005). Today, material mandates are being applied to a broader range of materials, products, and industries with arguably less direct health impacts. For instance, the European Union's Restriction on Hazardous Substances (RoHS) Directive is one of the more aggressive bans of materials in history (EU, "Directive on the Restriction of Hazardous Substances," 2003). The directive specifically targets the electronics industry and requires the phase out of lead, mercury, cadmium, hexavalent chromium and two groups of flame retardants in all products by 2008. This type of material mandate not only challenges the technical capabilities of product designers, but also the organizational capabilities across the electronics industry. Although materials for electronics are often selected far

(continued)

Table 8.3 (continued)

Category	Description	Application
		up the supply chain for commodity components, RoHS places responsibility for a complete bill of materials and certification on the final producers, requiring a level of information exchange and data management unprecedented in the electronics industry
Extended producer responsibility legislation	In an effort to reduce material waste, conserve resources, and prevent hazardous disposal, several countries have enacted the principle of extended producer responsibility (EPR) within statutory frameworks. EPR directives place financial responsibility for the collection and disposal of products at the end of their useful life on manufacturers, thereby aiming to create incentive to redesign products for reuse and recycling. EPR legislation, also referred to as "take-back," is attractive to policy-makers not only because it is a market-oriented instrument for environmental improvement, but also because it reduces the burden of waste disposal from individual municipalities (Walls 2005)	While deposit schemes for the recovery of aluminium cans and car batteries represent variations of "take-back" directives, EPR as discussed here has approximately a fifteen-year history beginning with packaging initiatives in Europe. The early efforts of several European countries were formalized in 1994 by the EU's Packaging and Packaging Waste Directive that stipulates national collection systems and recycling quotas (EU, "Directive on Packaging and Packaging Waste," 1994). A variety of public and private systems have developed in response, including Germany's Dual System which collects waste and coordinates recycling at a profit for producers who pay an upfront fee to display the "green dot" logo on their packaging. EPR directives have since targeted more complex products, including automobiles, appliances, and electronics. The more aggressive legislative efforts are coming out of East Asia and Europe, and include Japan's End-of-Life Vehicle Recycling Initiative (1996) and Home Appliance Recycling Law (2001), and the EU's Directive on End-of-Life Vehicles (2000) and Directive on Waste Electrical and Electronic Equipment (2002) although regulations have been adopted or proposed in Korea, China, India, Brazil, Venezuela, Chile, and some states within the United

(continued)

Table 8.3 (continued)

Category	Description	Application
		States as well. In order to comply with EPR requirements, companies must design, implement, and possibly operate comprehensive reverse supply chains (Toffel 2003). Representing no small endeavor, reverse supply chains may involve collection facilities, reverse logistics, partnerships with disassembly and recycling providers, integrated remanufacturing and reuse plans, and marketing initiatives to encourage consumer participation. Altogether, "take back" requires considerable organizational, technical, and financial commitment from the industry (Paquette, 2005)

Third, the optimal supply chain response to product-focused directives will be difficult to determine in the near future. Not only are global production systems increasingly complex, but such regulatory frameworks are relatively new, still evolving, and seemingly unclear about the ultimate environmental goals. For instance, it is unclear whether EPR legislation is intended primarily to minimize waste, reduce the toxic constituents of waste, encourage alternative waste disposal methods, or achieve a combination of these things (Paquette 2005).

Evidence from past governmental initiatives suggests that it is difficult to achieve multiple goals with one policy instrument (Walls 2005). For this reason, it may be presumed that future regulations will require multiple activities as an integrated response to multiple policy goals.

8.1.1.2 Taxes and Fees

Environmental taxes either "impose a tax cost on a product or activity that is environmentally damaging or they give a tax benefit to some product or activity that is environmentally beneficial" (Environmental Tax Policy Institute 2005). For example, in the United States, the federal government imposes an excise tax on ozone-depleting chemicals and offers a tax credit to people who buy electric vehicles. In this sense, environmental taxes do not replace regulatory directives, but rather help regulate the use of resources by visibly changing the purchase price. Environmental taxes, if applied aggressively and globally, may transform the way supply chains are designed and operated. For instance, suppose the United States levied a substantially higher gasoline tax, logistics systems might change dramatically in light of escalating transportation costs. This response could either foster regional supply chains and economic development or irreparably damage international markets. Environmental fees create the same effect, increasing the cost of select activities to environmentally preferable ends. Fees may be applied to landfill, hazardous waste, or raw material extraction, with ramifications that ripple along the supply chain.

While a large body of literature discusses the use of taxation to shape consumer behavior and raise government revenue, the direct impact of various taxation schemes on the management of global supply chains is not addressed. Environmental taxes and fees may be effective instruments for environmental progress, though arguably less effective for *supply chain* progress. In changing the visible price of a product or activity, supply chain decision outcomes may be different, but the decision framework and business processes in place may remain the same.

8.1.1.3 Liability

Another factor driving companies to improve their environmental performance is the risk of being held liable, or found negligent, for accidents or environmental damage. When a company experiences an accident or an incident with significant real or perceived environmental damage, the company may be held liable to pay for remediation. This is true even when the company is acting prudently and using state-of-the-art technology. If corporate action does not meet social requirements, companies also may face punitive actions for their behavior. To limit liability and negligence claims, a company may choose to implement strict risk reduction mechanisms.

To this end, IT can be used to continuously evaluate the impact on the environment from industrial processes. This can be done by monitoring emissions from facilities, tracking the transportation of goods and services, collecting information about the effects of certain chemicals, and monitoring the use of products by customers and throughout the product life cycle. In general, the lower the level of pesticides, biocides, and toxics (PBT) associated with a company's products, the lower the level of risk from such liabilities. Thus, many companies are now moving to assess the PBT content of their products and are attempting to find substitutes that lower this content. Reducing PBT content is also an effective tool to reduce perceived risks by customers. As the use of potentially harmful inputs is reduced, so are claims of damage.

Supply chain coordination plays a key role in limiting liability in three dimensions. First, to monitor a product's use throughout the supply chain, a company must be able to follow the product's flow. If a product is shipped to a customer who then uses it, perhaps mixing it with other products, and then passes it off to another party, it may be difficult to know how the product is used. If a client further down the supply chain uses the product and causes damage, the original producer still may be held liable. Proper design of supply chains is needed to eliminate such occurrences. Material Safety Data Sheets (MSDSs), mandated by the Occupational Safety and Health Administration (OSHA) for listed chemicals, have played an important role in explaining the proper use of chemicals in various production processes and ways to mitigate their environmental impacts.

Second, under the Comprehensive Environmental Response, Compensation, and Liability Act, companies face a joint and several liability standards. This means that even if only the final user of a hazardous substance disposes of it incorrectly, other companies that were involved in the process of production and distribution may be held liable to clean up the disposal site. This is especially problematic when the final user is unaware that a product may be hazardous and does not have the means to properly handle the substance.

Third, reducing liability in the supply chain can be achieved by not producing certain substances that have a high probability of being misused, or by choosing responsible supply chain partners.

Risk reduction activities may include training initiatives, product redesign, management of end-of-life products, and service offerings. For example, Greentech Assets, Inc. (2005) in Rhode Island offers recycling services specifically targeted at corporations aiming to limit the environmental and privacy risks associated with retired electronics. Ashland Chemical reduced their own liability and that of their customers by offering chemical services rather than sales. Ashland sells product on a "turn-key basis, taking on all the responsibilities of providing and disposing chemicals." In this sense, liability becomes an extremely effective regulatory instrument for several reasons. One, assigning liability to the most influential player creates incentive for the adoption and diffusion of environmental practices. Two, liability also invites pressure for environmental practices from insurance providers who underwrite industrial activities. Third and perhaps most importantly to supply chain processes, liability creates business opportunities to those companies that have invested in environmental literacy and services because they are able to reduce the risks associated with the activities of their customers' and the supply chain as a whole (Boyer and Porrini 2002).

8.1.2 Consumers and Ethical Responsibility

Markets create powerful venues for change since a savvy consumer base continually demands more value from products, services, and the organizations that offer them. In this sense, end consumers drive fundamental characteristics of the supply chain, including environmental performance.

8.1.2.1 Quality

Consumers demand quality products. As environmental awareness and expectations increase, so do demands for products with improved environmental qualities, including energy-efficient appliances, organic food and fabrics, recycled paper goods, and non-toxic cleaners. Past studies have shown that pinning down the exact status of environmental consumerism is challenging and subject to debate. Even as "79% of Americans consider themselves environmentalists and 67% state they would be willing to pay 5–10% more for environmentally compatible goods," (Roberts 1996) actual buying practices have *not* supported opinion polls. Consumers rarely accept environmentally preferred products with inferior performance, and very few are willing to pay a price premium for environmental attributes (Hoffman 2005). While environmental expectations may be high all around, many companies still view the green consumer as a niche market.

Regardless, the niche market has demonstrated consistent growth in recent years and currently comprises more products with improved environmental attributes than ever before. Sales in select product categories demonstrate this phenomenon:

- Organic: While the conventional food industry is generating a steady 2–3% per year growth, the organic industry has grown at rates between 17–20% annually for the past several years (Hansen 2004).
- Energy-efficient: Energy Star, a labeling program administered by the United States EPA since 1992 to reward the most energy-efficient products, has expanded to include 11,000 different models within 40 product categories, ranging from washing machines to light bulbs (US Environmental Protection Agency 2005).
- Non-toxic: Natural household cleaners, including laundry and dishwashing detergents, have risen in sales from $140 million in 2000 to $290 million in 2004.

Industrial sales mirror these trends. Purchasing Magazine reported in 2002 that "the most significant factor affecting supply, demand, pricing, and availability of solvents is the environmental issue." While demand for conventional solvents will be essentially flat at 0.2% per year growth, green solvents will post robust gains averaging 5.7% per year through 2005 (Atkinson 2002).

The issue of branding adds another element to managing consumer pressures for environmental performance. Research suggests that environmental expectations are higher when products are marketed with a strong brand. Since branding efforts essentially encourage consumers to develop an emotional attachment to a company's image and reputation, consumers in turn expect a relatively higher level of social and environmental performance. In fact, one of the most comprehensive surveys conducted in this area, covering 25,000 individuals in 26 countries, found that "more consumers base their impression of a company on its corporate social responsibility than do on (product) reputation or financial factors" (Roberts 2003).

The higher the profile of the brand, the more responsibility that company must take for environmental activities along its supply chain. Environmental activities, however, represent just one aspect of the broader corporate social responsibility (CSR) agenda which has gained wide appeal in the past 15 years. Also referred to as corporate citizenship, CSR involves the ethical treatment of employees, resources, the natural environment, communities, and nations in which companies operate. Non-profit advocacy organizations have evoked the concept of CSR to raise awareness and build pressure for more ethical corporate behavior. For example, Global Exchange launched an infamous campaign against Nike, Inc. for sub-contracting to "sweatshops" throughout South East Asia that employed children, required long hours, and maintained no environmental health and safety policies (Connor 2001). The Silicon Valley Toxics Coalition condemns brand name electronics manufacturers for toxic components and hazardous waste as a result of irresponsible disposal. Their seminal publication, "Exporting Harm: The High Tech Trashing of Asia," drew public attention to the practice of exporting electronic waste to be processed in parts of Asia (Puckett 2002). On the other hand, some companies such as Stoneyfield Farm and Aveda have built a name for themselves on a basis of CSR. The efforts of these companies may drive both consumer demand for environmentally advanced products and competitive pressure for more responsible behavior in general. In a time when marketing, media, and public relations define success for many high profile companies, pressure to project an image of corporate ethical responsibility is very high. While it may be relatively easy to pay tribute to CSR in annual reports, it appears considerably more challenging to implement and enforce practices along the supply chain that yield measurable environmental benefits.

8.1.2.2 Community Relations

Improved relations with local communities and other external stakeholders are becoming increasingly important for companies, as a matter of both law and of best practice. This has led companies to improve their environmental practices and the information they make available to the public concerning these practices. By doing so, a company can maintain its social franchise while enhancing its economic franchise. In communities, public interest groups are increasingly pressuring companies to practice environmental prudence and to prove those actions. The availability of easily accessible information about a company's environmental objectives and measurable performance criteria therefore can be expected to play an increasingly important role in assuring stakeholders that a company or facility is adhering to stated objectives.

A frequently cited example of the interaction of information and performance is in the area of toxics release inventory (TRI) reports. Parallel to the requirement to file TRI reports under the Superfund Amendments and Reauthorization Act Title III in 1986, many companies formed community advisory councils that played an important part in reducing toxic emissions. The fact that stakeholders had access to information regarding emissions, and the fact that this information was required by law, led companies and their communities to take action to reduce emissions.

A further example of the growing importance of public information systems related to a company's environmental performance is the requirement, under section 112(r) of the Clean Air Act Amendments of 1990, that risk management plans (RMPs) be developed by a number of facilities that store certain chemicals and that summaries of these RMPs be made available to the public.

8.1.3 Resources

Escalating global population and affluence create demand for more and more products. The corresponding rates of production inevitably place strains on the natural environment's ability to supply resources and absorb wastes.

An examination of the global supply and demand for fish illustrates this point well. The World Resource Institute reports that consumption of fish and fishing products has doubled in the past 30 years and has increased fivefold since 1950 (Kura et al. 2004). "Fish supply has become one of the major natural resource concerns, as seventy-five percent of commercially important marine and most inland water fish stocks are either currently being over-fished, or are being fished at their biological limit" (Kura et al. 2004). This situation bodes poorly for those in the fish business, including global corporations such as Unilever that sells fish and uses fish products as raw materials. Unilever is one of the world's leading suppliers of food, home care, and personal care consumer goods. In the mid-1990s, Unilever launched a comprehensive effort to secure a sustainable supply of fish. First, they provided seed money to the World Wildlife Foundation to research the situation and establish the Marine Stewardship Council as an independent organization to certify sustainable fish supplies. Then, they initiated discussion with competitors and national regulatory bodies in support of the Council's standards. Finally, Unilever publicly endorsed the work of the Stewardship Council and committed to purchasing only certified fish (Unilever's annual report 2005).

The availability of energy and water resources for manufacturing also presents a challenge to supply chain management. Water shortages are increasing worldwide as the demand for drinking and irrigation grows. The United Nations Environmental Program reports that one-third of the world's population lives in countries where consumption exceeds 10% of the total supply and more than 2.7 billion people will face severe water shortages by the year 2025 (United Nations Environmental Program 2002). Supply chain managers must consider resource constraints when locating facilities and planning operations, since energy and water shortages may dramatically affect business.

While it may be easy to take for granted the availability of natural resources to support industrial activities, resource constraints represent a systemic environmental pressure.

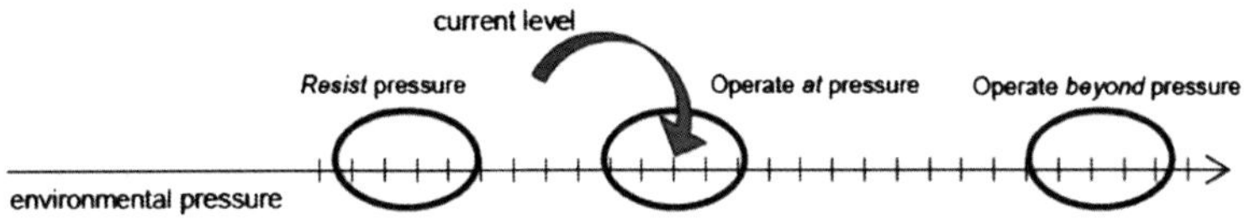

Fig. 8.3 A response to environmental pressures requires an environmental operating model. (*Source* Paquette 2005)

8.2 The Supply Chain Response

As environmental pressures increase and require action at the supply chain level, a company must choose (1) to operate beyond environmental pressures, (2) to operate at environmental pressures, or (3) to resist environmental pressures (See Fig. 8.3).

This categorization of environmental operating models is not a new concept. Several researchers have described various corporate environmental orientations in a similar way. Kopicki (1993) presents three approaches in environmental management: the reactive, proactive, or value seeking. Walton et al. (1999) offers a comparable model in characterizing the purpose of environmental activity as either "comply with the letter of the law," "clean up," or "be proactive." Klassen (2004) describes the continuum of behavior from reactive to proactive orientations in several publications. Ron (1998) designates environmental strategy as following, market-oriented or sustainability-oriented. Finally, Murphy (1996) introduced a survey tool that classifies companies across industries as environmental progressives, moderates, or conservatives.

Paquette (2005) proposed "*Balanced operational objectives*" in order to respect environmental pressures. Classic supply chain objectives are described by the Supply Chain Council to include reliability, responsiveness, flexibility, cost, and asset utilization. A "balanced set" may include only one or two of these operational objectives depending on the designated operating model. For instance, a corporation may focus on supply chain efficiency and may employ metrics such as line-items-picked-per-hour or cash-to-cash-cycle-time to indicate performance. With regard to the environment, operational objectives may be developed for each environmental *operating model* in response to each type of environmental *pressure* as proposed in Table 8.4:

Suppose a corporation elects to operate *at* regulatory pressure. This corporation's operating objective, therefore, is to comply with all regulatory directives that affect its activities with the least disruption to other business processes. Metrics such as number-of-non-compliance-incidents, or fines-for-non-compliance may be selected to indicate direct environmental performance. Metrics such as cost-to-compliance and time-to-compliance may be used to indicate efficiency and environmental performance. A large body of research discusses the application of metrics to indicate direct environmental performance, such as energy use or total

Table 8.4 Environmental operational objectives (modified by Paquette 2005)

		Environmental pressures		
		Resources	*Regulations*	*Markets*
Environmental operating models	*Operate beyond pressure*	Substitute Expand	Obviate the need for Exceed	Drive Create
	Operate at pressure	Conserve Secure	Comply	Meet Satisfy
	Resist pressure	-	Breach Relocate	Exit Ignore

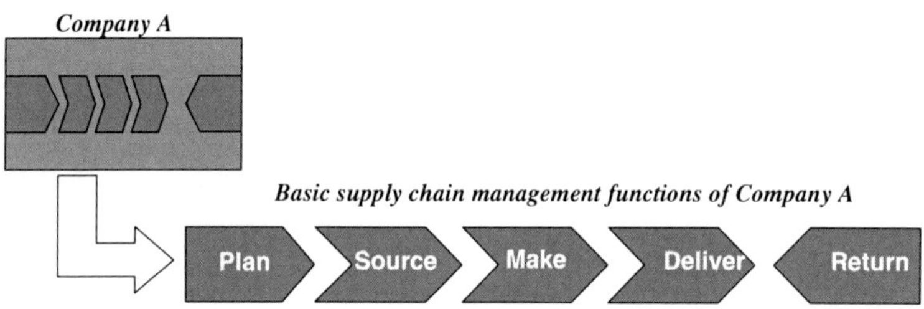

Fig. 8.4 Basic supply chain management functions as defined by the Supply Chain Council

waste generated (Gregory et al. 2004). An interesting extension of this research involves the development of metrics to indicate environmental performance of an entire supply chain (Clift 2003).

While comprehensive supply chain management may require hundreds of processes to be performed in a structured manner, the greatest operational and financial benefits result from concentrated efforts on a relatively small number of unique business processes. The same may be said about environmental benefits: an excellent supply chain with respect to environmental performance focuses on a small number of processes that are aligned with environmental operating objectives. Consider environmental processes arranged by the most basic functions of supply chain management as defined by the SCOR model (Fig. 8.4).

Each function provides a contribution in response to environmental pressures.

According to Paquette (2005) the chief variables that influence the environmental performance of a product or system are determined during the **planning**

phase. A number of processes may be used to aid environmental decision-making while planning the supply chain.

- *Environmental cost accounting* is a technique to identify and assign discrete costs to environmentally harmful activities within a broader system. This topic has been carried out in the chapter "*The costs of a sustainable supply chain*".
- *Environmental life cycle analysis* is a method used to identify and evaluate the environmental impacts associated with a product or service throughout its entire life from material extraction to eventual disposal and assimilation into the environment. This topic has been carried out in the chapter "*Environmental regards in product life cycle*".
- *Design for environment* is an approach to reduce the environmental impacts of a product by introducing specific design criteria during the product development phase, such as "design for recyclability" or "design for energy efficiency." This topic has been carried out in the chapter "*Design for Environmental concepts*".

Sourcing professionals may consider the environmental attributes of materials, components, and products, as well as the environmental performance of the suppliers' direct activities using the following processes.

- *Environmental auditing* is a procedure to verify the environmental performance of a material, component, product, or facility. Auditing may be conducted by a third-party organization or the buyer in accordance with previously established environmental guidelines. Many multinational companies, including Limited Brands, Inc., Texas Instruments, and General Motors have designated standards and routine audit suppliers for environmental performance. Internal auditing is also widely promoted as part of the ISO 14000 environmental management standards.
- *Environmental certification* is a guarantee that a product or facility meets environmental standards defined by a third party. Certification typically involves product labeling for consumer marketing in response to regulatory pressures or consumer demands for products with improved environmental attributes. Examples of prevalent certification programs include Green Seal (Green Seal Product Certification 2005), Germany's Blue Angel (Germany's Blue Angel Certification 2005), Certified Organic (2005) and the building industry's Leadership in Energy and Environmental Design certification (US Green Building Council 2005).

The **make** function is involved through:

- Pollution prevention: an approach to preemptively identify and alter activities that create waste. Prevention techniques including substitution, product modification, improved maintenance, and recycling have been successfully applied at several facilities following the Pollution Prevention Act of 1990 and several state-level regulatory directives.

- Environmental management systems: sets of processes that enable an organization to identify, monitor, and address the environmental impacts of its activities. Systems typically include guidance for employees in environmental health and safety procedures and facilitation tools for continual improvement of environmental performance.

The environmental implications from **Delivery phase** is growing, as materials, components, and finished products travel longer distances through production and distribution cycles. The total impact of delivery functions correlates to two variables that logistics professionals manage directly: transportation distance and mode. "Green" logistics is an approach that considers the environmental impacts of procurement, transport, inventory control, and distribution activities along with other considerations in order to minimize environmental costs. For example, in addition to considering monetary cost, time, and reliability of freight service, one may also consider the volume of carbon dioxide emissions. This aspect will be considered in the case study "*Carbon footprint in the textile supply chain*" of this book.

Return processes are gaining in strategic importance as companies compete further to maintain customers, recover assets, minimize liability, and meet extended producer responsibility regulatory requirements.

- Reverse logistics is a set of activities to collect, transport, and manage products and materials after sale and delivery to the customer. Reverse logistics has been typically used to facilitate unsold products and warrantee returns, and it is being further developed to address "take back" regulatory obligations and to pioneer concepts of closed-loop supply chains. This subject represents an important area of emerging research within supply chain management and has been studied in Sect. 3.3.4 of this book.
- Remanufacturing is a process to clean, repair, and restore used durable products to good condition for resale. Remanufacturing is typically integrated with reverse logistics processes because valuable products and components must be appropriately transferred from the consumer to the manufacturer. In addition to logistical challenges, remanufacturing involves serious technical, planning, and inventory management challenges.
- Recycling is a procedure to reuse materials, which may otherwise be considered waste, in a form other than primary use. Recycling is facilitated by return processes in part because existence of a secondary market depends on the quality of recycled materials. Whether recycling recovered materials or using purchased recycled content in production, processes require additional planning due to fluctuations in material timing and availability.

Box E: How Could SCM Deal with Sustainability? (Extract from Halldorsson et al. 2009).

The intersection of SCM and sustainability can be understood by three different approaches:

- An integrated strategy, where sustainability is fully consistent with SCM.
- An alignment strategy, where sustainability is complementary to the traditional SCM focus on costs and service.
- A replacement strategy, where the traditional SCM concept is replaced by an alternative approach to cope with the environmental and social aspects.

An integrated sustainability strategy for SCM

This is the approach taken in research streams: reverse logistics, product stewardship, and green SCM. On the supply side, suppliers are chosen, developed, and monitored based on their compliance with international codes of conduct, and this provides sufficient comfort to industrial buyers, who want to know about the origin of raw material and components (Maloni and Brown 2006; Mamic 2005). The products are designed for the environment by eliminating hazardous or harmful materials and making recycling and disposal easy (Srivastava 2007); the current systems and solutions can already cope with these circumstances. The production process is organized to reduce waste of materials, emission of gases and polluted water, and minimize the consumption of non-renewable energy resources. Transportation and distribution is organized to minimise total mileage, maximize capacity utilization by consolidation of shipments, and to use environmentally friendly transport modes when possible; this is seen as coherent with the logic of efficient distribution systems. The reverse logistics system is organized to maximize the value creation of the returned products, whether it is end-of-life products that are recycled or remanufactured or commercial returns, which are taken back to the market as soon as possible (Jayaraman and Luo 2007).

An integrated strategy is thus a strategy where the focus has changed to include not only the traditional costs and service considerations, but also the social and environmental impacts. Thus, the responsibility of greening the supply rests on everyone in the supply chain; from design of products, supply of materials, and components, through production processes and delivery to the customers, and finally the return recycling processes (Handfield et al. 2005). The measurement of supply chain efficiency such as costs, dependability, quality, reliability, and speed can be captured with measures of environmental and social impacts.

By integrating social and environmental objectives and performance criteria into the strategic and operational decisions in their supply chains, the firms can make the appropriate balance between costs, service, and environmental and social impacts.

An alignment sustainability strategy for SCM

Following an alignment strategy, economic, social, and environmental concerns will be balanced against each other. They are considered as complementary, meaning that they all have to be taken into account simultaneously, when companies are making important decisions regarding the design or operations of their products and supply chains. The triple-bottom line approach is an example of an alignment strategy (Elkington 1998), and tradeoffs between these three dimensions are made to meet the desired output of a particular process.

The same is true for CSR (Maloni and Brown 2006). The main difference between an aligned strategy and an integrated strategy is that the overall objective in the integrated strategy is to increase value creation to customers. If the customers ask for environmental and social concerns these are integrated in the planning, design, and operations decisions. If not, the company will comply with existing laws and regulations. An aligned SCM strategy puts equal weight on profit, people, and planet, irrespective of the customers' requirements. Social and environmental issues are part of the company's mission statement and the company accounts for all three aspects in its annual reports. On some occasions, measures such as speed and reliability must be replaced by, e.g., 'energy use, material use, carbon footprint and waste/landfill'. An aligned strategy focuses on cooperation and competence building in the entire supply chain. It is not enough to claim and control that the suppliers comply with the focal company's codes of conduct.

It is also necessary to cooperate with the suppliers and help them to improve the working and environment conditions.

A replacement sustainability strategy for SCM

It could be argued that the full implementation of the idea of SCM is contradictory to sustainability. If, for example, an agile supply chain design can increase the responsiveness to meet customer requirements, this may happen at the cost of resources (inventories, warehouses) and CO_2 emissions (e.g., by use of airfreight from China to Europe instead of container transport).

The main objective of SCM is to maximize customer value in the most effective and efficient ways possible. In order to survive in the global competition, companies extend their supply chains to further distant locations in order to reap the benefits of differences in labor costs and take advantage of efficient transportation networks. Outsourcing and offshoring to the Far East, South America, and Eastern Europe has increased dramatically during the past decade. This shift has increased the distance between production and consumption considerably with negative results on the use of non-renewable energy resources and emission of greenhouse gases.

From this perspective, the success of the extended enterprise is positive for the revenues of the companies, but negative for the environment. Therefore, a call for a paradigm shift is needed. Instead of continuing to squeeze every penny out of the total costs of products, firms have to reconsider how and where they produce their products, and customers have to reconsider their decision criteria for buying the products and the way they dispose them after use. A possible way to change the traditional SCM approach could be to measure the carbon footprint of the product throughout the supply chain over the product's life cycle, including the disposal and recycling (Grenon et al. 2007). A label on the product indicating the total amount of CO_2 per unit will tell the customer how green the supply chain of the product has been. However, calculating the carbon footprint of a product through the supply chain is not an easy task. First, a global standard for measuring carbon footprint is still lacking.

Second, carbon footprint is only one aspect of the environmental impacts of a product. Other aspects are related to water waste, air pollution, energy use, use of raw materials, etc. Third, the implication of focusing on the carbon footprint may be a major change in the design of the global supply chain in terms of location of production, choice of distribution channel and transport mode, selection of suppliers, etc.

One possible consequence might be a local-to-local approach instead of the current global-to-local approach. This is actually taking place in various countries; take for example the 'farmers markets' in the UK that are now competing with supermarkets. Local market squares in towns and cities around UK regularly (weekly or every fortnight) arrange a sales outlet for local farmers and food producers. Such community efforts focus heavily on the sustainable agenda.

First, the food must be produced locally, i.e., challenging the distance and volume of global supply chains. Second, packaging is at a minimum, where the extensive use of packaging is also due to long distances, dispersed responsibilities, and complex liability in the global supply chain. Third, products are fresh and preferably organically grown, i.e., challenging the artificial maturation process during long-distance transport in refrigerated ships and trucks. This is seen as complementary to local shops, and to the business prosperity of the local community; farmers and producers must operate within a 30–50 mile radius from the market outlet. Should global supply chains now turn local as regards other products? Pearce identifies thereby how difficult it is—from a consumer's perspective—to act in an 'environmental-correct' manner. He calls some of the relocalization strategies also patriotism-strategies and shows that in some cases it is better to buy green beans from Kenia instead of from England as most of the CO_2 emissions stem not from transport but from the production process (Pearce 2009).

8.3 The Costs of a Sustainable Supply Chain

Supply chain efficiency and cost reduction programs will not fully succeed unless managers consider critical environmental, health, and safety activities.

Companies of all sizes are further enhancing these fundamental supply chain changes by considering the environmental impact—and related bottom-line effects—of their decisions and actions.

They have increased their competitiveness by engaging in such environmental performance-enhancing activities such as:

- Reducing the obsolescence and waste of maintenance, repair and operating (MRO) materials through enhanced *sourcing,* and inventory management practices.
- Substantially decreasing the costs associated with scrap and material losses.
- Lowering the training, material handling, and other extra expenses associated with hazardous materials.
- Increasing revenues by converting wastes to by-products.
- Reducing the use of hazardous materials through more timely and accurate *materials tracking* and reporting systems.
- Decreasing the use and waste of solvents, paints, and other chemicals through chemical service partnerships.
- Recovering valuable materials and assets through efficient *product takeback* programs.

Yet, despite the potential for significant financial gains, most supply chain managers currently do not focus on environmental concerns. One reason for this is that cost accounting systems typically hide the frequency and magnitude of the "environmental costs" that companies incur. While raw material and labor costs are directly allocated to the appropriate product or process, other costs are accumulated into overhead accounts, which are allocated proportionally (e.g., based on the number of units manufactured) to all products, processes, or facilities. This allocation method might be appropriate for many overhead costs, such as rent and upper management salaries. However, this approach can lead to inaccurate costing and ineffective decisions when significant costs—waste disposal, training expenses, environmental permitting fees, and other environmental costs—are not allocated to the responsible products and processes.

For these reasons, supply chain managers often cannot achieve their overall objectives unless they tackle important environmental concerns.

Many companies have tackled this issue by using *environmental accounting* techniques to substantially reduce supply chain costs.

With these costing methods, companies can systematically identify environmental costs throughout the supply chain, e.g., costs associated with management of hazardous materials, which typically are not captured through conventional accounting methods. Once the costs (or potential benefits) have been identified, companies can analyze the cost drivers and evaluate alternative cost reduction opportunities.

A number of companies have successfully applied environmental accounting methods to supply chain management decisions. Some examples from well-known companies include:

- GM reduced its disposal costs by $12 million by establishing a reusable container program with its suppliers.
- Commonwealth Edison, a major electric utility company, realized $25 million in financial benefits through more effective resource utilization.
- Andersen Corporation (2000) implemented several programs that reduced waste at its source and had internal rates of return (IRR) exceeding 50%.
- Public Service Electric and Gas Company saved more than $2 million in 1997 by streamlining its inventory process to avoid product obsolescence and disposal. On a similar note, many companies are also benefitting from chemical service programs through which they outsource such responsibilities as chemical inventory management, distribution, training, and waste management. Companies in industries as varied as the semiconductor, automotive, and aerospace industries have achieved enormous materials management cost reductions, reduced production downtime, and significantly decreased solid waste and air emissions. Significant savings are possible because the costs to manage chemicals often range from 100 to 1,000% (and sometimes up to 1,500%!) of the costs to buy these materials (Bierma et al. 1999). With the help of the chemical service provider's expertise, environmental gains are coupled with substantial operating cost savings to achieve win–win improvements.

The key to achieving these improvements is changing the company decision-making processes to incorporate environmental information.

Materials management decisions affect many dimensions of company performance, including operating costs, investment requirements, product quality, and meeting delivery schedules. While materials managers typically address these business objectives during the decision-making process, environmental concerns are commonly overlooked. Unfortunately, failure to incorporate these "hidden costs" not only hinders a company's efforts to reduce a variety of environmental burdens, but also hinders efforts to improve financial performance. Materials managers typically do not address environmental concerns due to the structure of traditional cost accounting systems. While raw material and labor costs are directly allocated to the appropriate product or process, other costs are accumulated into overhead accounts. These overhead costs are in turn allocated proportionally to all business units, product lines, or facilities, typically based on gross sales or output. When significant costs are not allocated to the responsible products and processes, this approach may lead to inaccurate costing data and ineffective decision-making. Environmental costs are often misallocated in this manner. For example, take the case of a company's wastewater treatment facility. The costs of operating the facility are predominately caused by a few of the company's products whose

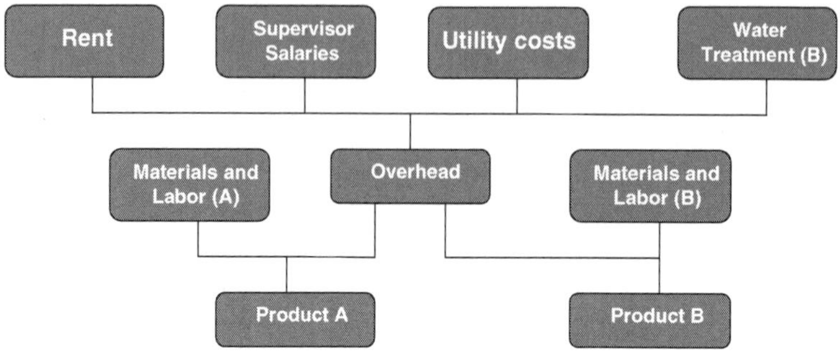

Fig. 8.5 Misallocation of environmental costs

production generates significant quantities of wastewater. If, as shown in Fig. 8.5, the costs of operating the treatment facility are accumulated into an overhead account and allocated equally to all of the company's products, the wastewater costs are obscured and product cost information is misleading. In this case, product B appears to be less expensive to produce than it actually is. Figure 8.6 demonstrates a correction to the allocation problem and depicts the primary goal of many environmental accounting efforts: to track environmental costs directly to the responsible product, process, or facility.

When environmental costs are hidden in overhead accounts, business decisions are made without sufficient consideration of the potentially costly environmental impacts downstream of the decision. For example, product design decisions that specify the use of hazardous materials inherently increase the risk of employee exposure or other incident. Lack of awareness can be financially detrimental to a company.

For example, the costs to landfill waste materials are generally accumulated into an overhead account and would therefore be "hidden" and not incorporated into product pricing or other decisions. However, several firms have applied activity-based costing methods to make these costs "conventional" and are allocating the landfill fees directly to the products or processes that generate them. In addition, interrelationships exist between the cost categories. For example, if a new production process required the use of hazardous materials, the expenses that a company might incur to clean up hazardous material spills would be classified as "contingent" costs. However, any future spills might also trigger "image/relationship" costs, such as concern among the company's employees or neighbors, and "external" costs, such as damage to a nearby aquatic ecosystem. The purpose of the cost framework is to help to identify and address the full set of consequences that might result from materials management decisions.

This cost framework is not limited to analyzing environmental costs and benefits but rather is universally applicable to any type of financial impact. For example, a purchasing agent might be able to lower the company's conventional costs by switching to another approved supplier that provides a lower priced

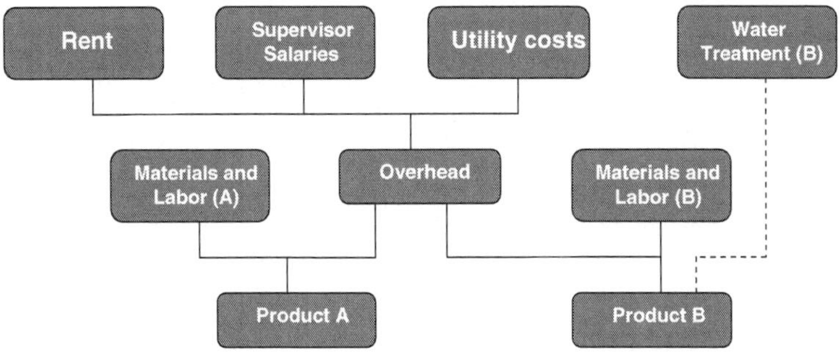

Fig. 8.6 Improved allocation

material. However, this switch could lead to increased waste disposal costs, a potentially hidden expense, if the new supplier shipped materials in containers with excess packaging. Similarly, delivery missteps by the new supplier could inadvertently cause a variety of contingent costs if these breakdowns resulted in significant production delays. It is easy for materials managers to overlook environmental costs and benefits during decision-making because they tend to occur upstream or downstream of the immediate decision, e.g., a purchasing action can have materials handling, storage, and disposition repercussions.

This focus on a single functional area is not limited to environmental considerations, and the resulting problems are a primary motivation for supply chain integration efforts. However, an increasing number of companies have discovered ways to reduce operating costs or otherwise improve performance by implementing practices that optimize supply chain, rather than functional area, performance.

EPA (2000) provided a four-step framework for identifying and using environmental information to improve financial performance. Box F reports the main steps of this method.

Box F: The Four Steps: An Overview

Step 1, Identify Costs, a systematic review of the facility or process is conducted to determine if and where significant environmental costs occur. This analysis enables the team to later focus where the probability for significant improvement is greatest (See Fig. 8.7).

The method proposes to conduct routine EH&S performance reviews to uncover sources of environmental costs. These reviews of facilities' processes can be carried out in order to determine levels of waste and pollution, health and safety risks, and effectiveness of EH&S management systems. These reviews consist of interviews with appropriate personnel, observations of day-to-day operational practices, and reviews of accounting and

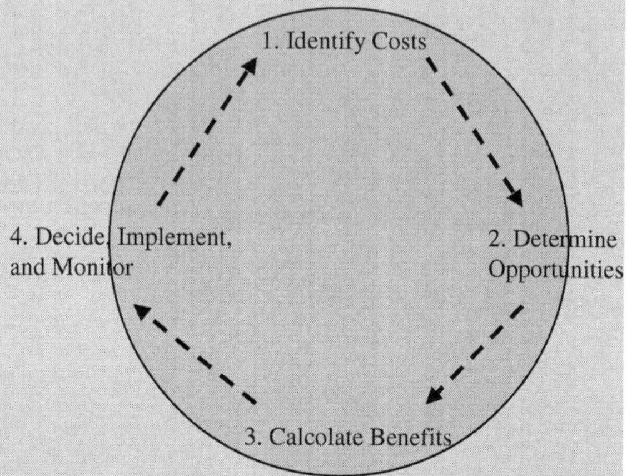

Fig. 8.7 Decision-making framework (*Source* EPA 2000)

manufacturing records. The abbreviated questionnaire in Tables 8.5 and 8.6 contains some key questions drawn from existing review checklists.

EHS Performance Review—Sample Questions
Purchasing

- *Have suppliers supported the company's efforts to reduce the facility's quantities and costs of waste?*
- *Are environmental, health, and safety performance criteria (e.g., flammability, biodegradability, toxicity, recyclability, and other environmental or regulatory requirements) clearly articulated in new product specifications to suppliers?*
- *Does the plant accept samples from chemical suppliers?*
- *Are suppliers required to take back unused samples they provide?*

Handling

- *Are all raw materials tested for quality before being accepted from suppliers?*
- *Are plant material balances routinely performed?*
- *Does the company's personnel training program include information about the safe handling of raw materials, spill prevention, proper storage techniques, and waste handling procedures?*

The questions and tables in this list are taken largely from a series of Guides to Pollution Prevention published by the U.S. Environmental Protection Agency. The Guides span a wide array of industrial sectors and each includes waste minimization assessment worksheets. (various publication numbers, all with the format EPA 625/7-9X/0XX). See www.epa.gov/ttbnrmrl/Guides.htm.

Storage

- *Are specialized containers or storage facilities required to prevent environmental damage?*
  Is inventory used in first-in-first-out order?
- *Is there a computerized inventory system to track inventory and material waste (e.g., bar-coding, MRPII, etc.)?*
- *What information do(es) the system(s) track?*
- *Are all storage tanks routinely monitored for leaks?*
- *If yes, describe the procedure and monitoring frequency for above-ground/vaulted and underground tanks:*

Disposition and Material Recovery

- *Is your solvent waste segregated from rinse-water streams and other wastes?*
- *Does the plant generate waste streams that contain valuable process chemicals or metals?*
- *Are all empty bags, packages, and containers that contained hazardous materials segregated from those that contained non-hazardous wastes? Are containers properly cleaned prior to disposal?*

Step 2, Determine Opportunities, the identified functional areas and processes are evaluated to determine which changes will likely yield significant cost savings and reduce environmental impacts. Potential changes are evaluated with criteria that can include the magnitude of potential cost improvement, the types of environmental burdens, and the barriers to change. Step 2 yields a set of possible alternatives with significant potential for improving costs savings and reducing environmental impacts.

Many companies have found that the Pareto (Fig. 8.8), or 80/20, principle applies, i.e., that a few supply chain improvements provide most of the achievable gains (Pojasek 1998).

These initial screening steps lead to Step 3, Calculate Benefits, where quantitative, and sometimes qualitative, analyses of the costs and benefits of a selected group of projects are conducted. Some of the analytical tools and methods used during this step are activity-based costing approaches, *net present value* (NPV) calculations, and risk evaluations. In EPA742-R-00-XXX (2000) a case study from Andersen Corporation was analyzed:

Andersen Corporation, the manufacturer of Andersen Windows and Patio Doors, applied several quantitative methods when evaluating an automated

Table 8.5 Input materials summary

Attribute	Description		
	Stream #	Stream #	Stream #
Material name/ID			
Source/supplier			
Hazardous component			
Annual consumption rate			
Purchase price, $ per _			
Overall annual cost			
Material flow diagram available (Y/N)			
Delivery mode			
Shipping container size & type			
Storage mode			
Transfer mode			
Control mode			
Empty montainer misposal/management			
Shelf life			
Supplier would			
accept expired material (Y/N)			
accept shipping containers (Y/N)			
Acceptable substitute(s), if any			
Alternate supplier(s)			

Table 8.6 Waste Sources

	Significance at plant		
	High	Medium	Low
Hazardous obsolete raw materials			
Non hazardous obsolete raw materials			
Spills and leaks (liquids)			
Empty container cleaning			
Container disposal			
Evaporative losses			
Off-spec materials			
Pipeline/tank drainage			
Laboratory wastes			
Contaminated wipes and gloves			
Other			

paint blending system. As the regulatory pressures and costs associated with its painting lines became more significant, Andersen began searching for ways to reduce emissions and material costs. After reviewing several alternatives, managers conducted an in-depth analysis of the most promising approach: a point-of-use paint mixing system. This "meter mix" system

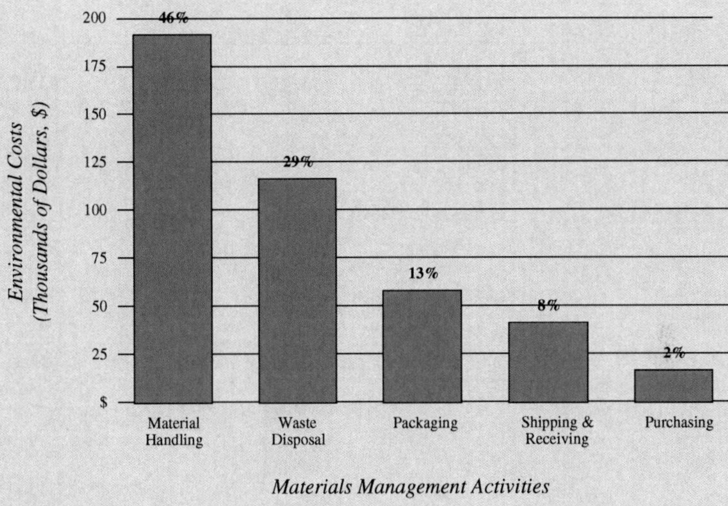

Fig. 8.8 Pareto Diagram

replaced the existing batch system. They evaluated several cost and material usage areas, including:

- Paint materials: purchasing and shipping costs.
- Waste: treatment, transport, and disposal costs.
- VOC emissions: associated fees.
- Solvent materials: purchasing and shipping costs.
- Solvent emissions: material losses and associated fees.

In addition, the managers factored in the labor and expenses associated with the following:

- Raw material handling and storage.
- Waste handling, storage, and disposal, as well as related training activities to ensure that waste materials were properly handled and disposed.
- Analysis, reporting, and record keeping associated with the paint line.
- Material obsolescence.

In each of the above four cases, the team determined material usage rates or described the financial impacts.

However, the team did not calculate actual dollar figures for these costs, but rather simply recognized that the total financial impacts were significant. This qualitative information provided important insights that supplemented

Table 8.7 Economical analysis

	Year 0 (Installation) ($)	Year 1 ($)	Year 2 ($)	Total (Years 0–5) ($)
Investment				
Equipment	(115.541)			(115.541)
Installation and other expenses	(14.559)			(14.559)
Total investment	(130.100)			(130.100)
Costs				
Operating costs		(109.355)	(115.302)	(623.197)
Total costs		(109.355)	(115.302)	(623.197)
Savings				
Paint use and waste reductions				
Paint purchase and shipping		110.374	113.685	585.991
Waste treatment, transport, disposal		14.387	15.106	79.497
VOC emissions and associated fees		162	170	895
Dilute solvent use and waste reductions				
Solvent purchase and shipping		58.710	60.471	311.699
Solvent emission losses and fees		560	588	3.094
Flush solvent use and waste reductions				
Solvent purchase and shipping		10.687	11.008	56.739
Solvent emission losses and fees		130	137	718
Total savings		195.010	201.165	1.038.633
Net Benefit	(130.100)	85.655	85.863	415.436

the more precise calculations. As shown below, installation of the meter mix system was attractive because the quantified costs yielded an 18-month payback and 58% internal rate of return. The qualitatively evaluated activities further strengthened this decision to proceed. The payback calculations are relatively straightforward, as shown below.

$$P = I/M$$

Where

P = Payback period (months)
I = Investment ($)
M = Monthly savings ($/month)

Based on the forecasts in Table 8.7, the initial investment (I) was $130,100 and the monthly savings (M) during the first 2 years averaged $7,146. With these values, the payback period was 18 months.

Notes and assumptions:

- "Operating costs" are additional costs required to operate point-of-use system.
- 3% annual increase in material and labor costs.
- 5% annual increase in all other costs, e.g., waste management.

In Step 4, Decide, Implement, and Monitor, the team shifts from evaluation to implementation. First, a decision is made to continue with the status quo or to pursue a new approach. Financial benefits and/or environmental improvements then occur as changes are put into action. The new practices institutionalized as information collection processes are integrated into the company's materials resource planning (MRP II), enterprise resource planning (ERP) systems, and other information systems. After implementation, a periodic review and continuous improvement effort allows decision-makers to evaluate their progress and pursue additional opportunities.

References

Alting L, Hauschild M, Wenzel H (1998) Elements in a new sustainable industrial culture environmental assessment in product development. Robot Comput Integr Manuf 14:429–439

Andersen Corporation, Ashland Chemical (2000) EPA742-R-00-XXX, forthcoming in 2000. www.epa.gov/opptintr/acctg

Atkinson W (2002) Demand for green solvents will boom. Purchasing Magazine Online, October 10. www.purchasing.com/article/CA250861.html. Accessed 20 May 2005

Bierma TJ, Waterstraat FL (1999) Chemical management: reducing waste and cost through innovative chemical supply strategies. In: Chemical Strategies Partnership, Tools for Optimizing Chemical Management Manual, forthcoming, Wiley, New York. www.chemicalstrategies.org

Boyer M, Porrini D (2002) The choice of instruments for environmental policy: liability or regulation? An introduction to the law and economics of environmental policy: issues in institutional design. Res Law Econ 20:1–41

Certified Organic Food Standards (2005) http://www.ifoam.org/about_ifoam/standards/pgs/PGSDefinitioninEngFrenSpanPort_web.pdf

Clift R (2003) Metrics for supply chain sustainability. Clean Technol Env Policy 5(3–4):240–256

Connor T (2001) Still waiting for nike to do it. A Global Exchange Report. www.globalexchange.org/campaigns/sweatshops/nike/stillwaiting.html. Accessed 5 June 2005

Daniel SE, Diakoulaki DC, Pappis CP (1997) Operations research and environmental planning. Eur J Oper Res 102:248e63

Ron Ad J de (1998) The ultimate result of continuous improvement. Int J Prod Econ 56–57(1): 99–110

Dyllick T, Hockerts K (2002) Beyond the business case for corporate sustainability. Bus strategy Environ 11(2):130–141

Elkington J (1998) Cannibals with forks. The triple bottom line of 21st century business. New Society Publishers, Tintown

Elkington J (2001) The Chrysalis economy. Capstone Publishing Ltd, UK

Environmental Tax Policy Institute. (2005) www.vermontlaw.edu/elc/index.cfm?doc_id=134. Accessed 17 May 2005

EPA U.S. Environmental Protection Agency (2000) A practical guide for materials managers and supply chain managers to reduce costs and improve environmental performance. www.epa.gov/wastewise

U.S. Environmental Protection Agency (2000) Enhancing materials management and supply chain performance with environmental cost information: examples from commonwealth edison, Andersen Corporation, and Ashland Chemical, EPA742-R-00-XXX, "http:// www.epa.gov/opptintr/acctg"www.epa.gov/opptintr/acctg.

Faucheux S, Nicolai I (1998) Environmental technological change and governance in sustainable development policy. Ecol Econ 27:243e56

Germany's Blue Angel Certification (2005) www.blauer-engel.de/englisch/navigation/body_blauer_engel.htm. Accessed 1 June 2005

Green Seal Product Certification (2005) www.greenseal.org. Accessed 1 June 2005

GreenTech Assets Inc. (2005) www.greentechassets.com. Accessed 31 May 2005

Gregory J, Atlee J, Isaacs J, Kirchain R (2004) Sustainability metrics for materials use at the system and operational level. Materials Systems Laboratory discussion paper

Grenon M, Joseph M, Turner M (2007) How big is your carbon footprint. CSCMP Supply chain quarterly

Halldorsson A, Kotzab H, Skjøtt-Larsen T (2009) Supply chain management on the crossroad to sustainability: a blessing or a curse? Logist Res 1:83–94

Handfield R, Sroufe R, Walton S (2005) Integrating environmental management and supply chain management. Bus Strategy Environ 14(1):1–9

Hansen N (2004) Organic food sales see health growth. MSNBC News Online, 3 Dec 2004. www.msnbc.msn.com/id/6638417/. Accessed 20 May 2005

Hoffman A (2005) Business decisions and the environment: significance, challenges, and momentum of an emerging research field. In: Brewer G, Stern P (eds) National Research Council, Decision Making for the Environment: Social and Behavioural Science Research Priorities

Hutchins MJ, Sutherland JW (2008) An exploration of measures of social sustainability and their application to supply chain decisions. J Clean Prod 16(15): 1688–1698

Jayaraman V, Luo Y (2007) Creating competitive advantages through new value creation: a reverse logistics perspective. Academy of management perspectives, pp 56–73

Klassen RD, Johnson PF (2004) The green supply chain. In: Westbrook R, New S (eds) Understanding supply chains: concepts, critique and futures. Oxford University Press, Oxford

Kopicki RJ, Berg MJ, Legg L, Dasappa V, Maggioni C (1993) Reuse and recycling: reverse logistics opportunities. Council of Logistics Management, referenced in Hoek, 1999

Kura Y, Revenga C, Hoshino E, Mock G (2004) Fishing for answers: making sense of the global fish crisis. World Resource Institute Report. pubs.wri.org/pubs_description.cfm?PubID=3866. Accessed 30 May 2005

Maloni M, Brown M (2006) Corporate social responsibility in the supply chain: an application in the food industry. J Bus Ethics 68(1):35–62

Mamic I (2005) Managing global supply chain: the sports footwear, apparel and retail sectors. J Bus Ethics 59:81–100

Murphy PR, Poist RF, Braunschweig CD (1996) Green logistics: comparative views of environmental progressives, moderates, and conservatives. J Bus Logist 17/1:191–211

Nagel C, Meyer P (1999) Caught between ecology and economy: end-of life aspects of environmentally conscious manufacturing. Comput Ind Eng 36:781e92

Paquette J (2005) Supply Chain 2020, a research initiative investigating the critical factors shaping supply chains of today and tomorrow. Engineering Systems Division, Massachusetts Institute of Technology, June

Pearce F (2009) Confessions of an eco sinner: travels to find where my stuff comes from, Eden Project Books

Pojasek RB (1998) Activity-Based Costing for EHS Improvement. Pollution Prevention Review, Winter 1998, pp 111–120. www.pollutionprevention.com

Puckett J (2002) Exporting harm: the high tech trashing of Asia. A Silicon Valley Toxics Coalition Report, February 2002. www.svtc.org/cleancc/pubs/technotrash.pdf. Accessed 5 June 2005

Rabs H, Bohn C (2003) Bæredygtig supply chain management: et studie i muligheden for at anvende supply chain management til at gøre en virksomhed bæredygtig. Unpublished M.Sc. thesis, CBS, Denmark

Roberts JA (1996) Green consumers in the 1990s: profile and implications for advertising. J Bus Res 36/1:217–231

Roberts S (2003) Supply chain specific? Understanding the patchy success of ethical sourcing initiatives. J Bus Ethics 44/2:159–170

Sarkis J (1995) Manufacturing strategy and environmental consciousness. Technovation 15(2):79e97

Shrivastava P (1995) Environmental technologies and competitive advantage. Strateg Manag J 16(3):183e200

Srivastava SK (2007) Green supply-chain management: a state of-the-art literature review. Int J Manag Rev 9(1):53–80

Tsoulfas GT, Pappis CP (2006) Environmental principles applicable to supply. J Clean Prod 14:1593–1602

Unilever's annual report describing efforts to build and sustain reliable fish supplies is available at www.unilever.com/ourvalues/environmentandsociety/default.asp. Accessed 5 May 2005

United Nations Environmental Program (2002) Vital graphics: an overview of the state of the world's fresh and marine waters. www.unep.org/vitalwater/. Accessed 30 May 2005

US Environmental Protection Agency (2005) Energy star program. www.energystar.gov/. Accessed 1 June 2005

US Green Building Council. Leadership in energy and environmental design is a national rating system to certify green buildings, administered by the www.usgbc.org/LEED/. Accessed 1 June 2005

Walls M (2005) The role of economics in extended producer responsibility: making policy. Res Energy Econ 27(4):287–305

Walton SV, Handfield RB, Melnyk SA (1999) The green supply chain: integrating suppliers into environmental management processes. Int J Purch Mater Manag 34(2):2–11, referenced in Hoek, 1999. www.rff.org/Documents/RFF-DP-03-11.pdf. Accessed 28 May 2005

Chapter 9
Environmental Aspects in Strategic Decisions

Global supply chains pose challenges regarding both quantity and value:

- Logistics, and Procurement have become strategic
- Organizations outsourcing, forming partnerships, alliances
- Product environment becoming more complex
- Time-based competition requires time compression
- Managing suppliers and customer relationships is necessary
- Competition shifting from company versus company to Supply Chain versus Supply Chain

Companies must carefully justify all environmental changes through either cost reduction or customer satisfaction issues. The focus of continuous improvement (used so effectively during TQM program implementation) can be applied quite effectively to improving environmental efficiency and effectiveness. The cross-functional relevance of environmental supply chain management is ensured by its direct impact on the supplier selection and management processes. Change should be viewed as a competitive tool and environmental efficiency should be viewed as a positive catalyst for change.

Large-scale adoption of environmental aspects within the framework of SCM is essential. The starting point as for any other major strategy is the top management commitment to the environmental issues. This is not only essential but is also critical for the overall success of the project. It should be the mission of the company's top management to sensitize everyone in the company toward environmental issues. It will also be a good idea to have a separate environmental department reporting directly to the top management. This department should look at all the aspects of the environment barometer and track its progress on a regular basis. Any discrepancy is immediately reported to the CEO.

Table 9.1 illustrates the major decisions made on the three levels of decision-making (strategic, tactical, and operational). These decisions are grouped by decision type (facility location, material flow, information, and customer service).

M. Bevilacqua et al., *Design for Environment as a Tool for the Development of a Sustainable Supply Chain*, DOI: 10.1007/978-1-4471-2461-0_9,

Table 9.1 Supply chain management and design decisions (modified by Ballou 1999 and Beamon 1998)

Type of decision	Strategic	Tactical	Operational
Facility location	Number of facilities Number of echelons Facility locations	Facility design	
Material flow	Transportation mode selection Supplier selection Distribution center-retailer assignments plant-product assignments Inventory positioning strategies Distribution strategies.	Inventory control policies Stock rotation policies	Production scheduling Replenishment quantities Replenishment intervals Carrier routing and scheduling Order expediency mechanisms
Information	Information system design Information protocol Selection information policies	Demand forecasting methods Periodic statistics reporting	Real-time information control
Customer service	Standards and objectives	Order priority rules	

The decisions in Table 9.1 have varying magnitudes of effect on the natural environment (Fig. 9.1).

For instance Table 9.2 focuses on two decisions that have a comparatively effect on the natural environment.

Therefore, in terms of internal supply chain management and design, the decisions most substantially affecting the environment are strategic facility location and material flow. These decisions are most substantial since they are both likely to disturb the external environment through emitted pollutants (noise, air, and water) and energy consumption.

In practicing ethical supply chain decision-making, one can adapt the steps of ethical decision analysis (modified from Harris, Pritchard, and Rabins):

1. Identify the facts associated with the decision.
2. Identify the ethical considerations (environmental concepts or principles) that affect the safety, health, and welfare of the public.
3. Take necessary action to ensure that the resulting decision holds as most important the safety, health, and welfare of the public.

Example 1 New facility location

- *Step 1:* Facts associated with the decision. Given a set of objectives and constraints, determine the best location for a new facility.

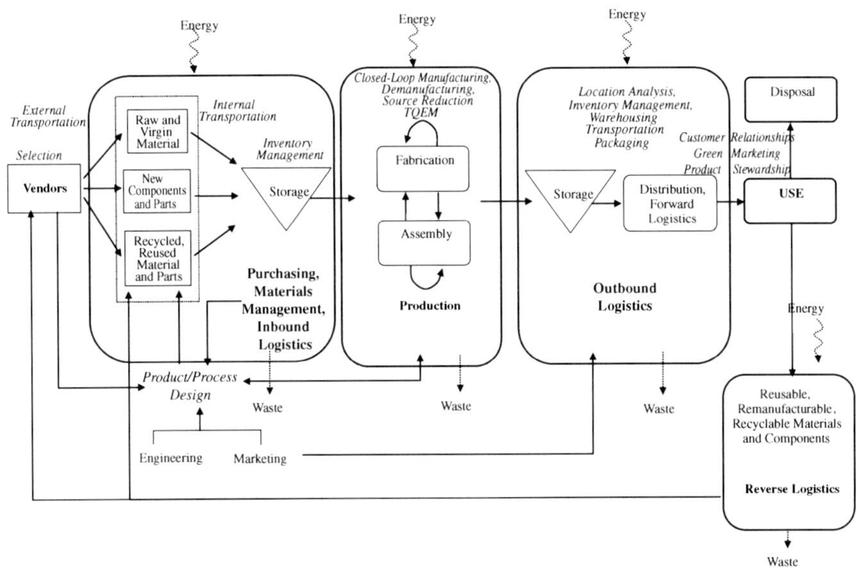

Fig. 9.1 Interaction between supply chain and environmental

Table 9.2 Supply chain decisions and their effects on the environment (*Source*: Beamon 2005)

Type of decision	Potential environmental effects
Strategic facility location	Strategic decisions pertaining to facilities affect the natural environment by affecting natural habitats (ecosystems), humans, and animals, primarily through habitat destruction, increased air, water, and noise pollution and energy consumption
Material flow (all levels)	Decisions pertaining to transportation modes and material movement have significant effects on energy consumption and motor vehicle congestion (and subsequently air and noise pollution)

- *Step 2:* Ethical considerations. The construction and operation of the new facility affects the natural environment by affecting natural habitats (ecosystems), primarily through habitat destruction and increased air, water, and noise pollution. As the engineering codes do not provide for broad protection of ecosystems, the ethical consideration in this case is to the public in terms of the effect of increased energy use, as well as increased air, water, and noise pollution.
- *Step 3:* Necessary action. In determining the location and construction plans for the new facility, take action to reduce the effects of increased air, noise, and water pollution on the public. Actions: take steps to minimize total material and personnel travel distances to and from the facility, make provisions to avoid runoff from construction activity and new pavement, consider noise-abatement strategies, and take steps to reduce the air pollution effects of the new facility, particularly with respect to population centres.

Example 2 Inventory control policy

- *Step 1:* Facts associated with the decision. Given a set of objectives and constraints, determine the best inventory control policy.
- *Step 2:* Ethical considerations. The selected inventory control policy has significant effects on ecosystems and the public, primarily through noise pollution, increased energy use, and increased motor vehicle congestion (and subsequently air pollution). As the engineering codes do not provide for broad protection of ecosystems, the ethical consideration in this case is to the public in terms of the effects of increased motor vehicle congestion and air pollution.
- *Step 3:* Necessary action. In designing the inventory control policy, take action to reduce motor vehicle congestion and air pollution. Actions: take steps to minimize total material movement to and from the facility by: order/delivery consolidation, accepting larger inventories by lengthening or shortening shipment intervals to avoid rush hours, reducing the number of shipments, or scheduling truck traffic density patterns to be acyclic to passenger car traffic. Other actions include considering noise-abatement strategies and taking steps to reduce the air pollution effects of the new facility, particularly with respect to population centres.

Direct interaction with supply chain partners can enable a company to reduce total **inventory** levels, decrease product obsolescence, lower transaction costs, react more quickly to changes in the market, and respond more promptly to customer requests. Essential to supply chain performance is improving the effectiveness of *materials management*—the set of *business processes* that support the complete cycle of material flows from *purchasing* and internal control of production materials, through planning and controlling work in process, to warehousing, shipping, and distributing finished products. Managers can improve their materials management performance by first understanding how their decisions affect the purchasing, storage, *handling*, and asset *recovery* activities throughout their organization.

Another key component of supply chain management is ***logistics***—the activities to obtain incoming materials and distribute finished products to the proper place, at the desired time, and in the optimal quantities. Companies can greatly improve business performance by working with suppliers, shippers, distributors, and customers to better coordinate logistics activities.

In this section four topics were considered strategic and they were studied in deep in the next sections:

- decisions on the locations of production and logistics facilities,
- suppliers selection,
- environmental Material Management,
- closed-loop supply chains and reverse logistics.

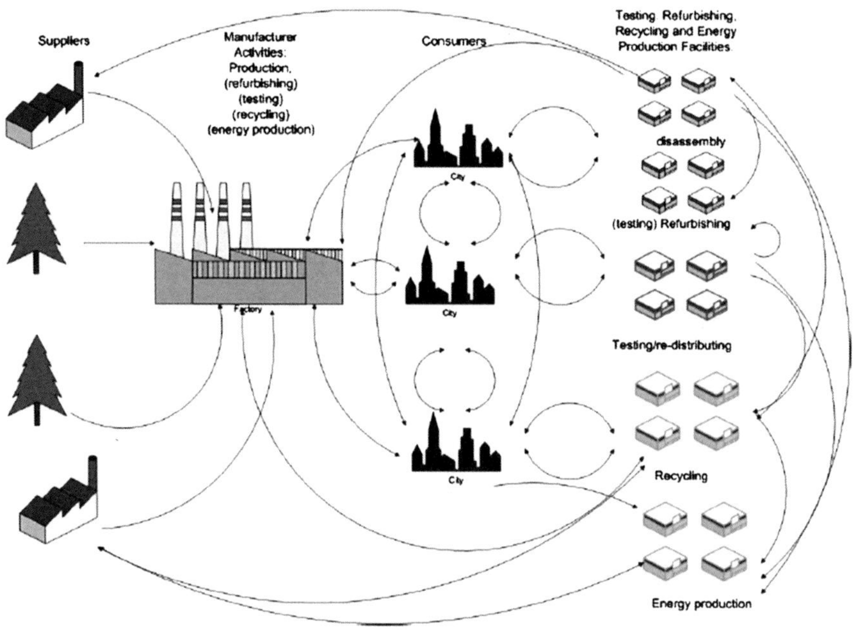

Fig. 9.2 Sustainable logistic network (*Source*: Neto et al. 2006)

9.1 Sustainable Logistic Network

In a Logistic Network, a number of actors will influence efficiency in terms of costs and the environment. Suppliers, manufactures, consumers, logistic operators, and third parties operating in testing, refurbishing, recycling, and energy production for the end-of-life products are the main players. These actors perform the majority of the activities impacting business and the environment.

In general terms, the activities performed in a logistic network are related to manufacturing, transportation, use, and end-of-life products' destination. Figure 9.2 pictures them.

The decisions regarding these activities will, therefore, determine the network costs and environment impact. These decisions are strategic (e.g. location of factories), tactical (e.g. products end-of-life destination) as well as operational (e.g. choosing suppliers, third parties in collection, refurbishing, etc.).

The most important trend in Logistic area is:

- Rapidly growing transport work
- Longer transport distances
- Larger vehicles and vessels
- Larger logistics service providers
- More consolidation

- More intermodality
- Tighter control of logistics systems
- Demand for: more advanced logistics services and more environmentally sustainable transport services

There are various approaches for cross-company logistics models that conform to the general network model of logistics. These models represent networks transporting rights, goods, finance and information where spatial, quantitative, informational and temporal differences as well as company boundaries are crossed. Parameters defining the structure of a logistics network are paramount:

- Number, locations, and functions of source points (= loading locations, making goods available),
- Number, locations, and functions of target points (= unloading locations, points of reception),
- Number, locations, functions of connections, or nodes between sources and targets.

The network nodes are called transhipment terminals. This implies that only transhipment but not storage in general (no inventory) is foreseen at these locations. Transshipment terminals serve as consolidation terminals where the flows of goods are collected and/or as break-bulk terminals where the flows are in turn distributed. The **basic structure** of transportation links can be represented either as **direct connection** ("point-to-point" transport) in its simplest form (single-stage, uninterrupted transport chain) or as a **multi-stage system** with preliminary leg, main leg, and subsequent leg with transshipment terminals where the network nodes serve as consolidation terminals where the flows of goods are collected and/or as break-bulk terminals where the flows are in turn distributed Fig. 9.3.

The mixture of logistics systems made up from the given basic structures is decided in the logistical network structure. The processes are designed when the logistical capacities are superimposed on this. The logistical capacity can be subdivided into transport capacity, warehousing capacity, and information capacity. In addition to the basic structure of the systems, the speed of traffic flowing between the individual points in the system must be taken into account. The network strategy is also based on geo-economic considerations such as the long-term development of customer demand or the development of the required delivery time. Summing up, the criteria logistics costs, supply service, adaptability, susceptibility to interference, transparency, and time for planning and establishment of the system are important in the moment of developing and evaluation logistics models.

The current economic development is not sustainable. The biggest and most challenging threat is the worldwide dramatically increasing greenhouse gas (GHG) emissions. In the absence of an effective climate policy this trend will continue and emissions will double until 2050.

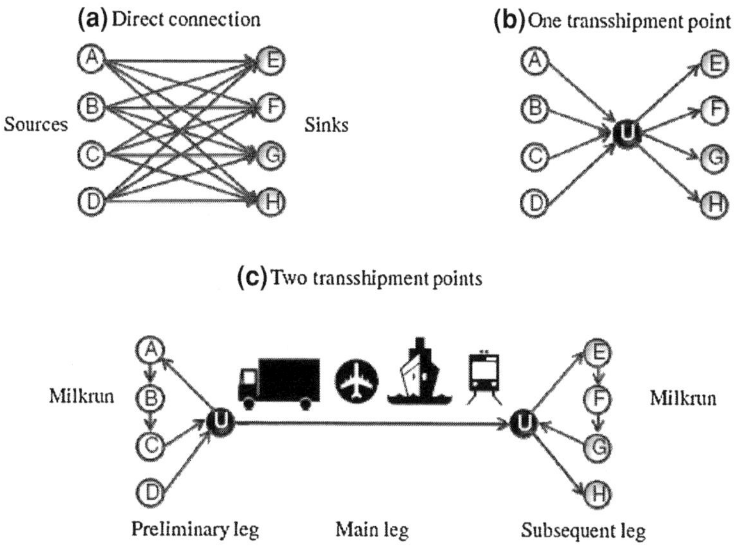

Fig. 9.3 Illustration of consolidation by means of transhipment points (*Source*: Brauer and Krieger 1982)

One of the most important European regulations regarding the environment is the directive Emissions Trading Scheme (ETS) based on the so-called Kyoto Protocol which is the world's first multinational emissions trading scheme covering greenhouse gases. This ETS, which has taken effect in 2005, caps emissions from 10,000 plants in the oil refining, smelting, steel, cement, glass, and paper sectors, and allows trading of their emissions allowances. The year 2008 corresponds to the Second commitment period which implies that it is the turn of the transportation companies, in addition to the industry sector to pay credits for their CO2 over-emissions. Companies across all the European Union must incorporate climate change into day-to-day commercial decisions, and assess what innovative steps they can take to reduce emissions (Lemathe and Balakrishnan 2005). In this respect decision makers have to face a brand new problem when designing their supply chain network. The environmental criteria have to be taken into account as well as economic feasibility.

Today's transport system considerably contributes to the current unsustainable development (Fig. 9.4). It accounts for approximately one quarter of global CO2 emissions. Road transport accounts for ca. 65% and is the major contributor (Fig. 9.5). Road transport emissions are split two-third to passenger transport and one-third to freight at present in the OECD (European Conference of Ministers of Transport 2007). It is one of the few industrial sectors where emissions are still growing because the continuously growing demand for freight transport is preferably met by road, the most CO2-intense land transport mode, while the use of the CO2-efficient mode rail not increases (Fig. 9.6).

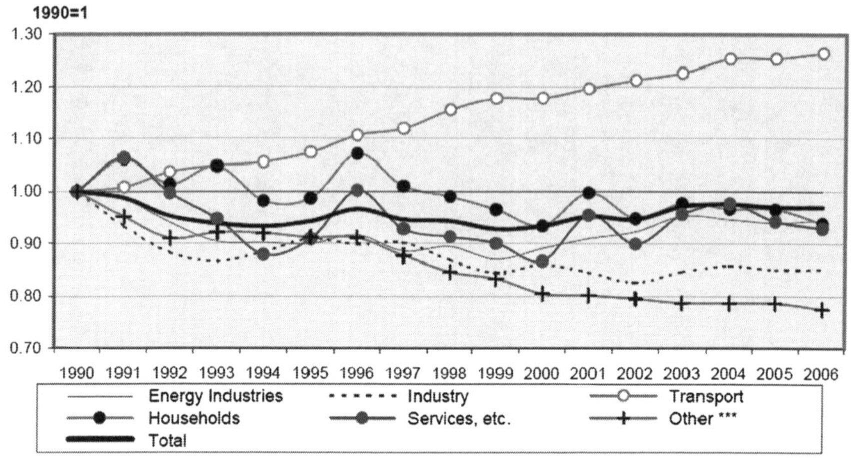

Fig. 9.4 CO2-emissions per sector, EU-27 (*Source*: Eurostat and the EU Commission 2009)

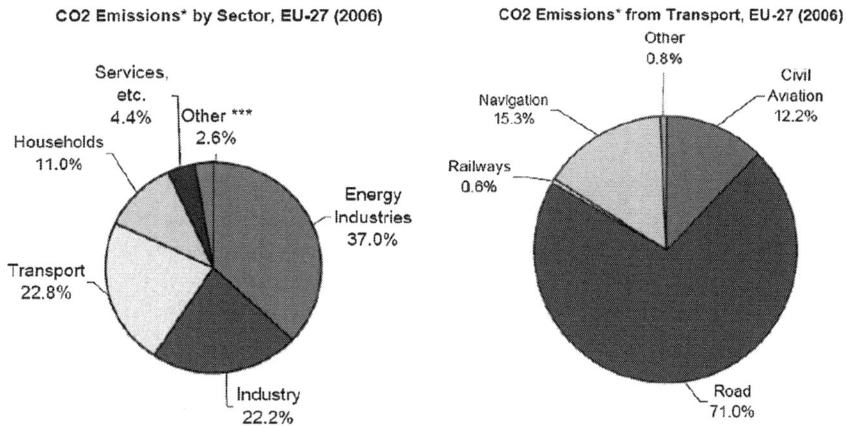

Fig. 9.5 Distribution of CO2-emissions, EU-27 (*Source*: Eurostat and the EU Commission 2009)

The growth of road freight traffic is the result of complex interactions derived from the economic activities that it serves (McKinnon 2007). Supply chain structures, improved logistics, alternative fuels, and other technological measures can contribute to reduce the emission intensity of road freight. Nonetheless, freight transport volumes are expected to continue to grow significantly (European Commission 2006b) and outweigh these potential reductions of the emission intensity in the road freight sector. Consequently, it is this expansion that needs to be targeted. If a return to sustainable levels is to be achieved, shifting transport volumes from road to more CO2-efficient modes like rail is inevitable.

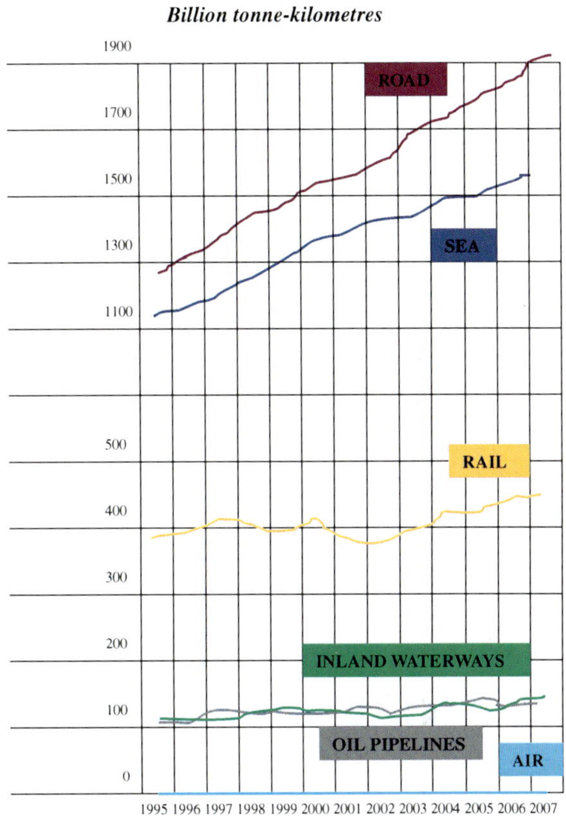

Fig. 9.6 Transport system (*Source*: Eurostat and the European Commission 2009)

Woxenius (1998) states that the suitability of rail transport for substantial transport market shares is limited by, among other things, the limited extension of the railway network and the high costs of shunting wagons into private sidings. Road transport, on the other hand, offers accessibility with maintained economy for smaller shipments over short distances. Hence, intermodal transport, the combination of pre- and post-haulage (PPH) on road and the long haul on rail, has the potential to offer the required accessibility and flexibility of road yet maintaining CO2 efficiency of rail transport.

Woxenius (1998) continues that using more than one transport mode in the transport of goods between origin and destination requires a transhipment of part-loads between two modes which is time-consuming, costly, and also involves a risk of damage. To limit these problems the goods are loaded into standardized units which can be transhipped automatically and be handled by all modes of transport. This method is called the principle of unit loads and the transport arrangement is commonly referred to as intermodal transport. Basically, intermodal transport is a combination of at least two modes, where road transport is used for pick-up and delivery of the standardized load unit in order to secure broad accessibility and flexibility. The by far biggest distance is performed by large-scale transport modes such as rail, inland waterways, short sea shipping, or ocean shipping where the units are consolidated with other shipments and economies of scale are being achieved.

Macharis and Bontekoning (2004) define intermodal transport by the following characteristics:

- Task division between transport modes with respect to the short-haul and long-haul of the transport chain. Road transport is assigned to the short-haul while rail is assigned to the long-haul.
- Transhipment to enable the division of task between short- and long-haul.
- The use of standardized load units, which increase the efficiency in the transport chain.
- Multi-actor chain management with many organizations which each of them control a part of the transport chain.
- Synchronized and seamless schedules between different modes without storing and handling of freight during its journey from origin to destination.

In order to run the intermodal transport chain smoothly, a lot of actors need to work in collaboration. These actors are (a) PPH operators, who take care about the pick-up and delivery, (b) terminal operators, who take care of the transhipment operations, (c) network operators, who take care of the infrastructure planning and organization of the long-haul transport (rail, barge, ocean vessel), and (d) forwarders, who can be considered as users of the intermodal infrastructure and services, and take care of the route selection for a shipment through the whole intermodal network (Macharis and Bontekoning 2004). Figure 9.7 provides a simple depiction of intermodal freight transport where the long-haul is represented by rail.

PPH operations accessing consignors and consignees to and from terminals often take place in urban areas, where they are likely to increase the local external effects compared to all-road transport (Woxenius 2001). At the same time, urban congestion makes haulage operations more costly and less reliable, which reduces the competitiveness of the whole intermodal transport chain.

Different consolidation networks for intermodal transport services exist, i.e. point-to-point, line, hub-and-spoke, and collection-distribution network (Fig. 9.8). In these networks, different production models for operating the trains can be applied, including frequency of services and train lengths. For obvious reasons, the choice of consolidation network and production model has a significant impact on costs and quality of the intermodal transport service.

Furthermore, the design and organization of intermodal transport services require decisions from more than one actor and more time horizons. Currently, each actor is striving for optimizing its own operation, but what is best for each single actor is not necessarily for the chain. As a consequence of the high costs of rail and terminal equipment, the actors aim at economies of scale and minimizing the costs of intermediate transhipments. This has lead to a high concentrated

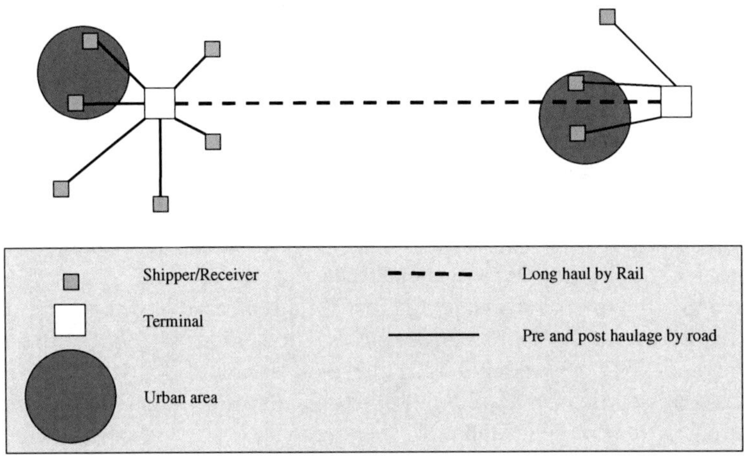

Fig. 9.7 A typical road-rail intermodal transport

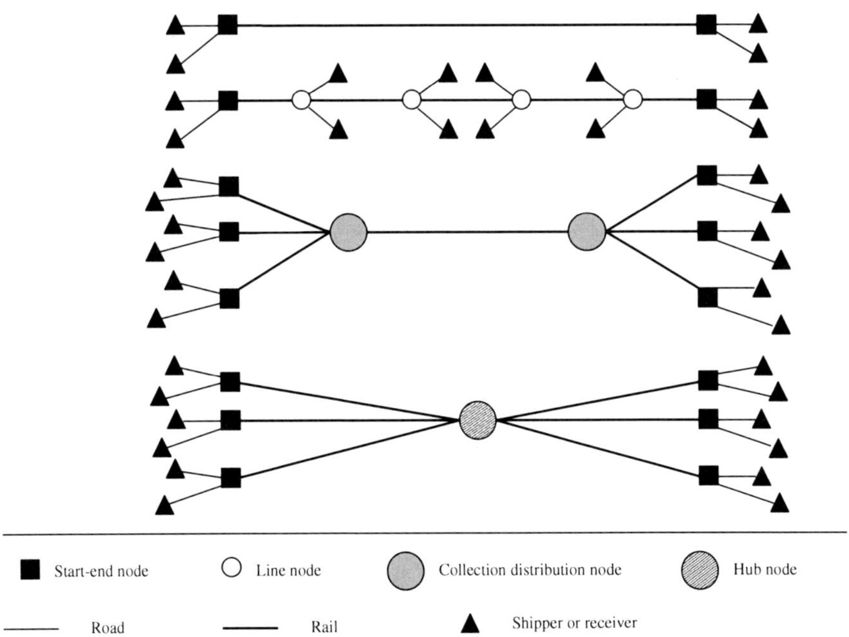

Fig. 9.8 Four basic consolidation networks (modified by Macharis and Bontekoning 2004)

intermodal network with a relatively small number of nodes and a strong focus on a limited number of high-volume corridors. The trains stay at the terminal throughout daytime and are operated overnight as full trains between terminals.

High volumes of load units are required in order to distribute the costs of the terminal between a large number of transhipments.

The break-even distance of intermodal transport compared with all road transport is most sensitive to PPH for accessing consigners and consignees. Despite its relatively short distance compared to the rail line haul, it accounts for a large fraction (between 25 and 40%) of total expenses for the transport chain (Macharis and Bontekoning 2004). Generally, traditional intermodal transport is competitive only at distances in excess of 300–500 km on high volume corridors (Bärthel and Woxenius 2004). To compete successfully with all-road transport for dispersed goods flows over medium distances, an alternative design and organization of intermodal transport services are required, addressing rail, terminal and PPH operations. One possibility is a line train design with high transport frequencies and additional small-scale stops between the large-scale origin and destination terminals. By combining less dense flows over long distances and dense flows over short distances, this service can attract the amount of freight needed for high frequencies and utilizing economies of scale (Woxenius 2001).

The stops at the small-scale terminals need to be short in order not to prolong the total transit time which requires a fast transhipment technology with a low costs per move. Hence, innovative systems do not only require an operational shift of consolidation network and production model, but, they also have to cope with a technical challenge since a cheap, flexible, and scalable transhipment technology is required that does not rely on economies of scales. Bärthel and Woxenius (2004) state that a wide range of transhipment technologies have been proposed by inventors and evaluated by researchers, but with very view exceptions, they have not been explored commercially.

The operational design of line trains can also improve the preconditions for PPH in the intermodal transport chain. PPH usually takes place in urban areas where congestion and most environmental impacts are most critical. A larger number of terminals allow a greater chance of shorter local road haulage which allows restricting operational costs. Kreutzberger et al. (2003) therefore conclude that both spatial and network policies are crucial for the efficiency of intermodal transport.

Hence, besides the actors that actively run the intermodal transport chain, there are a lot of other actors and stakeholders which influence its competitiveness. As Woxenius' model illustrates (Fig. 9.9), these actors are those which influence (a) infrastructure, (b) demand for transport services, (c) competition from single-mode transportation, (d) laws and regulations, and (e) political and economic decisions. Thus, also the actors representing these factors need to be considered when strategies for improving the competitiveness of intermodal transport are to be developed. However, research and studies focus usually on single actor categories with operational problems in a certain time horizon. There is a lack of studies that address decisions from more than one operator and/or time horizon (Macharis and Bontekoning 2004).

Road freight transport is a major contributor to the unsustainable impacts of the freight transport sector. If a return to sustainable levels is to be achieved, shifting

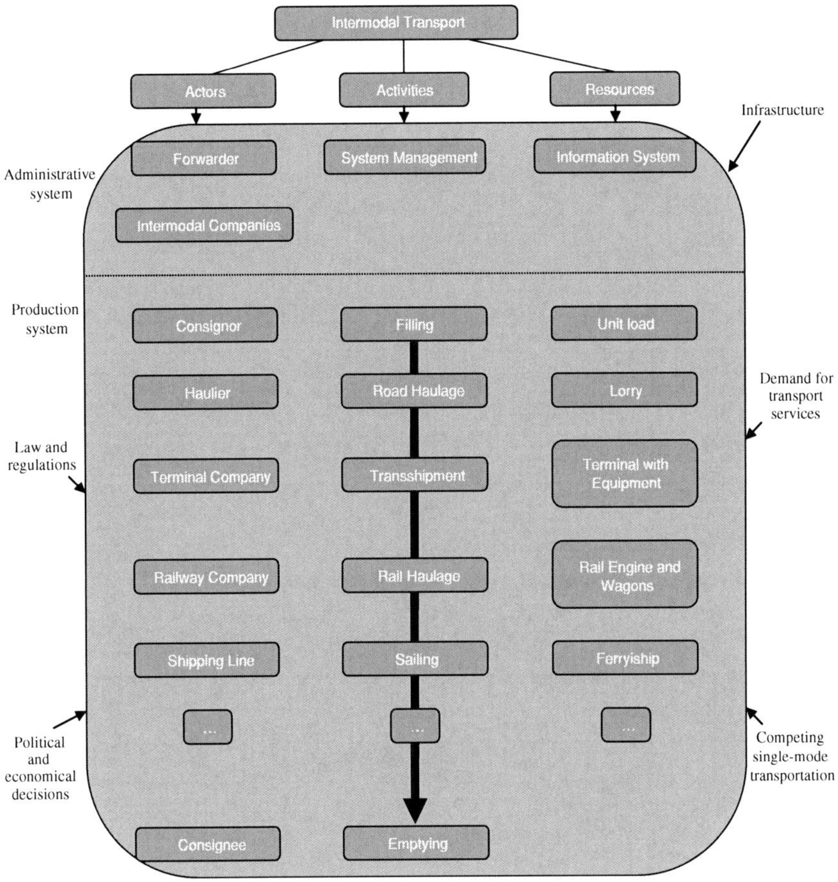

Fig. 9.9 A reference model of intermodal transport (modified by Woxenius 1998)

transport volumes from road to more CO2-efficient modes like rail is inevitable. A modal shift from road to rail requires more competitive intermodal transport for dispersed transport flows over short and medium distances.

Box G: Application of the Modelling Approach to a Romanian Automotive Cluster

In a research project funded by the Federal Ministry for Transport, Innovation, and Technology (BMVIT) as well as the Austrian Research Promotion Agency (FFG) the developed logistics models are demonstrated

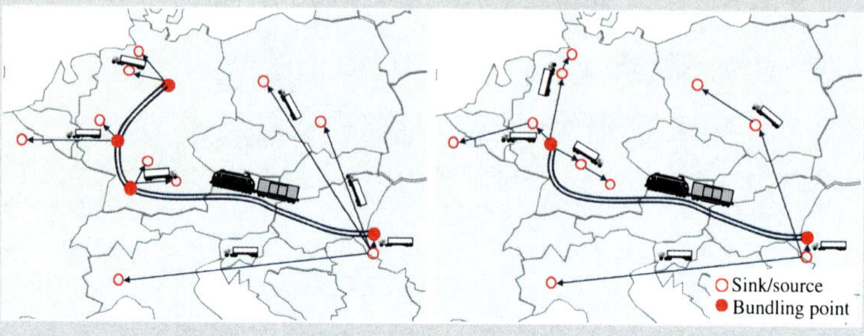

Fig. 9.10 Scenarios 1 and 2

by means of the region Timis in Romania (Sihn et al. 2010). Focusing on 7 automotive companies volumes of outgoing transports were analyzed. Starting from the current state of individual transports, different scenarios were defined. The scenarios are aimed at cost reduction and sustainability in using modes of transport for high volumes like rail traffic.

Figure 9.10 shows 2 defined scenarios of transport bundling for the Timis Region. Scenarios 1 using block train with 3 stops and direct relations from the end of the train not considering locations in Poland and Italy that cannot profit from consolidation with the block train. Scenario 2 limits the block train to one stop but bundles transports further leaving the train to their final destinations. Destinations not considered in the main leg bundling were consolidated as well. Shifting the main leg to railway and optimizing the collection and distribution of goods from and to transhipment points logistics costs could be reduced by 15% in the given case. The ecological impact in reduction of CO2 emissions by 40%, cutting fuel consumption in half, shows the success in more than one target dimension. The main deficit of the models is overcoming the doubled lead time coming from the ceteris paribus inspection of transports.

In addition to the simulation and evaluation of scenarios a sensitivity analysis was executed to cover the ecological and economical results. Therefore the evaluated scenarios indicated were simulated with lower basic loads keeping all other factors stable. At a level of 70% of the load, block train concepts as well as the transfer of 66% of transports to railroad could be maintained. Negative effects of the change in basic loads were determined in the capacity utilization of transport capacities and the flexibility especially for block trains. Nevertheless the developed transport concepts and cross-company models can stand up to the actual transport handling. Economic considerations show lower costs of scenarios compared to the actual situation.

Table 9.3 Supplier selection criteria

Prequalification of suppliers	
	Require or encourage environmental criteria for approved suppliers
	Require or encourage suppliers to undertake independent environmental certification
Environmental requirements at the purchasing phase	
	Build environmental criteria into supplier contract conditions
	Incorporate EHS staff on sourcing teams
Supply base environmental performance management	
	Supplier environmental questionnaires
	Supplier environmental audits and assessments

9.2 Suppliers Management and Selection

Environmental supply chain management recognizes the crucial role to be played by the purchasing and the function's involvement in activities that include reduction, recycling, reuse and the substitution of materials. Working with suppliers on environmental issues not only generates significant environmental benefits, but also opportunities for cost containment, improved risk management and enhanced quality and brand image. This will also help companies streamline their supply base and develop more co-operative, long-term relationships with key suppliers, a practice that has fostered greater opportunities to work together on environmental issues.

Managing vendor relationships is moving away from an emphasis on negotiation and cost reduction to become a key managerial and technical challenge. Experience shows that more long-term, collaborative arrangements mutually benefit both the customer and the supplier (Burt and Soukup 1994). A manufacturer's eco-performance is largely determined by ''upstream'' environmental impacts. Already, many writers have commented on the need for the monitoring/auditing/assessment of suppliers in the management of the value chain (Lloyd 1994) and have published guidelines to assist procurement departments to work in partnership with their suppliers to include environmental factors in their supply operations (e.g., CIPS 1995).

Table 9.3 shows some important aspect to take into consideration for supplier selection.

Supplier audits on environmental issues are an absolute must and this process should be a cross-functional initiative. This should involve employees from quality assurance, environmental affairs, and purchasing. Similarly, teams should include financial analysts who decide whether suppliers will be the most productive from the perspective of maximizing Environment Value Add (EVA) make-or-buy decisions. Adherence of suppliers to quality and environment standards is a necessary prerequisite for achieving a company's various objectives. Early sourcing

Table 9.4 Supplier management criteria

Build environmental considerations into product design	
	Jointly develop cleaner technology with suppliers
	Conduct life cycle analysis in cooperation with suppliers
	Engage suppliers in design for environment (DFE) product innovation
	Coordinate minimization of environmental impact in the extended supply chain
	Develop tools that assist in the DFE effort
Cooperate with suppliers to deal with end-of-pipe environmental issues	
	Reduce packaging waste at the customer/supplier interface
	Reuse/recycle materials in cooperation with the supplier
	Launch reuse initiatives (including buy backs and leasing)
Reverse logistics	
	Give supplier an incentive to reduce the customer's environmental load

and early supplier involvement in basic product developmental work is required and while doing so it will be easy for the companies to strictly adhere to the norms it has prescribed for itself. These sourcing decisions must also take into consideration safety issues, capacity of suppliers, and ability to treat compounds and effluents. Products and processes should be subjected to continual critical analysis at every stage of the value-added process. The early integration of suppliers into all decisions affecting them is critical to environmental effectiveness.

Table 9.4 shows some important aspect to take into consideration for supplier management.

The close alignment of supplier capabilities with buying firm's environmental goals is critical to program success. This alignment can be achieved through an alliance supporting organizational and informational framework and the benchmarking of performance with environmental, quality, and cost parameters.

Table 9.5 shows some more Environmental Supply Chain practices for supplier management.

Purchasing with its proximity to its suppliers needs to play a broader role in the company's environmental agenda. It needs to conduct timely supplier evaluations using criteria such as risk and environmental capability. Because of the emphasis on environmental value-added performance, companies turn to suppliers for use of their waste treatment facilities. Companies also seek to develop suppliers who can collect, clean and reship process waste back to the company.

In the Box H and I Dell and Hitachi policies for suppliers management and selection were presented.

Table 9.5 Environmental supplier management

Influence legislation to facilitate better SCEM policies	
	In cooperation with suppliers, lobby to strengthen environmental regulation
	Lobby on behalf of SCEM initiatives
Work with industry peers to standardize requirements	
	Create inter-firm procurement group to collaborate on environmental issues
	Standardize supplier questionnaires
Inform suppliers of corporate environmental concerns	
	Issue statements of EHS priorities to suppliers
	Draft and distribute comprehensive SCEM policy
Promote exchange of information and ideas	
	Sponsor events to facilitate discussions between customers and suppliers on environmental issues
	Host training and mentoring programs.

Box H: Dell Supplier Management System

The core components of Dell's global supplier management program are:

- *Certification and Standards*
  ISO 14001 certification. Dell requires their suppliers to be compliant with ISO 14001, the most widely recognized standard for environmental management systems, by Jan. 31, 2004 or submit a schedule for achieving certification and obtain Dell approval.
  OHSAS 18001 certification. Dell requires their suppliers to be compliant with OHSAS 18001, a prominent European standard for workplace health and safety management systems, by Jan. 31, 2004 or submit a schedule for achieving certification and obtain Dell approval.
- *Training and Communication*
  Training. Dell provides training on its supplier expectations during its annual supplier conference as well as updates in quarterly business reviews. In addition, Dell provides training in a variety of relevant areas, such as environmental practices, to suppliers.
  Supplier social responsibility and environmental responsibility agreements. As part of every new supplier contracting process, Dell requires suppliers to sign an agreement acknowledging they are aware of and will abide by Dell requirements and principles.

- *Reviews and Compliance*
  Business reviews. In order to embed socially responsible behavior into business activities, Dell includes a review of requirements and principles in quarterly business reviews that are held with key suppliers.
  Self audits. Dell asks key suppliers to conduct a self-audit to review with Dell management on an annual basis, using a standard scorecard format. These audits are required to be signed by a member of senior management of the supplier.
  Dell executive oversight and review. Dell's supply-chain management system is overseen by chief procurement officers and senior vice presidents of Dell Inc. Issues of concern related to supply chain practices are raised in regular operations reviews with these executives and, if appropriate, also raised in Dell's business conduct committee and to the chief executive officer.
  Board of directors oversight. Issues of concern related to supply-chain practices will, as appropriate, be raised with the Dell Board of Directors and/or its various committees.
  Engagement with third parties and non-governmental organizations (NGOs). Dell will engage with third parties and NGOs as it deems necessary in order to ensure the effective implementation and oversight of its supplier principles.
- *Correction and Enforcement*
  Correction. In recognition the complexity of the world in which Dell does business, when suppliers fail to meet these principles, Dell and the supplier will create an action plan to ensure future compliance. Performance against the plan to adhere to Dell's standards shall not take more than one year.
  Enforcement. Dell reserves the right to terminate at any time, even before corrective plans are developed and implemented, agreements with suppliers that fail to comply with their Supplier Commitment Policy or their Supply Chain Management Requirements.

Box I: Hitachi Green Procurement Guidelines

Commitment to Corporate Social Responsibility (CSR) policies of Hitachi group was formulated in March 2005. This policy aims to foster common awareness of social responsibility with their business partners as well as to promote environmental protection activities. Since the environmental protection activities involve a whole society, their suppliers' cooperation is requested and appreciated to promote the items included in "Hitachi Action Guidelines for Environmental Conservation".

The main requests to their suppliers for cooperation in investigation of green procurement concern:

Investigation overview

(1) Investigation categories

The investigation is to be conducted for each of the following categories:

(i) The status of the supplier's environmental activities
(ii) The status of environmental load reduction of the procured products
(iii) Information about chemical substances contained in the procured products

(2) Investigation replying method

Suppliers are kindly requested to provide the Hitachi Group with information by means of its Green Procurement System based on the Internet (A Gree'Net). To access A Gree'Net, user registration is required in advance. Contact the procurement department, to which you deliver your products, or Environment CSR Support Center.

(3) Investigation frequency

Suppliers are requested to review the following items periodically (once a year) and enter updated data into
the Green Procurement System (A Gree'Net): (i) the status of their environmental activities, and
(ii) the status
of reducing environmental loads of the products delivered to Hitachi
Investigation related to (iii) information of chemical substances contained in the products, will be requested when necessary.

Contents of investigation

(1) The status of environmental activities of suppliers

The investigation of the following items will be made for each supplier (or each business place of a supplier):

(a) Items related to environmental certification

- Acquisition of ISO 14001 certification or other external certifications that Hitachi approves

(1) Already obtained ISO 14001 certification.
(2) Already obtained other EMS certification.
(3) Facilitating or finalized the plan to acquire external certifications including ISO 14001.

(b) Items related to endeavors for "Green Procurement"

- Status of planning green procurement

(1) Implement Green Procurement.
(2) Have a plan to implement Green Procurement.

(c) Items related to environmental activities (19 items)

- Corporate philosophy and policy

  (1) Have a corporate policy regarding environmental protection
  (2) Have set up an environmental policy and committed continuous improvement and pollution prevention efforts.
  (3) The company's environmental policy commits to observe legal restrictions.
  (4) Company environmental policy is known to all employees and available to any third party.

- Plan and organization

  (5) Have a goal/target for environmental protection.
  (6) Assign specific organizations/persons to carry out relevant responsibilities toward the goal/target.
  (7) Have an implementation plan to achieve the goal/target.

- Environment assessment/system
  Control and assess the following items in the manufacturing process in an effort for improvement:

  (8) Reduce water pollution.
  (9) Reduce air pollution.
  (10) Reduce noise and vibration.
  (11) Treat waste properly and reduce the amount of waste disposal.
  (12) Reduce energy consumption (electricity, gas, fuel, etc.)
  (13) Reduce the use and discharge of hazardous chemical substances.
  (14) Have a product assessment program.
  (15) Have a systematic plan for emergency.
  (16) Have any internal environment audit program.

- Provision of education, training, and information

  (17) Implement an environmental education program.
  (18) Implement training for personnel engaged in work that may significantly affect the environment. Have a list of such personnel.
  (19) Provide information related to environmental protection.
  (d) Manufacturing process information

- Use or non-use of ozone layer depleting substances in the manufacturing process

  (1) Used in the product manufacturing process.
  (2) Not used in the product manufacturing process.
  (3) Under survey.

  (2) The status of environmental load reduction of the procured products

  (a) Environmental load reduction of products to be delivered (10 items)
  With regard to the products the Hitachi Group procures from suppliers, suppliers are requested to make efforts in accordance to the items below.

Suppliers are also requested to make the same considerations for the raw materials and parts that they procure themselves.

- Resource saving

(1) Make an effort to reduce weight and size.
(2) Use recycled parts or resources (Recycled material content rate).
(3) Product durability improvement is considered.

- Energy saving

(4) Energy saving during use/stand-by time is considered (Reduction rate of energy).

- Recycling

(5) Collect and recycle products (Recycling rate).
(6) Uniform and standardize materials.
(7) Consider ease of disassembling and sorting.

- Packing materials

(8) The supplier is reducing packing materials and giving consideration to collecting, reusing, and recycling.

- Provision of information

(9) The supplier is providing environmental information related to products.

- Chemical substances

(10) The supplier is endeavouring to properly use chemical substances.
(3) Information about chemical substances contained in the procured products
(a) Information about contained chemical substances to be input to A Gree'Net.

(i) Basic product information
(ii) Product composition information
(iii) Information about inclusion or noninclusion of regulated chemical substances
(iv) Information about the submission or nonsubmission of a Warranty of Non-inclusion
(b) Investigation format for chemical substances contained in products
For the format of controlling chemical substances contained in the products, A Gree'Net flexibly supports the formats widely adopted in the industrial world, primarily considering usability for suppliers. The Hitachi
Group conceives that the format released by Joint Article Management Promotion consortium (JAMP) is currently the most reliable and rational tool for chemical substance information transfer.

9.3 Environmental Material Management

The materials management activities within most companies are undergoing fundamental changes. Companies are, among other things, reducing their supplier base, forming partnerships with key suppliers, and implementing lean manufacturing methods. The purpose of these supply chain initiatives is to reduce both costs and wastes. Conventional decision-making approaches commonly overlook or inadequately represent the costs and benefits that may occur in departments outside the decision-maker's area. For example, a purchasing agent can often reduce material procurement costs by switching to a new, lower-price supplier. In some cases, this change increases overall production costs, even though the engineering group may approve the switch. For example, if the new supplier uses excess packaging, the increased disposal costs could exceed the procurement savings. In contrast, proactive purchasing managers are reducing costs by conducting *value analyses* of materials to minimize the firm's *overall* costs, forming cost reducing supplier partnerships, demanding less or reusable packaging, and influencing the material specification processes. These expenses are critical because the costs to purchase, handle, store, and dispose of materials represent some of the largest operating expenses for most manufacturing companies.

Companies can increase their competitiveness by developing more efficient, responsive supply chain processes. This is evident from the massive efforts that automotive, chemical, and other companies are expending to implement enterprise resource planning (ERP) systems and to change their relationships with both suppliers and customers. Many companies are striving to improve their logistics and materials management processes, but important *environmental burdens* are usually not addressed. While these considerations may seem to be a more appropriate focus for the Environmental, Health, and Safety (EH&S) group than for materials managers, a number of companies have saved millions of dollars by addressing difficult-to-spot materials management expenses such as inefficient material handling steps that unnecessarily lose excessive materials to air and water. In fact, the bulk of the activities that are considered the responsibility of EH&S are actually due to the operational activities under the purview of supply chain managers. Those managers who overlook the EH&S elements of their activities are probably overlooking opportunities to increase the responsiveness and efficiency of the supply chain. As highlighted in Table 9.6, Intel, Andersen Corporation, 3M, Commonwealth Edison (ComEd), and other leading companies have improved their purchasing, material handling, storage, *material recovery, disposition*, and product take back processes through environmental accounting initiatives. While these success stories are drawn from several industries, the common thread is a proactive, rigorous effort to incorporate environmental considerations into decision-making processes.

Substantial improvements are feasible because environmental management problems are, to a large degree, material-driven. For example, the types and costs

Table 9.6 Supply chain improvement

Innovation	Results
Purchasing	
Several electronics companies, including Nortel and Intel, have moved away from purchasing *materials* toward receiving chemical *services* via chemical management programs. These services can include procurement, inventory management, data tracking, and waste management	By providing incentives for suppliers to reduce material quantities and by leveraging the suppliers' expertise, these companies have achieved substantial savings and reduced wastes. Chemical management providers are no longer compensated based on the volume of chemicals they sell to their customers, but on value-added services instead. With appropriate incentives, providers are rewarded for reducing chemical usage (and costs), increasing productivity, or decreasing waste
Materials handling	
A number of companies are switching to reusable packaging systems. 3 M recently launched an Eco-Efficient packaging business, and GM has a well-established reusable pallets and containers program	By working closely with suppliers, GM successfully switched to reusable packaging systems and reduced its disposal costs by $12 million between 1987 and 1992. Additionally, reusable containers can reduce solid waste, decrease product damage during shipping, and eliminate ergonomic and safety problems (e.g., cuts while slicing open boxes)
Storage	
Several companies have changed their inventory storage procedures for maintenance, repair, and operating (MRO) materials by consolidating storage areas and requiring suppliers to adhere for stringent material return policies	Public Service Electric and Gas Company streamlined its purchasing and storage processes and saved more than $2 million in 1997. The changes significantly decreased the disposal of obsolete paint and other materials, reduced storage space requirements, and lowered carrying costs. Previously these costs had been hidden in overhead accounts

of wastes, the significance of regulatory constraints, and the risks of improper handling are all dependent upon which materials are purchased and used within a facility.

A number of companies have worked with suppliers to eliminate unnecessary packaging and reduce hazardous materials quantities. In fact, some companies are significantly improving both their environmental profile and profit margins by taking a strategic approach to purchasing. They are asking suppliers (and their suppliers' suppliers) to evaluate and lower environmental costs. These approaches show that environmental insights can improve core business processes. Table 9.6

Table 9.7 Materials management improvement

Innovation	Results
Materials recovery	
By focusing on their high volume material flows and striving to eliminate wastes, several companies have justified material recovery projects by applying environmental accounting methods	Andersen Corporation (manufactures of Andersen Windows and Patio Doors) developed a composite material from wood wastes generated during its manufacturing processes. This innovation is expected to yield 50% return of investment and has already enabled Andersen to decrease solid lumber purchases by 750,000 board-feet
Disposition	
Many companies are saving money by ensuring that major waste disposal costs are made explicit and attributed directly to the responsible product or business unit. Companies use this cost innovation to identify more financially attractive alternatives to disposal	Environmental accounting techniques enabled Commonwealth Edison, a Midwestern electric utility company, to greatly reduce its landfill disposal volume. A *life cycle* accounting approach highlighted the indirect costs created by a variety of activities including disposal. Once these costs were made explicit, the company began developing a cost-effective method for grinding three limbs. Com Ed netted $2 million annually making and selling tree mulch instead of sending the tree limbs to landfills
Product take back	
A variety of companies have developed cost-effective ways to recover products from their distributors or costumers. By working with products designers and other functions, supply chain managers can establish systems that enable them to recover these assets and reduce manufacturing costs	Kodak's logistics system currently recovers 70% of the cameras sold to consumers. Since its inception, over 200 million FunSaver cameras have been returned to Kodak facilities. Kodak reduces operating costs by recycling or reusing 77 to 86% of each camera's materials. One of its significant accomplishments has been developing a system that overcomes the financial and non financial barriers that initially limited product returns from photo processors

illustrates how strategic innovations in several core business practices have improved both the financial and environmental performance of their supply chain.

In each of the Table 9.6 illustrations, the roles of purchasing, inventory, or other supply chain managers were pivotal (EPA 1998).

While this guidebook focuses on environmental improvements, the framework and tools are equally applicable to health and safety issues, e.g., injuries and subsequent workers' compensation claims that result from materials handling (Table 9.7)

Box L: Materials Management at Public Service Electric and Gas (PSE&G)

At Public Service Electric and Gas (PSE&G), the Materials Management Team led the effort to reduce disposal and other costs. The team determined that the utility company was over purchasing numerous products and inefficiently storing them in several locations within each facility. The net result was unused, excess materials that wasted storage space and eventually required disposal. The PSE&G Materials Management Team first analyzed its inventory supply chain, particularly for its chemical commodities and paints. The inventory analysis revealed that each facility purchased materials separately from many suppliers. Since most suppliers encouraged purchasing in bulk, the facilities often had excess product.

Leftover inventory was eventually sent to PSE&G's central resource recovery facility where materials were sorted, sent to disposal facilities, or, if possible, sold. The company realized that it could avoid the costs of sorting, disposing, and finding buyers for excess inventory by simply improving its purchasing practices.

The Materials Management Team narrowed PSE&G's list of suppliers from over 270 to only nine. As part of the long-term agreements that the company established with these select suppliers, the suppliers agreed to track the inventory that each PSE&G facility purchased. Thus, whenever purchasing agents from one plant called to order products, the vendor would check to see if another PSE&G facility already had the material in stock. The suppliers also committed to take back any extra, obsolete, or discontinued products. (Usually they are able to sell these materials to other customers.) By placing the responsibility on its suppliers, PSE&G eliminated the disposal of unused, excess products and, in doing so, reduced operating costs and environmental burdens.

Improved materials management practices can also improve environmental, health, and safety performance. For example, converting to reusable containers can reduce the solid waste burdens associated with primary and secondary packaging materials, and can minimize the number of injuries caused by using razor knives to open packages. Similarly, chemical management service providers closely track material consumption and can quickly spot production floor inefficiencies that lead to material losses and resulting waste disposal.

Tables 9.8 and 9.9 list a number of financial and non-financial environmental considerations associated with supply chain activities. As shown, purchasing, material handling, and several other core materials management functions can substantially affect environmental performance. Tables demonstrate how better materials management decisions can result from a more complete understanding of the costs that a company incurs throughout the life cycle of its products, processes, and facilities.

Table 9.8 Examples of environmental costs and benefits

Purchasing	Material handling	Storage
Volume of production material purchased	Purchase price of packaging materials	Specialized storage space for hazardous materials
Potentially hidden		
Activities to certify suppliers	Efficiency gains from automated handling of reusable containers	Costs of cleaning up spills
Contingent		
Reduced risks of environmental incidents by working with suppliers that have product stewardship programs	Ergonomic and safety issues, including cuts from razor knives	Employee exposure to hazardous material and subsequent workers' compensation claims
Image/relationship		
Positive media coverage because of partnership with "green" suppliers	Positive public image from using reusable containers	Improved employee satisfaction from reduced exposure to hazardous materials
External		
Improved ecosystems through eliminating the use of hazardous materials	Accidental emissions to air and water due to spills	Decreased releases to ecosystems because of fewer and less severe spills

Table 9.9 Examples of environmental costs and benefits

Material recovery	Disposition	Product take back
Revenues from recovered materials	Labor and fees associated with manifesting, hauling and dumping solid waste	Reduced material costs because components recovered from returned products
Ecoefficiency gains, e.g., reduction of materials "lost to scrap"	Decreased environmental insurance premiums by decreasing quantity and hazard of disposed materials	Increased shipping costs due to product returns
Decreased remediation liability due to lower volume of waste sent to landfills	Potential liability for cleaning up hazardous wastes that leak out of landfills	
Ability to attract investors and insurers because of saving from material efficiencies	Positive public image by avoiding the community backlash related to leaks from inadequate disposal sites	Increased costumer concern about the quality of products
Reduced exposure-related medial expenses for local residents	Decreased landfill burden	Decreased mining or harvesting of raw materials due to lowered overall material requirements

9.3.1 A Revised Economic Order Quantity: Improving the Inventory Management Model

As companies establish just-in-time and other lean inventory systems, they reevaluate the lot sizes of purchase orders and production runs. This hypothetical example demonstrates how a company could increase its overall efficiency by changing a common inventory practice— determining the economic order quantity.

Commonly hidden in inventory management decision-making are the costs to dispose of obsolete, hazardous materials. Several factors, especially ordering costs and inventory carrying costs, are typically considered when companies determine what quantity of specific materials to order. In contrast, companies rarely incorporate the risk of obsolete materials.

This example illustrates how two environmental considerations, material losses, and waste disposal, can be included in this common inventory management model. This example does not address other hidden and contingent costs, such as those related to spills and worker exposure. The basic equation for determining the order quantity that minimizes total costs is:

$$Q = \sqrt{\frac{2SD}{HC}}$$

where

Q = Optimal order quantity (units)
D = Annual demand for material (units)
S = Procurement and setup costs per order ($)
H = Inventory holding cost rate, often 10–35%
C = Cost of inventory item ($/unit)

Holding costs are the costs to maintain inventory and include cost of capital (that could be invested elsewhere), warehousing costs, insurance, and other expenses. For hazardous materials, the costs of disposing obsolete materials should be considered because disposal costs, on a per unit basis, can be comparable to the initial purchase costs.

For example, many industrial paints cost approximately $3/lb, while disposal costs for these materials can exceed $1/lb. If a company analyzed its hazardous waste disposals and observed that ~5% of its paint was eventually disposed of instead of used, then the company should increase its holding costs considerably (Table 9.10).

For this hypothetical case, the economic order quantity decreases by 23% when material obsolescence and the accompanying costs of lost materials and waste disposal are considered. With this approach, a company could reduce the environmental burdens and decrease the overall costs associated with paint disposal.

Table 9.10 Example

	Conventional	Revised
Annual demand (D)	4,000 lb	4,000 lb
Setup costs (S)	$50	$50
Holding cost rates (H)[h,i]	10%	15% (+ 5% for disposal)
Item cost (C)	$3/lb	$3/lb (+ 1/lb for disposal)
Order quantity (Q)	1,155 lb	895 lb

Several options, including recycling and energy recovery, can reduce these costs. In this case, assume that disposal is required. Losses include the purchase cost of the lost material (5% x material price) and the disposal cost due to obsolescence (5% × disposal cost). The disposal ratio (5%) could be determined empirically and would depend upon the types of materials, purchase quantities, and consumption rates. Periodic reevaluation of this ratio would be warranted. For a conventional case, assume that material price = $3/lb and holding costs = (10% × material price).For a revised case, assume that disposal cost = $1/lb and revised holding costs = [(10% + 5%) × material cost] + [5% × disposal cost].

Box M : Materials Tracking

In 1993, ComEd recognized that its total cost of managing materials was greater than its initial purchasing expenses. In particular, a significant portion of the company's waste disposal costs were caused by inefficient purchasing, storage, and other materials management practices. Based upon this realization, the company initiated a materials tracking effort to identify and characterize the waste and by-product streams from its generating facilities. At several facilities, engineering teams used a waste accounting software package to evaluate five high-priority material categories: chemicals and oils, coal by-products, PCBs, recyclables, and solid waste. The teams did not conduct extensive mass balances, but rather addressed some key questions:

- Which types of wastes were being generated at the various facilities?
- What quantities of each material were being disposed?
- What were the waste disposal costs or recycling revenues for each material?

The goal of the first step, Identify Costs, was to identify the waste streams that were significant enough to justify additional evaluation. Thus, the initial evaluation was limited to identifying the waste streams and completing preliminary analyses at a few facilities. The teams did not track emissions, spills, or other types of environmental burdens, nor did they estimate purchasing, storage, or other costs. The approach pinpointed three high volume and high-cost waste streams: coal ash, contaminated soil, and solvent waste. In particular, the ComEd team discovered that the facilities annually disposed of over 500,000 pounds of solvent-contaminated materials. Since these waste solvents were regulated and accordingly disposed of as hazardous wastes, the disposal costs were substantial.

Team Approach

The waste stream evaluation revealed a number of significant opportunities for reducing waste at its source. ComEd then focused on the company's use of solvents and alternative cleaning materials, and broadened its scope of analysis to include the complete set of materials management activities. Action committees were formed with members from all of ComEd's key materials management processes: procurement and contracting, receiving and testing, warehousing and distribution, operations, and recovery and distribution.

The team began by evaluating requirements that could constrain its ability to replace a solvent with a less costly, less burdensome cleaning material. Three questions were answered for each cleaner:

- For which applications is the solvent best suited?
- Does the cleaner leave a residue?
- Is the cleaner corrosive to metals, vinyls, plastics, or insulations?

These questions helped the team determine which cleaners could be easily substituted and which ones could not, e.g., obviously, a cleaner that corroded metal could not be used on steel or other metallic surfaces. After determining which cleaning materials were viable substitutes for each other, the team then evaluated the product performance of each cleaner. A cross-section of personnel who used or managed the solvents rated the materials in five categories:

- Operating
- Purchasing and Supply
- Environmental and Regulatory
- Safety and Health
- Analytical.

The survey participants rated each cleaner on several criteria within each category. For example, the Purchasing and Supply criteria were

- Cost
- Vendor Performance
- Shelf Life
- Packaging Safety
- Availability of Various Sized Packages
- Storage Ease
- Dispensing Ease.

The solvents were ranked from 1 to 10 (1 being the worst rating, and 10 the best). Since some factors were considered significantly more important than others, the team assigned a weighting factor to every criteria. A final product rating was determined by calculating a sum of the weighted criteria scores.

The screening questions enabled the identification of substitute cleaners, and the product ratings revealed which cleaning materials had superior operating and environmental performance.

Thus, during the second step, Determine Opportunities, the team uncovered numerous opportunities to replace solvents with equivalent or superior performing cleaners.

Life Cycle Cost Analyses

The third step of the Inventory Minimization Program, Calculate Benefits, was a life cycle cost evaluation. The team quantified three categories of cost:

– Purchase Costs
– Inventory Carrying Costs
– Waste Disposal Costs.

Since the team had already determined the last category of costs, the team focused on the first two categories and obtained data by reviewing purchasing and warehouse records. The analyses revealed that the cost reduction opportunity was spread across several of ComEd's materials management processes. In fact, disposal cost savings were less than one-third of the total cost savings obtainable.

The team was also concerned about hidden and contingent costs. Since these costs were more difficult to

estimate, the team qualitatively evaluated several additional cost categories, including:

– Solvent Specification Cost
– Record Keeping Cost
– Cost of OSHA and EPA Citations
– Training Costs
– Cost of Solvent Misuse.

These cost evaluations enabled the team to address the full spectrum of solvent performance, from purchasing cost to environmental considerations to operating effectiveness.

The life cycle cost analyses confirmed the opportunities to replace ComEd's current cleaning materials with lower cost, better (or at least equivalent) performing alternatives.

Results!

The Inventory Minimization Program proved to be quite successful at the two facilities that first implemented it. The teams' recommendations led to cost savings and environmental gains. ComEd changed its procurement decision process and reduced the number of solvents from fifteen to three.

Based upon this initial success, the program was implemented and transferred to other generating facilities. The overall results were:

- Replacement of over 100 solvent products with nonhazardous materials
- Reduction of hazardous solvent waste by 88% during a two-year period
- Five-year estimated cost savings of over $1,000,000
- Considerable reduction of difficult-to-estimate costs, including those for record keeping and training.

This initial improvement effort has since evolved into a continuous improvement effort— the New Product Evaluation Process. The purpose of this process is to ensure that ComEd continues to reduce the life cycle costs and the environmental concerns of cleaning activities. After some initial screening steps (e.g., determining if the material has special handling or storage requirements), new solvents are evaluated as described in Steps 2 and 3 with a performance evaluation and life cycle cost evaluation. This effort has enabled ComEd to continue reducing its hazardous material use, accompanying waste, and overall costs, while helping operating personnel perform their jobs quickly and effectively.

Since this initial success, ComEd has monitored its progress and successfully expanded its program to address several other materials management activities.

Box N: Collaborative R&D; The Case of the Smart Flow of Goods Project (Source: Romsdal 2009)

In 2008, the British Standards Institute (BSI 2008) introduced the PAS 2050 methodology; a specification of assessment of GHG goods and services. The methodology is based on the view that not only the direct burning of fossil fuels but also the consumption of goods and services give rise to GHG emissions—referred to as indirect or "embodied" emissions (BSI 2008; Minx et al. 2008). This is particularly relevant in supply chains of perishable products such as food, where an overwhelming proportion of GHG emissions arise in the primary production stage and much less in distribution and packaging (Nereng et al. 2009). This implies that any assessment of GHG emissions for a product provided to the customer should include emissions at all the stages of the supply chain. The embodied emissions perspective thus highlights the importance of efficient logistics and operations in the food supply chain in order to speed up the product's journey through the supply chain to increase the likelihood of the product fulfilling its function, i.e. being consumed before its use by date.

Supply chain planning and control

In order to meet some of the challenges facing today's complex supply chains, new concepts for supply chain planning and control are emerging. Manufacturing planning and control (MPC) tasks in a supply chain involve determining what, who, when, and how to act in order to meet customer demands with the exact supply in a coordinated chain (Vollmann et al. 2005). Since supply chain actors affect each other, they cannot be managed in isolation (Shi and Gregory 1998). Today, most of the planning and control systems used in supply chain operations are based on traditions of make-to-stock (MTS) and material requirements planning (MRP) where forecasts and expectations of future demand are the main inputs. The consequences are that a number of supply chain operations are decoupled from actual end customer demand, and that inventories are used as a buffer against uncertainty and fluctuating demand. The needs for increased supply chain integration and collaboration has resulted in the development of several collaborative models for orchestrating supply chain and network activities, such as vendor managed inventory (VMI), collaborative planning, forecasting and replenishment (CPFR), and automated replenishment programs (ARP). The aim of such models is to achieve seamless inter-organizational interfaces by specifying control principles and models for the flow of materials and information, where network operations are tied and adjusted to customer demand and more make-to-order (MTO) strategies.

Information and supply chain transparency

In their 2008 literature review on sustainability, Carter and Rogers (2008) identified transparency as one of the supporting facets of supply chain sustainability. Increased transparency is in part being driven by the rapid speed of communication and improvements in software, and can be improved through vertical coordination across supply chains, as well as horizontal coordination across networks (Carter and Rogers 2008). Access to real-time demand and event information in supply chains is a critical element in the implementation of the demand-driven control concepts described above. Advances in technology within areas such as RFID, sensor technology, and electronic product code information services (EPCIS) will enable access to more real-time information than the existing technology solutions, supporting a shift in the planning and control concepts towards purer demand- and pull-driven supply chains.

Project background

The project *Smart flow of goods* was initiated in 2006 by nine major players in the Norwegian grocery industry; three food manufacturers, two manufacturers of packaging material, two wholesaler–retailer dyads, and two logistics solution providers. The background for the project was threefold; there was a need for track and trace solutions in the food sector, a wish to increase the competitiveness of actors through responsive and efficient logistics solutions in the supply chain, and a need for development and testing of intelligent and automated logistics solutions enabled by RFID and EPCIS technology.

The fresh food supply chain

Food supply chains are complex and form large networks, and the grocery industry's solution to its logistics challenges has tended to set a standard for other industries. The logistics structure is often centrally coordinated, enabling use of cross-docking and terminal facilities in distributing goods to retailers in parallel with direct shipments from manufacturers. The Norwegian grocery market is dominated by a handful of large actors within each stage of the supply chain and there is a near-full consolidation into four chains of wholesaler–retailer dyads which in total control 98% of the market. Control concepts in the industry are dominated by traditional push and forecasting-based control. Information on consumer demand is in many cases not accessible due to lack of, or infrequent, information sharing, and those who have access to information do not necessarily utilize it for control purposes. Some other typical challenges and problems associated with the current operations and control of Norwegian food supply chains include:

- A high number of stock points or buffers along the supply chain
- Large amounts of waste/scrapping due to long lead times and temperature sensitive products
- Forecasting and planning in each node based on forecasted demand from subsequent node
- Forecasting based on historic sales and extensive manual parameter setting in forecasting software
- Limited information sharing and use of available information for operations planning and control

Since fresh food is highly perishable, efficient production planning and SCM is crucial. The amount of food waste generated along the supply chain and in consumers' households today is substantial, and a recent British study found that 61% of household food waste is avoidable and could have been eaten if it had been managed better (Ventour 2008). Even though cause-effect chains are complex and further mapping of these are required, it is reasonable to claim that a substantial share of this waste can be attributed to long lead times and resulting short shelf life of perishable food products. Thus, Manufacturing planning and control (MPC) systems in food supply chains should support cross-company processes in a manner that avoids increasing demand amplifications, stock levels, and lead and response times.

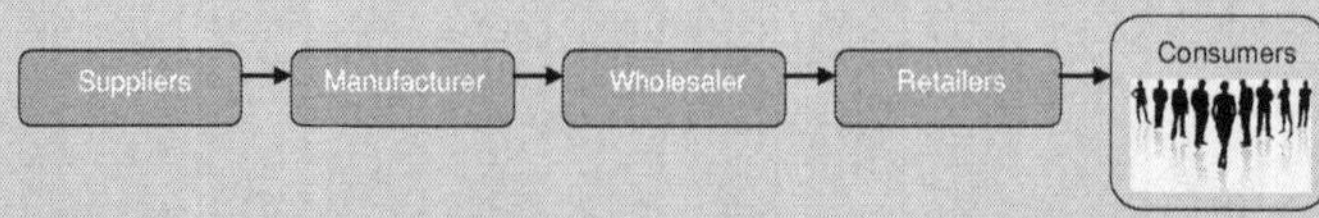

Fig. 9.11 Typical food supply chain

Organisation and working method

A typical supply chain involved in the project is depicted in Fig. 9.11.

New solutions in the project were developed using the *control model methodology*. A control model is a description of the material and information flows in a supply chain documenting the material and information flows between the various supply chain actors, as well as processes, transportation modes, and the detailed principles and rules used to control material flows. Initially, an AS IS control model describing the starting point for each supply chain was developed. The main purpose of this AS IS model was to make all involved actors aware of and agree on the structures and policies that were currently used to control the supply chain. Information was collected through workshops, meetings, interviews, observation, written documentation, databases, etc. The supply chain actors were responsible for providing the information requested by researchers, who then systematized the information into an illustrated, structured AS IS document that all the involved participants agreed on.

After an analysis of the AS IS control model and a mapping of improvement opportunities, several TO BE control models were developed introducing new aspects such as RFID and collaboration through real-time information sharing. The TO BE models specified how the individual supply chains should be controlled in the future. The future models were developed in a creative collaborative process consisting of workshops, meetings, discussions, etc. among key supply chain actors and the involved researchers. The solution development process drew heavily upon existing theoretical knowledge and concepts, best practice principles, researchers' experience, and decision makers' detailed knowledge of the supply chains. Finally, an implementation plan for the TO BE control model will be developed, mapping out the requirements and prerequisites for achieving increased collaboration and improved operations in a supply chain.

The Smart flow of goods concept

The core idea of the project is to investigate how the use of RFID and EPCIS technology integrated into packaging can increase supply chain sustainability. The concept builds on development and testing of the following key elements:

– New concepts for the flow of food products through the supply chain using RFID and EPCIS integrated in the packaging material

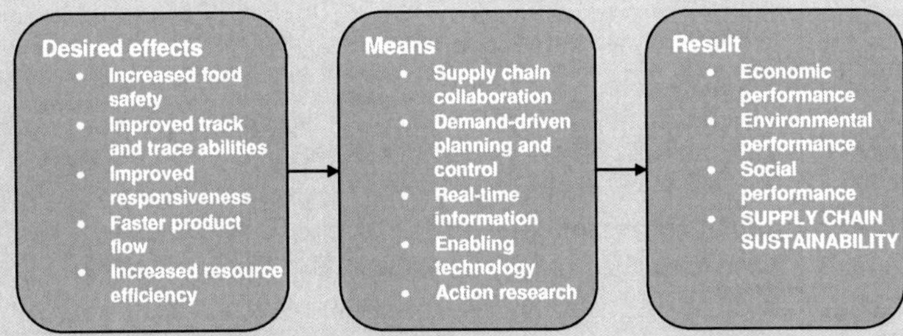

Fig. 9.12 Desired effects, means and result of the

- "Intelligent packaging" and returnable transport items (RTI) based on RFID and EPCIS as standard for data capture and information exchange in the supply chain
- New ICT architecture and systems that support SCM and eTraceability
- Control models for demand-driven supply chains based on real-time and transparent information and streamlined business processes
- New business and supply chain collaboration models

Figure 9.12 illustrates the desired effects of the project and the means used to achieve them in order to contribute towards the three pillars of sustainability.

Smart flow of goods project.

Based on literature, experience, and discussions within the project, a number of expected positive effects on supply chain performance from the implementation of the concept have been identified. Sharing of real-time information and more pull-driven MPC principles are expected to lead to better forecasts and reduced demand variability (the Bullwhip effect), improved manufacturing and inventory control, better capacity utilization, and reduced inventory levels. Other effects include better planning of transport and distribution, reduced response time and total lead time to demand satisfaction, fewer stock-outs, earlier warning of potential supply problems, and more responsive supply chain actors. Expected environmental effects are mainly related to the reduction in product loss/wastage due to improved planning and control in manufacturing and distribution, thus reducing energy, GHG and embodied GHG emissions, and increasing reuse of packaging resources.

Action research for sustainability

The origins and basic ideas of action research can be traced back to the psychologist Kurt Lewin, who has been attributed with the coining of the term "action research". However, action research is not so much a methodology as a collection of approaches that aim to: "*...contribute both to the practical* *concerns of people in an immediate problematic situation and to the goals of science by joint collaboration within a mutually acceptable ethical framework"* (Middel et al. 2006). The three crucial elements of action research are thus *action, research*, and *participation*. Research projects aim to contribute both to the specific organization and academia, and are in essence change processes in which one seeks to develop a holistic understanding of a complex dynamic system (Coughlan and Coghlan 2008). Another key characteristic is that action researchers are actively involved in the change initiative in the organization and thereby directly contribute to the results of the project. An action research project can therefore be said to have two parallel objectives: an improvement objective to solve a specific problem and a research objective to contribute to the generation of new knowledge. An important characteristic of action research is that it is situational and thus does not generate universal knowledge, and that theory generated through action research therefore is very hard to replicate or test. Theory emerges incrementally during the project—based on the theoretical understanding that grows through the reflection on the planning, implementation, and evaluation phases of the action research cycle(s) (Coughlan and Coghlan 2008). Generalization from qualitative studies such as action research and case studies takes place toward theory and not toward samples and universes. Thus the value of the findings will lie in the ability to achieve "extreme relevance" and practical applicability, leaving the question of generalization up to the practitioners' evaluation of whether or not the findings apply to their particular situation. Action research can contribute to increased supply chain sustainability.

Joint meaning construction: Action research facilitates joint meaning construction and problem definition. By spending considerable amounts of time talking to and interacting with each other, both the researchers and the organizations' members were better able to understand the situations they were facing and develop appropriate solutions.

Building trust and commitment: An environment of trust and commitment is essential in any collaborative project. Since action research is built on participation and close interaction between practitioners and researchers it can build trust and commitment not only within a project but amongst the supply chain partners as well, having a mitigating effect on some of the disturbing effects for instance power issues potentially have on supply chain collaboration efforts.

Multiple methods and disciplines: Action research is not synonymous with a purely qualitative methodology. Instead, it is a research strategy that combines qualitative and quantitative methods in analysis and problem solving where appropriate depending on the issues at hand. In addition, action research is not restricted to any particular scientific field and can easily incorporate theories, perspectives, and tools from other research disciplines, thus supporting the multi-disciplinarity required within sustainability.

Gap between industry and academia: The dual project objectives of improvement and knowledge generation contribute to closing the gap between industry and academia. The action research approach can ensure the relevance and applicability of solutions for increased sustainability and knowledge generated in an R&D project.

Action research for social change: Action research can be understood as a set of collaborative ways of conducting research that promotes democratic change through its basis of democratic inclusion. This facilitates the combining of technical and organizational measures that were identified in Chap. 2 as necessary to build sustainability.

Limitations of action research

Despite the advantages of action research described above, there are also some challenges associated with its use, particularly within a supply chain context. In general, action research puts fairly high requirements on the involved researchers in terms of skills in addition to the typical skills required of any researcher (e.g. diagnosis, intervention, learning in action, social skills, and dealing with uncertainty). Therefore, action research teams should always involve experienced action researchers to ensure knowledge transfer to lesser experienced researchers. Other challenges are related to the need for spending an extended period of time within the context under study—which in supply chain terms will involve researchers closely interacting with a number of companies. This puts resource constraints on projects in terms of time, cost, and personnel. In addition, gaining access to companies is frequently an issue in any action research project and might be particularly difficult in supply chain projects where success of the project depends on obtaining the commitment of and gaining access to a number of companies simultaneously. Also, the fact that action research does not generate universal knowledge might limit the usefulness and transferability of results to other settings. However, inherit in the action research strategy is the understanding that the judgement of the applicability of knowledge generated in one context to another context is left up to the practitioners' assessment.

Findings on sustainability

In terms of the economic aspects, both the academic and industrial partners agree that the control models for real-time demand-driven supply chain planning and control developed in the project are likely to impact positively on operational and logistical efficiency, thus improving economic performance.

The project has established a collaborative environment, which has enabled the development of new collaborative processes and collaboration models. Hopefully the industrial actors will be able to build on this in their long-term relationships, thus supporting their future efforts toward sustainability. The fact that the project was conducted in Norway must also be taken into consideration when discussion the project's perhaps somewhat lacking focus on social aspects. Norway has a long tradition of strong socio-democratic influence which has impacted on the culture and the business environment. There is extensive legislation in place regarding health, safety and environment (HSE), working conditions, pollution, gender equality, etc., and there has long been a fairly cooperative relationship between trade unions and employers' associations. These facts are likely to have reduced the need for explicit focus on social aspects in sustainability projects. However, for companies which outsource e.g. manufacturing to other countries and cultural contexts, social issues will be highly relevant and critical aspects that must be incorporated into any sustainability efforts. Technology has by some authors been suggested as a fourth pillar of sustainability. Investigation of the potential of RFID technology was the main trigger for the *Smart flow of goods* project. RFID has been found to provide a good starting point for successful information sharing by improving data quality and capturing data at multiple points in the supply chain, making it possible to share more relevant information. However, preliminary findings also indicate that the project's most important contributions toward sustainability not necessarily stem from the application of the technology itself, but rather from the increased ability and willingness to share information with supply chain partners. POS data and inventory information can currently be shared without RFID. However, the focus on RFID seems to have motivated the supply chain partners to work on supply chain transparency and information sharing—thus enabling improved supply chain planning and control.

9.4 Closed-Loop Supply Chains and Reverse Logistics

Closed-loop supply chains and recovery of used products, in particular, recently have received much attention. While traditional logistics are perceived as managing the supply of goods and/or services from the producer to the (end) customer as well as internal logistics, and input and output to the company, reverse logistics is the process of planning, implementing, and controlling the efficient and effective inbound flow and storage of secondary goods and related information opposite to the traditional supply chain (SC) direction for the purpose of recovering value or proper disposal.

Reverse logistics is defined by Rogers and Tibben-Lembke (2001) "the process of planning, implementing, and controlling the efficient, cost effective flow of raw materials, in process inventory, finished goods, and related information from the point of consumption to the point of origin for the purpose of recapturing value or proper disposal". The closed-loop supply chain is a wider concept, and includes, besides reverse logistics, the return product process of acquisition, test, sort, and disposition, including distribution and marketing.

Closed-loops consist of two supply chains: a forward and a reverse chain, whereby the recovered product either reenters the primary forward chain or is shunted to a secondary market (Fig. 9.13).

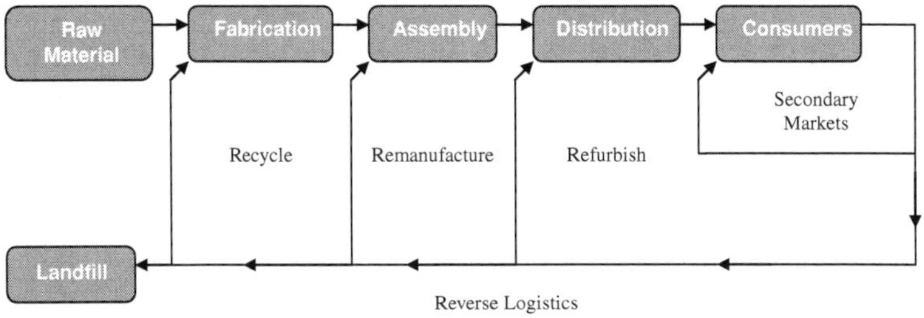

Fig. 9.13 Closed-loop supply chain scheme (modified by Kumar and Putnam 2008)

End-of-life management involves those options available to a product after its useful life. There are five product recovery operations aimed at recapturing value:

Repair and reuse: The purpose of which is to return used products in working order. The quality of the repaired products could be less than that of the new products.

Refurbishing: The purpose of which is to bring the quality of used products up to a specified level by disassembly to the module level, inspection, and replacement of broken modules. Refurbishing could also involve technology upgrading by replacing outdated modules or components with technologically superior ones.

Remanufacturing: The purpose of which is to bring used products up to quality standards that are as rigorous as those for new products by complete disassembly down to the component level and extensive inspection and replacement of broken/outdated parts.

Cannibalization: The purpose of which is to recover a relatively small number of reusable parts and modules from the used products, to be used in any of the three operations mentioned above.

Recycling: The purpose of which is to reuse materials from used products and parts by various separation processes and reusing them in the production of the original or other products.

Typically reverse (or closed-loop) supply chains span the following five groups of activities along with constraints shown in Fig. 9.14.

Collection: All activities rendering used items (product, component, or material) available and physically moving them to some point for further treatment. This may involve product acquisition, transportation, and storage.

Inspection/separation: Results in splitting the flow for various recovery and disposal options. This may involve testing, disassembly, shredding, sorting, and storage.

Re-processing: Reusable flows undergo the actual transformation of a used item into a reusable item of some kind. Depending on the recovery option chosen, this comprehends various activities such as disassembly, shredding, repair, and replacements.

Disposal: The non-reusable flows are disposed of to incinerators and landfills.

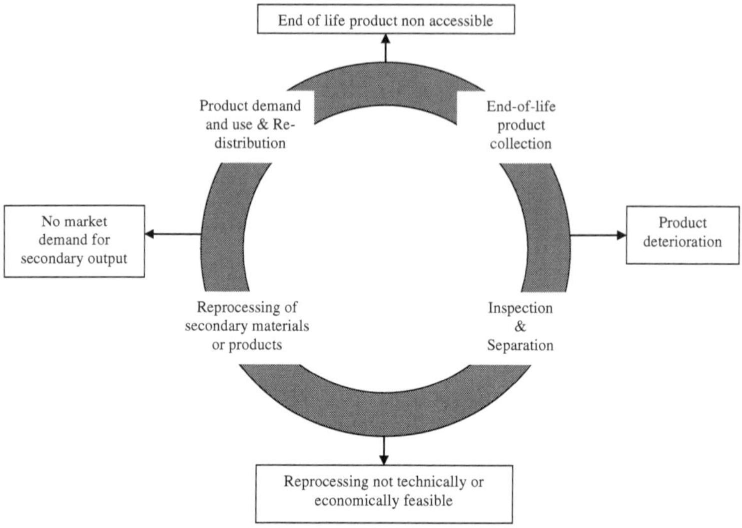

Fig. 9.14 Closed-loop supply chain: main processes and constraints (modified by Kumar and Malegeant 2006)

Re-distribution: Directing reusable items to be marketed to new markets, and physically moving them to potential new users. This involves sales activities, transportation, and storage.

Examples of closed-loop supply chains are Xerox-Europe (Guide VDR 2003), Kodak's Single-Use Cameras (Guide VDR 2003), IBM's spare parts (Fleischmann et al. 2003), Caterpillar's remanufacturing of diesel engines (www.planetark.com), and the European Recycling Platform founded in 2002 by Braun, Electrolux, HP, and Sony (www.erp-recycling.org).

Brodin and Flygansvær (2006) have empirically identified three types of coordinators within the EEE-industry: The dominant coordinator (closed-loop supply chains), the implicit coordinator (a recycling specialist of an open reverse logistics system), and the mediating coordinator (e.g. a TPL provider).

Stock et al. (2006) argue that effective product returns strategies and programs can enhance the competitive advantage. Similarly, Johnson (2005) states that "product returns have long been a necessary evil, but top companies today are managing their reverse supply chains as a source of value". In many cases procedures, policies, and processes related to the firm's reverse logistics are embedded in the operational routines of value-chain activities. Therefore, reverse logistics belongs to the firm's distinctive capabilities that are difficult to imitate, transfer, or substitute.

The variations in timing, quality, and quantity of product returns make it difficult to forecast requirements and allocate resources to return systems on other than an ad hoc basis. Only a few companies have a formalized information system and standard operating procedures for handling returns. An important problem is related to the fact that products returned by end-users are often unpacked, without barcodes, or other product identifications. When the products are returned to consolidation or return centres, it is a time-consuming task to identify the product and re-labelling it with a barcode.

Time-to-remarket is essential for time-sensitive returns, e.g. clothes, books, mobile phones, and electronic equipment.

Blackburn et al. (2004) use the term "preponement" as a strategy to make the reverse supply chain responsive by reducing time delays and promote early collection, sorting, disposition, and disassembly rather than late (postponement) process and product differentiation. Cannibalization is a problem for companies, which take back used products or new products returned from the end customer to be returned to the market. In some cases the products are repacked and returned to the primary market at the same price. In other cases, the products are sold on a secondary market, e.g. via a broker or an electronic auction (e-bay.com, lauritz.com, amazon.com).While performance measurements can be routine in the case of forward flows of products in the supply chain, return flows are rarely measured in a systematic way. However, it is also important to set up performance measures for the reverse supply chain, e.g., time from consumer complaint to replacement of new product/repaired defect product at the customer premise, time to pay-back the customer, quantity and quality of returns, causes of returns, costs involved in returns, etc. The responsibility of the reverse supply chain is often fragmented among different actors, and as a result, no one takes overall responsibility (see for instance Nike case, Box P).

Most developed countries have regulations focused on preventing and managing waste streams such as municipal waste, industrial waste, and hazardous waste. These regulations are the result of societal, consumer, and environmental values spurred by a variety of interests such as population density, land limitation for landfills, sustainable resources for natural resources, clean water, and air (van den Bergh and van Veen-Groot 2001). Additionally, the Basil Convention on the Control of Transboundary Movements of Hazardous Wastes and their Disposal (1992) prohibits the international shipment and trade of hazardous wastes—particularly to protect developing countries from polluting their own environments due to undeveloped environmental protection measures.

The European Union has been a leader in developing regulations such as End-of-life Vehicles Directive (ELV) (see "**Automobile closed-loop supply chains**" Box O), Waste Electrical and Electronic Equipment Directive (WEEE), Restriction of Use of certain Hazardous Substances Directive (RoHS), and the Packaging and Packaging Waste Directive.

Box O: Automobile Closed-Loop Supply Chains

Automobile recycling has long been a profitable business in the United States. The Steel Recycling Institute reported the 2003 recycling rates to be 103%. The automobile recycling rate is calculated by comparing the total

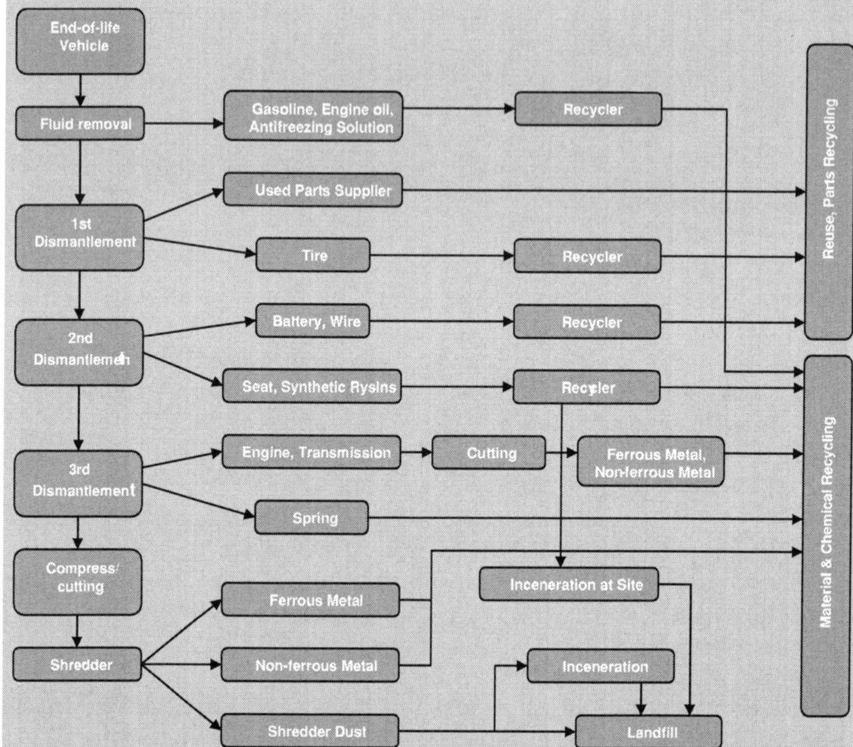

Fig. 9.15 An example of materials flow of end-of-life vehicle treatment system

steel used to produce new cars versus the total steel recovered from old cars (Steel Recycling Institute 2006b). There are over 10,000 dismantlers in the US and the infrastructures of dismantlers and shredders are strongly intertwined in the US (Johnson and Wang 2002). In the European Union, it is estimated that 8–9 million vehicles are discarded annually, which result in recycling of approximately 75% by weight of the vehicle and 9 million ton of waste created per year.

Due to the value of steel and other components of automobiles, recycling practices are attractive. ELVs contain hazardous substances such as spent oils, solvents, heavy metals, organic toxics, and ozone-depleting substances. Contaminated shredder residues are a significant issue around the world. See Fig. 9.15 for an example of a materials flow for an ELV treatment system (Kim et al. 2004).

A Strengths, Weaknesses, Opportunity and Threat (SWOT) analysis (see Table 9.11) has been selected to summarize the issues the automobile manufacturers face in taking on producer responsibility for their product's ELV regulatory drivers.

Table 9.11 SWOT analysis. *Source*: Johnson and Wang (2002); Kim et al. (2004); Steel Recycling Institute (2006a, b)

Strengths :

- There is a successful and strong reverse supply chain in existence for automotive recovery. Recovery, recycling and reuse technology and market exist
- There is market competition for significant wastes streams, particularly the metals, tires and recyclable solvents
- Some manufacturers successfully remanufacture or reuse parts. They have integrated these materials back into their supply chain
- Spare parts are a profitable business
- Volvo, Saab and BMW appear to have taken their producer responsibilities more seriously. They have redesigned their cars so that components can be dismantled more efficiently. Consequently they have lowered their producer allocation fees by 35-40% to ELV programs. Design paybacks have been demonstrated

Weaknesses :

- Each EU country has a slightly different structure for their ELV program (i.e., producer responsibility, first/last owner pays portion of scrapping fee, etc.)
- Dismantlers are not always incentivized to remove all hazardous materials before metal recovery occurs. Contamination of shredder material and landfills is a significant liability
- In the EU, approximately 25% of vehicle weight is not recycled today. Eight to 10% of the vehicle consists of plastic (polyvinyl chlorides which are difficult to recycle). Improved technology and markets are needed for glass and plastic waste streams
- In the EU regulatory management and subsidies re-incentivize and remove free-market forces by increasing complexity and cost of disassembly and recovery. Manufacturers are being forced to pay for the recovery and recycling
- Sustainable financial mechanisms have not been developed for all waste materials to enter into a perfectly competitive market place
- Information systems and monitoring systems are not in place
- Manufacturers do not have perfect control over end user's final disposition of the vehicle. Some country's scrapping fee structure causes last owners to abandon cars. There is a strong secondary market for used cars in Eastern Europe, South America, and China

Opportunities :

- Investment in R&D re-engineering could improve component design and design for disassembly. Others in the industry have already reduced their recycling fees by 35–40%
- Remanufacturing of some components could create profit growth in reuse markets or service. (This segment probably does not need to search for new market segments for remanufactured goods)
- There already multi-lifecycle use of some components in the industry

Threats :

- Abandoned vehicles are a problem in the countries that place some responsibilities on the end user of the vehicle (i.e., Germany and the Netherlands). End users abandon cars in order to avoid paying waste disposal fees. More EU countries are reorganizing their programs to place the responsibility fully on manufacturers. This will ultimately require R&D on redesign
- Foreign producers who do not have disassembly information or reverse supply chain networks may have to transport vehicles back to the country of origin or pay higher fees

(continued)

Table 9.11 (continued)

- Johnson Wang's research found that plastic, polymers and glass materials cannot be economically recovered and are expected to drain the recycling infrastructure unless markets for these materials are developed. These researchers did not believe the 2015 recovery rates were feasible within the industry
- EU monitoring programs are expected to increase to ensure compliance; noncompliant manufacturers will be fined
- US and Asian vehicle producers may be at a disadvantage compared to Volvo, Saab and BMW for efficient dismantling designs. Their producer allocations may be higher and those costs would be transferred to customers

Box P: Nike Case

Nike company has created a strategic alliance with an eco-non-profit organization ''National Recycling Coalition'' in order to collect used tennis shoes.

Both Nike and NRC have different goals:

– Nike goals are to generate profits and to build an environmental friendly image.
– The NRC's main goal is the environmental protection.

Nike and NRC have created an alliance around a common goal: reprocessing used tennis shoes. The NRC takes care of the logistics of collecting the shoes, leaving Nike to focus on the end uses for the resulting Nike Grind material. Their short-term goal is to have at least one athletic shoe collection center in every state in the continental US and to recycle 125,000 pairs/year (Nike Environmental Responsibility, 2004).

Nike does pay the NRC for its service. The company simply provides each organization with communications tools to promote its collection effort, including customizable radio spots, media releases, posters, and print ads. Three $20,000 grants will be awarded to participating recycling organizations that successfully apply to the Nike reuse-a-shoe and NRC grant program in 2004. Figure 9.16 illustrates the Nike closed-loop supply chain.

Collection of used products. The used products collected are the shoes:

– Post-consumer, non-metal-containing athletic shoes of any brand.
– Nike shoes that are returned due to a material or workmanship flaw.

Nike gives the different drop-off locations on its website. The collection process offers different options as shown in Fig. 9.17. The process of

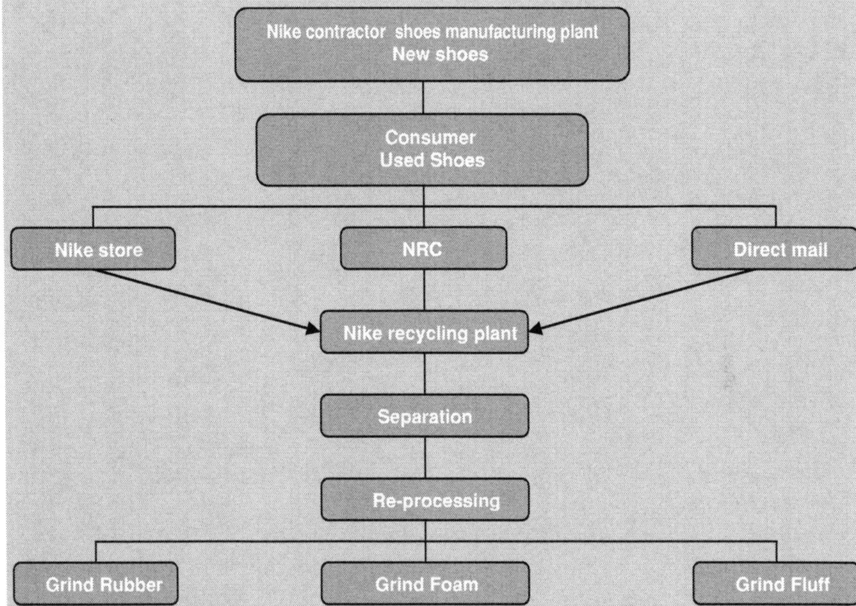

Fig. 9.16 Nike closed-loop supply chain organization (modified by Kumar and Malegeant 2006)

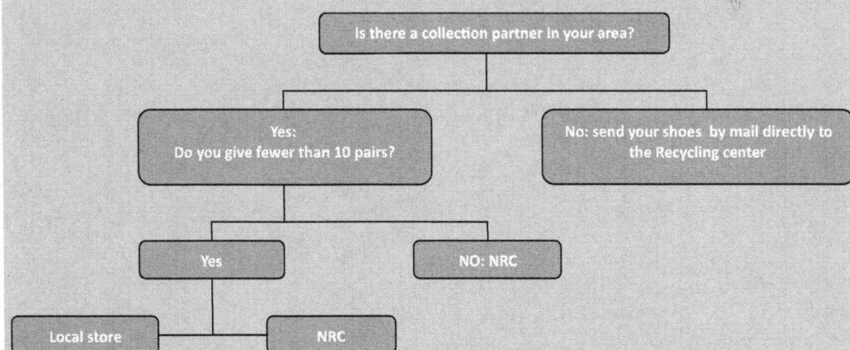

Fig. 9.17 Collection options

collection depends on the geographic location and number of items. The potential donor can send his used shoes by mail directly to the recycling plant, or give his shoes to some Nike local stores or to an eco-non-profit organization that is a member of the NRC.

Kumar and Malegeant (2006) used the Nash theory of game, in order to define the best option (see Table 9.12):

Table 9.12 Benefits and disadvantages of the three collection options offered by reuse-a-shoe program (*Source*: Kumar and Malegeant 2006)

Collection option	Nike		Donor	
	Benefits	Disadvantages	Benefits	Disadvantages
1. Mail	No transport or storage costs	None	None	Hassles: packaging
2. Local store	None	Transport and storage costs	Convenience No cost	None
3. Recycle partner	No product acquisition cost	Limited transport cost	Convenience No cost	None

- The first option, sending the shoes by mail, is a win–lose strategy for Nike. The company just focuses on the remanufacturing issues while the potential donor has to pay to send his used tennis shoes directly to the recycling plant. There is no disadvantage for the company; however, there are packaging hassles and expenses incurred by the donor.
- The second option is a lose–win strategy for Nike. This option is convenient for the donors but it increases the production acquisition costs for the company. That is why the company has limited this option to few local stores and to a maximum of 10 shoes/donor. This option is convenient for the donor while it is complicated to manage for Nike.
- The third option would be a win–win strategy for Nike. The company relies on its partner network to acquire the tennis shoes. Once a member of the NRC has collected 5,000 pairs of shoes, Nike will arrange for the shoes to be picked up by roadway and shipped free of charge to its reuse-a-shoe recycling facility in Wilsonville, Oregon. So, it is the best strategy for both.

Separation of material. NRC organizes separation of three main materials, and arranges to grind them up. Nike recycles and reuses 100% of the materials from post-consumer and defective athletic shoes that are currently able to be recycled.

Reprocessing of material. Through a chemical process in their Wilsonville recycling plant in Oregon, they obtain three distinct types of Nike Grind material (upper fabric, midsole foam, and outsole rubber). Each is used in a different way to make new sports surfaces such as soccer and football fields, basketball and tennis courts, tracks, and playground surfacing. The new products are:

Nike grind rubber, from outsoles and manufacturing by-products, helps make football, baseball, and soccer fields, as well as golf products, weight room flooring, and running tracks. (Nike licensee Field Turf uses Nike grind material from manufacturing by products exclusively.)

Nike grind foam, from midsoles, is used in synthetic basketball courts, tennis courts, and playground surfacing.

Nike grind fluff, from textile and leather uppers is used for padding under hardwood basketball floors.

In 2004, in the continental US, 125,000 pairs of shoes are expected to be recycled. From 1993 to 2004, more than 15 million pairs of shoes have been processed (Nike Inc. 2004). This remanufacturing process allows the multinational company both to create more profits and to improve its public relations: reuse-a-shoe plays an important role in Nike's long-term commitment to help increase the physical activity of young people to improve their lives. In addition, reuse-a-shoe is also a key component of Nike's long-term commitment to reduce its impact on the environment by helping to close the loop on the life cycle of literally millions of pairs of old, worn-out or otherwise unusable athletic shoes. Since the reuse-a-shoe program began in 1993, Nike has helped to donate more than 150 sport surfaces to communities around the world.

References

Ballou RH (1999) Business logistics management: planning, organizing, and controlling the supply chain, 4th edn. Prentice Hall, New Jersey

Bärthel F, Woxenius J (2004) Developing intermodal transport for small flows over short distances. Transp Plann Technol 27 5:403–424

Beamon BM (1998) Supply chain design and analysis: models and methods. Int J Prod Econ 55:281–294

Beamon BM (2005) Environmental and sustainability ethics in supply chain management. Sci Eng Ethics 11:221–234

Blackburn JD, Guide VDR Jr, Souza GC, Van Wassenhove L (2004) Reverse supply chain for commercial returns. Calif Manage Rev 46(2):6–22

Brauer KM, Krieger W(1982) Betriebswirtschaftliche Logistik, Berlin

Brodin MH, Flygansvær BM (2006) Taking control in reverse distribution systems: coordinators playing games of power and trust. In: The second ''Northern Lights of Logistics'' seminar, special track NOFOMA, Norwegian School of Management, pp 12–24, June 2006

BSI (2008). Specification for the assessment of the life cycle greenhouse gas emissions of goods and services: BSI British Standards

Burt DN, Soukup WR (1994) Purchasing's role in new product development. In: Clark K, Wheelwright SC (eds) The product development challenge. Harvard Business School Press, Boston, pp 333–345

Carter CR, Rogers DS (2008) A framework of sustainable supply chain management: moving towards new theory. Int J Phys Distrib Logist Manag 38(5):360–387

CIPS (1995) Supply chain—the environmental challenge. Chartered Institute of Purchasing and Supply, Stanford (UK)

Coughlan P, Coghlan D (2008) Action research for operations management. In: Karlsson C (ed) Researching operations management. Taylor & Francis, New York

EPA U.S. Environmental Protection Agency (1998), WasteWise Update: Extended Product Responsibility, EPA530-N-98-007, October 1998, www.epa.gov/wastewise

European commission (2006b) Keep Europe moving—sustainable mobility for our continent. Brussels

European conference of ministers of transport (2007) Cutting transport CO2 emissions—what progress? Paris, OECD

Eurostat and the EU Commission (2009) http://epp.eurostat.ec.europa.eu/portal/page/portal/eurostat/home/

Fleischmann M, van Nunen JAEE, Gräve B (2003) Integrating closed-loop supply chains and spare-parts management at IBM. Interfaces 33(6):44–56

Guide VDR Jr, Van Wassenhove LN (eds) (2003) Business aspects of closed-loop supply chains. Carnegie Mellon University Press, Pittsburgh

Jayaraman V, Luo Y (2007) Creating competitive advantages through new value creation: a reverse logistics perspective. Acad Manag Perspect, pp 56–73

Johnson LK (2005) Reversing course, Harvard Business School Publishing newsletters: Supply Chain Strategy. Harvard Business Review

Johnson MR, Wang MH (2002). Evaluation policies and automotive recovery options according to the European union directive on End of-life vehicles (ELV). Proceedings of Institution of Mechanical Engineers, 216 Part D. J Automobile Engineering

Kim, Ki H, Joung, Hyun T, Nam, Hoon, Seo, Yong C, Hong, John H, Yoo, Tae W, Lim, Bong S, Park, Jin H (2004) Management of status of end-of life vehicles and characteristics of automobile shredder residues in Korea. Waste Management 24(6), 533–540

Kreutzberger E, Macharis C, Vereecken L, Woxenius J (2003) Is intermodal freight transport more environmentally friendly than all-road freight transport? A review. 7th NECTAR Conference. Umeå, Sweden

Kumar S, Malegeant P (2006) Strategic alliance in a closed-loop supply chain, a case of manufacturer and eco-non-profit organization. Technovation 26(2006):1127–1135

Kumar S, Putnam V (2008) Cradle to cradle: Reverse logistics strategies and opportunities across three industry sectors. Int J Prod Econ 115(2008):305–315

Lemathe P, Balakrishnan N (2005) Environmental considerations on the optimal product mix. Eur J Oper Res 167(2):398–412

Lloyd M (1994)How green are my suppliers? Buying environmental risk. Purchasing Supply Manag. p 36, 9 October

Macharis C, Bontekoning YM (2004) Opportunities for OR in intermodal freight transport research: a review. Eur J Oper Res 153:400–416

Mamic I (2005) Managing global supply chain: the sports footwear, apparel and retail sectors. J Bus Ethics 59:81–100

MCKinnon A (2007) CO2 emissions from freight transport in the UK. Edinburgh, Logistics Research Centre, Heriot-Watt University

Middel R, Coghlan D, Coughlan P, Brennan L, McNichols T (2006) Action research in collaborative improvement. Int J Technol Manag 33(1):67–91

Minx J, Wiedmann T, Barrett J, Suh S (2008) Methods review to support the PAS process for the calculation of the greenhouse gas emissions embodied in goods and services. Report to the UK Department for Environment, Food and Rural Affairs. London, UK: Stockholm Environment Institute at the University of York and Department for Biobased Products at the University of Minnesota

Nereng G, Semini M, Romsdal A, Brekke A (2009) Can innovations in the supply chain lead to reduction of GHG emissions from food products? A framework. Paper presented at the Joint Actions on Climate Change

Neto JQF, Bloemhof-Ruwaard JM, van Nunen JAEE, van Heck HWGM (2006) Designing and evaluating sustainable logistics networks. Rotterdam School of Management (RSM), Erasmus University, The Netherlands. January 13, 2006

Nike Environment Responsibility (2004) October. Available at http://www.nike.com/nikebiz/nikebiz.jhtml?page=27

Nike Inc. (2004) Nike annual report. Available at www.nike.com

PAS 2050:2008 Specification for the assessment of the life cycle greenhouse gas emissions of goods and services, British Standards Institution

Rogers DS, Tibben-Lembke R (2001) An examination of reverse logistics practices. J Bus Logist 22:129–148

Romsdal A (2009) The appropriateness of action research to achieve increased supply chain sustainability—a case study. Proceedings of XIV Summer School 'Francesco Turco'

Shi Y, Gregory M (1998) International manufacturing networks—to develop global competitive capabilities. J Oper Manag 16(2, 3):195–214

Sihn W, Hillbrand C, Meizer F, Leitner R, Prochazka M (2010) Development of a simulation model for multimodal, Cross-company logistics networks. 8th International Heinz Nixdorf Symposium, IHNS 2010 Paderborn, Germany, April 21–22, 2010

Srivastava SK (2007) Green supply-chain management: a state of-the-art literature review. Int J Manag Rev 9(1):53–80

Steel Recycling Institute (2006a) Steel recycling holds strong despite inventory crunch. Available/http://www.recycle-steel.org/PDFs/2006RatesRelease.pdfS

Steel Recycling Institute (2006b) The inherent recycled content of today's steel. Available/http://www.recycle-steel.org/PDFs/Inherent/2006.pdfS

Stock J, Speh T, Shear H (2006) Managing product returns for competitive advantage. Sloan Manag Rev 48(1):57–62

Van den Bergh JCJM, van Veen-Groot DB (2001) Constructing aggregate environmental—economic indicators: A comparison of 12 OECD countries. Environ Econ Policy Stud 4:1–16

Ventour L (2008) The food we waste. Waste and resources action programme, London

Vollmann TE, Berry WL, Whybark DC, Jacobs RF (2005) Manufacturing planning and control for supply chain management. McGraw-Hill, Boston

Woxenius J (1998) Development of small-scale intermodal freight transportation in a systems context. Doktorsavhandlingar vid Chalmers Tekniska Hogskola

Woxenius J (2001) Intermodal freight transport—urban impact of new network operation principles and transhipments. Cities of tomorrow: Human living in urban areas. Transportation of people and goods. Göteborg

Chapter 10
Case Study: A Carbon Footprint Analysis in Textile Supply Chain

This study was proposed by the leader company itself compelled by the increasing attention of the market to environmental issues. The main aim of the project was to evaluate the application of the life cycle assessment (LCA) methodology in order to obtain a tool to support the company environmental policy. The first step was to define the object of the analysis: one of the basic items produced by the company was chosen as a case study. A simple wool sweater without buttons, laces, zippers ,or other accessories, with standard production processes for the yarn, weaving, and finishing treatments was taken into account. The result of the study was the evaluation of the carbon footprint of such product to assess its impact on the green house effect. The audiences to which the results were addressed were the textile leader company that promoted the research and his suppliers. Moreover, other than for external communication, the purposes of the management in applying the LCA methodology is related to the product improvement, to the support for strategic choices and benchmarking.

The Monte Carlo analysis has been used in order to obtain, for each calculated impact, not only the average value but also the distribution curve of the results characterized by uncertainty parameters.

Moreover a sensitivity analysis was carried out to evaluate the impact of management choices such as:

- a change in the transportation modality, from airplane to boat;
- a combination of road and rail transportation;
- a selection among suppliers that allows the firm to cut environmental impacts.

The following sections present the results of this research work: after an introduction section regarding carbon footprints in the supply chain, in Sect. 10.2 techniques utilized to define the input and the output of the system and to perform the uncertainty analysis will be described. In Sect. 10.3 the definition of the system

M. Bevilacqua et al., *Design for Environment as a Tool for the Development of a Sustainable Supply Chain*, DOI: 10.1007/978-1-4471-2461-0_10,

boundaries will be presented. Finally Sects. 10.6 and 10.7 present the results of the sensitivity analysis and conclusions.

10.1 Carbon Footprints in the Supply Chain

The European Commission has recently announced that Member States are to reduce their emissions of greenhouse gases by at least 20% before 2020 as compared to 1990 levels, the reduction possibly reaching 30% if other industrialized countries, such as USA, China,and India, commit themselves to a similar effort in connection with the climate conference in Copenhagen in 2009. This decision has an impact for all European Union Member States as well as all included stakeholder groups such as globally organized companies that have offshored a lot of their production capacities to low-cost countries and have set up long-linked intermodal transport chains in order to serve their markets (Halldorsson et al. 2009).

An initial step toward the reduction of greenhouse gas emissions has been the introduction of the carbon footprint as a measure, which is “the total amount of carbon dioxide (CO_2) and other greenhouse gases emitted over the entire lifecycle of a product or service” (Gereffi 2001). The carbon footprint is typically measured in tons of CO_2 and it can be used to understand the relative amount of damage, which a product or service causes to the environment.

Large UK-based retailing companies such as Tesco, Marks & Spencer, Boots, and Sainsbury have already started to label some of their products with carbon footprint-related information. Brand-owners, such as Walkers (snack food), Innocent Drinks (fruit smoothies), and Botanics (shampoos) have measured their products’ carbon impact and committed themselves to reduce the carbon footprint of their products. The British and Austrian governments are considering mandatory carbon footprint information labeling on all products. Another example is Timberland, manufacturer of the ikonic walking shoes, which has committed to become a carbon neutral enterprise by 2010 by using more renewable energy, incorporating more recycled and renewable materials, generating less waste, manufacturing with fewer chemicals, and planting more trees (www.interface-sustainability.com 2009). The Danish–Swedish dairy giant Arla Foods has decided to reduce their global CO_2 emissions from food production, transport, and packaging by 25% before 2020. When the British Standard for CO_2 emission is accepted, Arla Foods will start to affix a CO_2 label on their products in large UK retail stores (www.arlafoods.dk 2009).

10.2 Materials and Methods

The application of the LCA methodology was carried out by following the international standards and by using the Simaprò software and the Ecoinvent database (Pre Consultants 2006; Swiss Centre for Life Cycle Inventory) applied to the production phases of the chosen item (wool sweater). In accordance with ISO 14041 the stage of life cycle inventory assessment (LCIA) involves the collection of the data concerning the processes and the various calculation procedures. The relationship between the item produced and the environment are defined.

The focus of this project was the calculation of carbon footprint measured in CO_2 equivalent. For this purpose, the IPCC 2007 method was selected (Cambridge University Press 2007). This method developed by the International Panel of Climate Change, allows to quantify the greenhouse effect by measuring the equivalent CO_2, and lists the climate change factors of IPCC with a timeframe of 20, 100, and 500 years. In this work a timeframe of 100 years was considered. Normalization and weighting are not a part of this method.

The English guide lines (PAS 2050:2008)[1] were taken into account for the following aspects of the boundaries definition:

- the modeling of the transportation phase to the stores was considered and an average transportation model was used;
- the exclusion of the capital goods from the analysis;

10.2.1 Related Research Works

Many companies focused their attention on environmental impact of their products. Levi Strauss & Co (2009a, b) commissioned a life cycle assessment of two of their core products. By taking a product-lifecycle approach to their work, they were able to develop a set of strategies to address the greatest impacts of their business on the environment.

Wiendmann and Mix (2007) calculated textiles and clothing are responsible for around four per cent of the secondary carbon footprint of an individual in the developed world. The problem has been largely addressed at European level, the EU COST Action 628 (Nieminen et al. 2007) was established to produce first hand industrial environmental data of textiles in Europe, as well as to suggest tools for comparisons of present technologies and practices with cleaner applications, including the economic effects. LCA was used to set up criteria for an

---

[1] "Where products are distributed to different points of sale (i.e. different locations within a country), emissions associated with transport will vary from store to store due to different transport requirements. Where this occurs, organizations should calculate the average release of GHGs associated with transporting the product based in the average distribution of the product within each country, unless more specific data is available" (publicly available specification PAS 2050:2008, par. 6.4.6, note 4).

"the GHG emissions arising from the production of capital goods used in the life cycle of the product shall be excluded from the assessment of the GHG emissions of the life cycle of the product" (publicly available specification PAS 2050:2008, par. 6.4.3).

Environmental Product Declaration for textile products. Unique, first hand industrial data were collected from five European textile industries.

Mc Curry (2009) described a study about future trend in textile industries carried out from Freedonia Group, Ohio-based consultancy firm. The study said that the industries will pay attention to a host of environmental issues that aim to reduce their carbon and environmental footprints.

An ecological footprint study in the textile field has been proposed by Herva et al. (2008). They analyzed a textile tailoring plant with the overall purpose of developing a tool for evaluating the environmental impact evolution due to the performance of the plant, as well as for comparing the environmental behavior of different tailoring processes. Therefore, the selected data were those from the manufacturing work. Data were divided into three main categories: energy, resources, and waste. The principal contribution to the final EF (expressed in hectares of land) was the resources category, mainly due to the high value associated to the cloth. The consumed energy was the second contributor, while the waste category remained in third place. The final outcomes were divided by the production rates to obtain a comparable relative index, easy to be interpreted by the different stakeholders.

Regarding woollen products, Barber and Pellow (2006) presented a research that produced a detailed inventory of resource inputs for New Zealand merino wool and assessed its total energy use profile relatively also to other textiles. Results of this study show that New Zealand merino fiber production and early stage processing uses significantly less energy than synthetic fibers.

Brent (2004) proposed for a wool product a modified LCIA procedure, which is based on the protection of resource groups. A distance-to-target approach is used for the normalization of midpoint categories, which focuses on the ambient quality and quantity objectives for four groups: Air, Water, Land, and Mined Abiotic Resources. The case study establishes the importance of region-specificity, for LCIs and LCIAs.

In comparison to other works in textile sector this research carried out a carbon footprint analysis of woollen product involving all the supply chain stakeholders and evaluating the impact of new management choices, such as the transportation modality or the criteria for suppliers' selection.

10.2.1.1 Uncertainty Analysis

Uncertainty is a measure of the 'goodness' of results. Since the LCA is a model, there can always be errors that cause a certain level of uncertainty in the results. The uncertainty may depend on several factors: poor correspondence between the software models and the reality and between geographic and temporal aspect, non-representative measurement samples, non- complete information, poor reliability of data, etc. (ISO 14040 2006). Traditionally, the LCA (inventory and impact assessment) is a deterministic model used for estimating the potential impacts associated with a product. However, the LCA's primary weakness lies in its improper treatment of the uncertainty resulting from the sparse and imprecise nature of available information and the simplified model assumptions (Lo et al. 2005). This fact is demonstrated by Jimenez-Gonzalez and Overcash (2000) that, comparing the life cycle inventory (LCI) results for refinery products among several available databases, have shown that the variability of estimated emissions to the atmosphere, waterborne, and solid waste are approximately 50–150, 1000, and 30%, respectively.

Variability and parameter uncertainty of unit process' inputs and outputs, e.g. measurement uncertainties, process specific variations, temporal variations, etc., are expressed in quantitative terms on the level of individual inputs and outputs of unit processes. This type of uncertainty has been treated consistently and in a quantified way within the ecoinvent project (Frischknecht and Rebitzer 2005). In this work lognormal distribution has been assumed for all unit processes of ecoinvent data. In fact several reports in the field of risk assessment and impact pathway analysis have shown that the lognormal distribution seems to be a more realistic approximation for the variability in fate and effect factors than the normal distribution.

The Monte Carlo analysis has been used in order to obtain, for each calculated impact, not only the average value but also the distribution curve of the results characterized by uncertainty parameters.

The statistical principle of Monte Carlo method consists in repeating calculation many times. Each time a random value is chosen for each flow, for example an emission or raw material input. In this work the uncertainty analysis was based on 1000 calculations. According to (Langevin et al. 2010) this number of iterations was a compromise solution between simulation time and precision of results. Uncertainty information can be derived with basic statistical methods from the distribution of the calculation results.

The values chosen in the Monte Carlo analysis are within a specified distribution. In this study the uncertainty estimation for each emission was calculated by using the following equation (1) developed by Weidema and Wesnaes (1996). The Pedigree Matrix (see Table 10.5 in Appendix) was used to calculate the Ui terms (i = 1, …, 6) shown in the standard deviation equation.

$$SD_{95} = \sigma_g^2 = \exp^{\sqrt{[\ln(U_1)]^2+[\ln(U_2)]^2+[\ln(U_3)]^2+[\ln(U_4)]^2+[\ln(U_5)]^2+[\ln(U_6)]^2+[\ln(U_b)]^2}}$$

Eq. 1: standard deviation equation
with:
U1: uncertainty factor of reliability
U2: uncertainty factor of completeness
U3: uncertainty factor of temporal correlation
U4: uncertainty factor of geographic correlation
U5: uncertainty factor of other technological correlation
U6: uncertainty factor of sample size
Ub: basic uncertainty factor (see Table 10.6 in Appendix)

In order to reduce uncertainty as much as possible the maximum level of detail in the input data has been reached by obtaining the main data directly from the different parties involved in the process.

10.3 Case Study

The LCA methodology has been used to define the carbon footprint of a wool sweater made by a leader company in the textile sector. The entire production chain has been examined and all the single contributions to the environmental impact have been evaluated. Hypothesis for improving the resources use and management have been proposed, again using the LCA approach.

The idea with which many decisions were made was to maintain a quite generic point of view since the main goal was to assess if the LCA methodology could be applied continuously and systematically to the entire garments collection.

10.3.1 System Boundaries

It was decided to define the boundaries of the system starting from sheep breeding at the farms in South Africa, from which the greasy wool is obtained, then considering the other production phases, the distribution to the final stores all around the world and, finally the user phase (washing and final disposal) as shown in Fig. 10.1.

Those boundaries comprehend all the phases related to the sweater production, distribution, and use: wool scouring, dyeing, spinning, knitting, transportation to the distribution centers and then to the selling stores, washing, and final disposal.

All the packaging, especially the ones used for sending the sweater from the production site to the store, were considered.

Regarding the environment boundaries, it was decided to consider the agricultural systems as part of the environment and this means, for example, that pesticides are viewed as emissions.

10.3.2 Functional Unit

In order to maintain a general point of view and to consider an average sweater, all the different characteristics were included in the functional unit.

The sweater chose as case study had the following features:

- 100% Merinos wool;
- 4 color;
- 2009 winter collection.

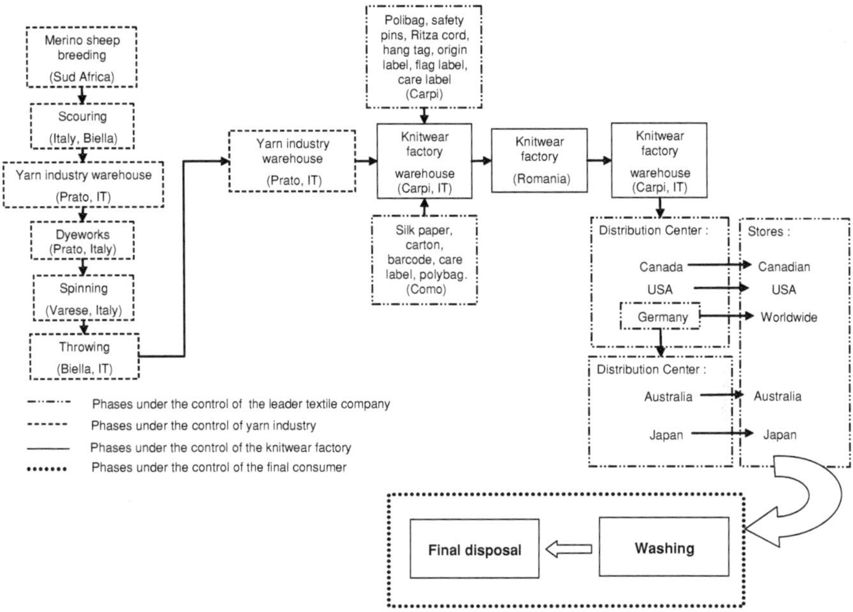

Fig. 10.1 System boundaries

Table 10.1 Percentage distribution for each size

	S	M	L	XL	XXL	XXXL	Tot
Distribution pieces-size (%)	4,71	20,59	25,29	24,71	17,65	7,06	100
Pieces for size (n)	141,18	617,65	758,82	741,18	529,41	211,76	3000
Weight tot for sizes (g)	32654.12	150798.53	195017.65	201911.29	151025.29	63131.29	794538.18

The medium-weight method was used to calculate the average sweater. The parameters included in the average sweater, representative of the functional unit were:

- Distribution size: each size assumes a different weight as shown in Table 10.1.
- Colour: each colour has a slightly different dyeing process. Based on the color distribution for the examined sweater in the entire winter 2009 collection, it was possible to calculate the contribution of each color in the medium sweater in terms of weight (Table 10.2).

Net weight: 264.85 g (without accessories).
Color and accessories:

Table 10.2 Average sweater characteristics

Accessories Number	Description	Weight (g)				
1	Main label (regular fit)	0,31 g				
1	Flag Label	0,11g				
1	Origin label	0,11g				
1	Hang tag	1,83g				
1	Safety pins (hantag)	0,2g				
1	Cord ritza (hang tag)	0,3g				
1	Care label	0,3g				
2	Barcode	0,1g				
1	Silk Paper	4,25g				
1	Polybag	15,65g				
Color	*001*	*032*	*038*	*521*	*522*	*Tot kg*
Weight distribution %	111,81	3,22	82,25	32,00	35,58	264,85

Table 10.3 Data sources

Supplying company	Production processes	Primary data	Secondary data		
			Software model	BREF	ARPA
A	Sheep breding	X	X		
B	Scouring			X	X
C	Dyeing	X		X	X
D	Spinning	X	X		X
E	Trowing e vaporizing	X			X
F	knitwear factory	X	X		

10.3.3 LCIA

The data collected have been divided into two different groups: primary and secondary data as is shown in Table 10.3.

The primary data were collected directly from the companies involved in the different phases of the production process by a questionnaire. Secondary data were extracted from:

- the software models (Yarn production bast fibres/IN U 2007)
- "BREF"(Best available technology reference document) of the European Community (2003). This document provides information about the best techniques for the textile sector.
- "Analysis of the production cycle in the textile and wool sector" of the Arpa of Piemonte Region (2005).

For the final phases of the life cycle: use and final disposal, data were obtained from the literature, market analyses, company data, and ENEA database.

10.4 The Production Process

According to Byoungho (2004) the textile industry represents one of the most various and complicated productive processes of the entire manufacture system. Generally speaking it is possible to distinguish two main different processes: mechanical (spinning and weaving) and chemical (washing, dyeing, and finishing).

The complete processing cycle of the analyzed item involves several steps. First comes shearing, followed by sorting and grading, making yarn, dyeing, finishing, making fabrics, making up the sweater, and distributing.

In synthesis 7 companies are involved in the production of the analyzed item with 17 transportation phases for a total of more than 10,000 km, 5 Kwh of electrical energy, 18 MJ of thermal energy, 60 g chemicals and detergents, 200 g of packaging, and 350 g of waste.

Regarding the use of the sweater, data were obtained from company data and from ENEA guidelines (2003). A medium life of 5 years with 15 washes per year was considered, 30°C of washing temperature, 10 l of water for each cycle, and 130 ml of chemicals[2] (soap and conditioner).

For the last life phase, the final disposal, from the literature data it was established that 49% was disposed of, 49% was burned, and the last 2% was reused.

The transportation scheme for the distribution centres and the relative percentage are shown in the following Fig. 10.2.

10.4.1 Impact Calculated by IPCC 2007

By using the IPCC 2007 it was possible to assess the contribution of each phase to the carbon footprint measured as CO_2 equivalent. The following diagram (Fig. 10.3) shows that the total amount of CO_2 produced is 1,947 kg for the single item for the phases that go from the breeding to the final disposal.

It is also evident that four phases give the main contribution to the CO_2 production:

(1) The transportation from the DC to the stores (0.470 kg CO_2).
(2) The sweater realization (0.384 kg CO_2), since the material reaches the knitwear factory in Romania by truck (Euro 3), and it goes back to Carpi in the same way.
(3) The breeding phase (0.376 kg CO_2).
(4) The washing phase (0.280 kg CO_2).

[2] Data obtained from ENEA.

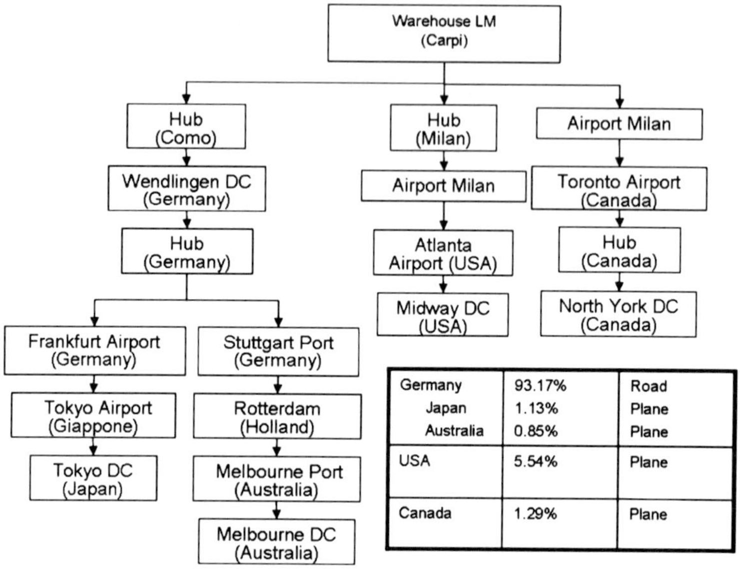

Fig. 10.2 Scheme of the transportation from the warehouse to the DC

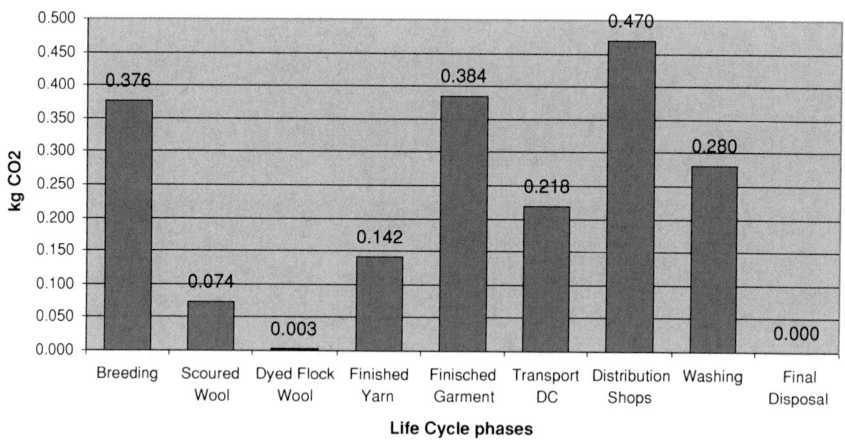

Fig. 10.3 Percentage of CO_2 in the different phases

10.5 Uncertainty Assessment IPCC 2007

The Monte Carlo analysis was used to determine the uncertainty in the calculation of the CO_2 equivalent by setting the boundaries from the breeding to final disposal. It was calculated an average value of 1.947 kg with a standard deviation Sd of 0.179. The range goes from 1,614 kg to 2,331 kg with a 95% confidence interval. (Fig. 10.4)

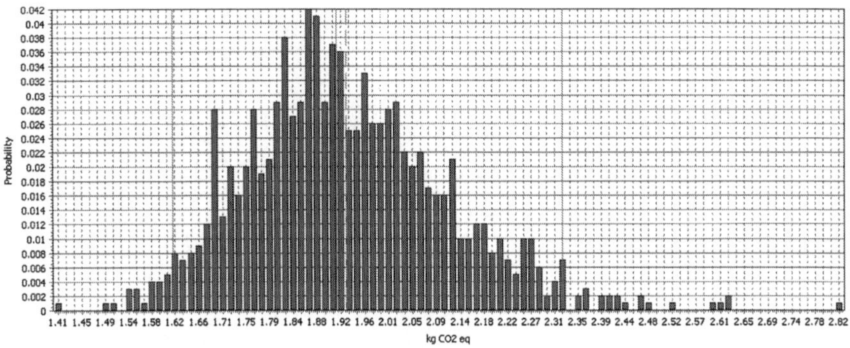

Fig. 10.4 Uncertainty analysis Kg CO_2 eq100 (IPCC 2007)

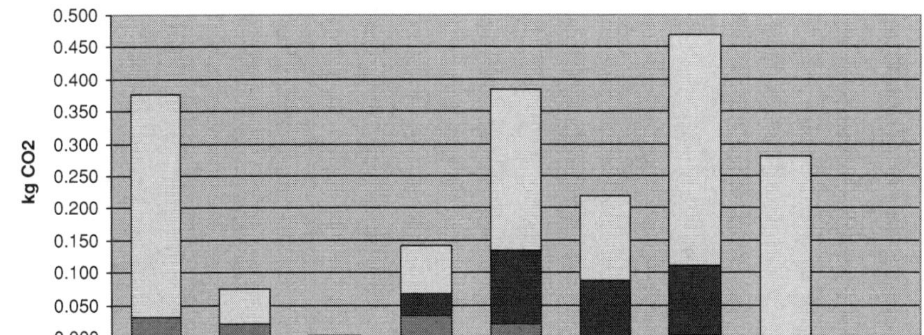

	Breeding	Scoured Wool	Dyed Flock Wool	Finished Yarn	Finisched Garment	Transport DC	Distribution Shops	Washing	Final Disposal
transp good	0.344	0.053	0.003	0.074	0.250	0.130	0.360	0.280	0.000
transp pack	0.000	0.000	0.000	0.035	0.113	0.087	0.108	0.000	0.000
other	0.032	0.021	0.000	0.033	0.021	0.001	0.002	0.000	0.000

Fig. 10.5 Contribute of transportation for each phase

10.5.1 Life Cycle Interpretation

In the research reported here, the LCI results were analyzed and processed by means of a contribution analysis and an uncertainty analysis based on Monte Carlo simulations. The accuracy obtained using such method is acceptable since the range of the confidence interval is about 20% of the steady-state mean.

The research demonstrated that the CO_2 production related to the chosen sweater depends mostly on the complexity of the supply chain and on the distribution system.

The transportation is the main CO_2 generator in the sweater production process; in fact, the CO_2 equivalent produced in each phases depends mainly on the goods transportation modality and the travel length. The following diagram (Fig. 10.5) shows the contribution of the transports (goods and packaging) in all the garment life cycle.

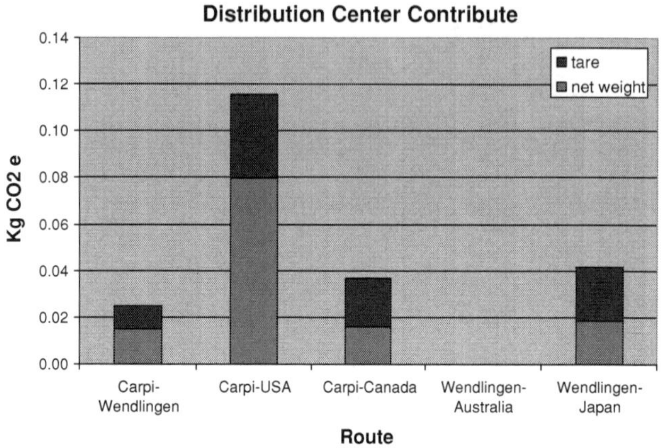

Fig. 10.6 CO_2 equivalent for the different destinations

Figure 10.5 also shows that the packaging gives an important contribution of nearly 40% to the CO_2 emission, especially in the final phases of the life cycle.

For this reason a model for designing a new packaging system has been created. New boxes of different sizes (small/large sized) and less weight have been tested in order to insert more sweaters into a single box. However, those attempts were often refused from sales managers mostly not only for esthetic reasons, but also because the solution proposed did not optimize the truck saturation. In fact, the results demonstrate that the packaging was already been conceived for minimizing the volumes transported and that a new design provided irrelevant improvements (less than 1%). Moreover, the production and disposal of the packaging do not contribute to the CO_2 production.

The model of the transportation to the distribution centers represents a particularly interesting field for the leader company for two main reasons: first it is under his direct control and second it is a phase that involves all the goods produced by the company. This means that the model represents big volumes of items and, as a consequence, an improvement in this phase can affect widely the entire production chain.

The Carbon Footprint for each destination was calculated considering not only the impact related to the transportation and his modality but also the volume of the goods. The results attributed the main impact, in terms of CO_2 equivalent produced, to the Carpi-USA destination as shown in Fig. 10.6. The following graph shows clearly that a variation for the USA transportation modality can have the greatest impact in terms of CO_2 saving.

Even if Australia and Japan have a similar volume, the goods travel on freight ship for Australia, and Airplane for Japan and this justifies the greater impact of the last route.

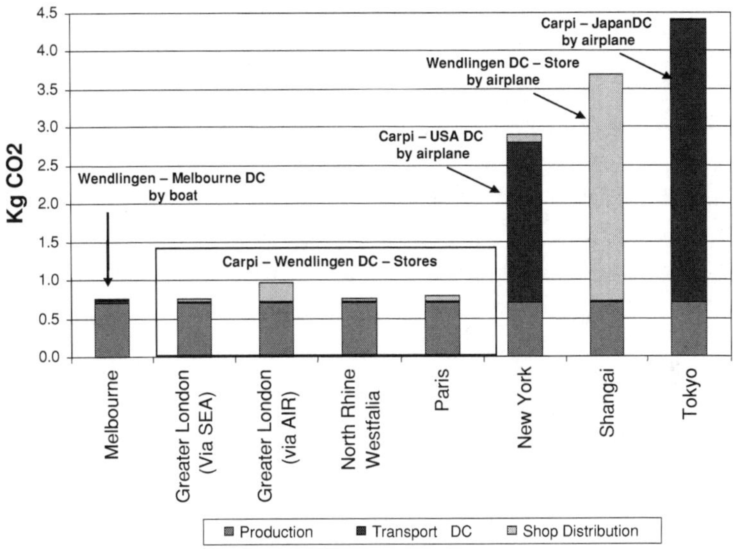

Fig. 10.7 Carbon footprint related to the selling point location

10.6 Sensitivity Analysis

A sensitivity analysis was carried out in order to evaluate the impact of management choices such as:

- a change in the transportation modality, from airplane to boat;
- a combination of road and rail transportation;
- a suppliers selection that allows the firm to cut down the environmental impacts;
- a change in consumer behavior.

10.6.1 Transport

It has already been demonstrated that the main impact in terms of CO_2 production is given by transportation. In this paragraph the sensitivity of this parameter to the distance, volumes of goods, and transportation means will be shown.

The calculated impact of 1.947 kg CO_2 is a global average value, the real value for the single sweater depends strongly on the selling point localization. The following diagram (Fig. 10.7) shows how the CO_2 equivalent produced depends on the transportation to the selling points but it does not give indications about the priorities to address in order to reduce it, since, for this purpose, an analysis of volumes is required.

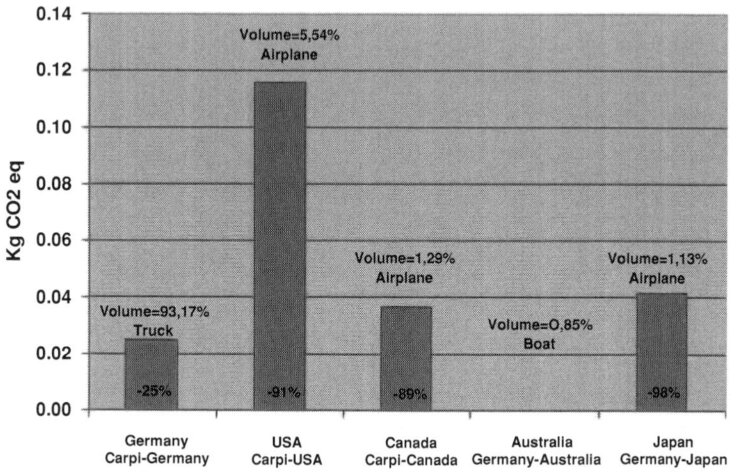

Fig. 10.8 Contribution of transportation modality related to the volumes of goods

The carbon footprint produced during the transportation phase to the stores is about 34%; by changing the transportation modality (from road to rail and from plane to boat), it is possible to cut down the CO_2 production of 20–30%.

The transportation of the goods to the Distribution Centers contributes 0.22 kg to the total Carbon Footprint. The contribution of each DC weighted over the volumes is shown in the following graph (Fig. 10.8).

In order to reduce the CO_2 equivalent, two different approaches are required. First it is necessary to look upon the destinations that give the higher contribution considering distances, volumes, and modality. Second the impact of a change in the transportation modality has to be modeled: from airplane to boat and from road to rail.

By changing those parameters, a total reduction of 0.18 kg that corresponds to 84% in this phase has been calculated.

10.6.2 The Combination of Road Rail Transportation

In the research project the transportation distance between Carpi and Wendlingen was analyzed with an "accompanied combined transportation" model. From Carpi the trucks travel by road to the hub in Como, from which they reach Chiasso where they can be loaded directly on the trains and arrive in Basilea. From there the trucks go back to the road until they arrive to the final destination in Wendlingen.

The difference, in terms of grams of CO_2 equivalent produced for each sweater, considered for the different transportation modalities is shown in the following diagram (Fig. 10.9).

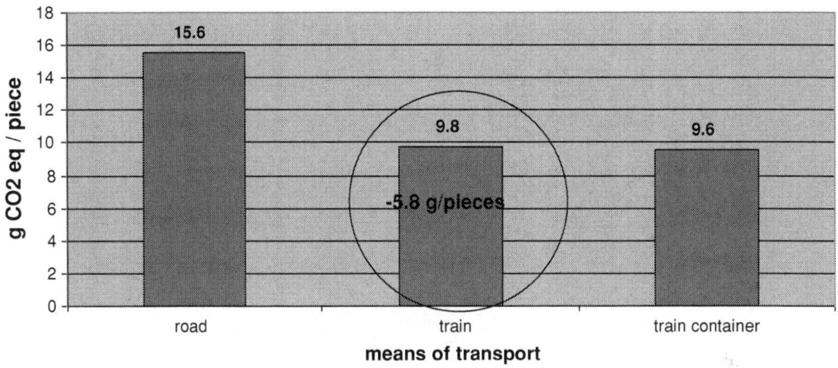

Fig. 10.9 CO_2 production along the road Como-Wellingen

The adoption of the combined transportation methods would allow the company to reduce of about 37% the carbon footprint along this road. The total reduction (considering the entire process from sheep breeding to the stores) would be only of 0.4%. This is due to the fact that the Como-Wendlingen road provides a small contribution to the global impact. Even if the benefit on the single sweater seems to be low, nevertheless, the improvement obtained with this operation on the total flow of goods transported by the leader company along this road is significant. Considering that the average gross weight of the sweater is 476 g and the annual flow is 500.000 kg the benefit, in terms of CO_2 saving, is of 5.3 tons per year.

10.6.3 Suppliers Selection

The selection of suppliers that guarantee good standards in terms of environmental impact with reference to the European Commission Standard (BREF) is a key factor in terms of CO_2 equivalent reduction. The sweater production processes has been accounted an amount of 0,066 kg CO_2 per piece. Referring to the BREF this is a fair low contribution and it is due to the company environmental awareness in the suppliers selection.

From the study it has been assessed that a high percentage of CO_2 derives from the transportation between suppliers. By hypothesizing that all the suppliers (the dyeing mills, the spinning, and the knitwear factory) are located in a nearest area, it is possible to save more than 0.4 kg of CO_2 for garment (28%). In the following Table 10.4 it is shown how the reduction of distances between suppliers could decrease the CO_2 produced.

Moreover the use of Euro 3 trucks to travel back and forth from the warehouse to the knitwear factory in Romania increases the effect. If using Euro 4 trucks the benefit would be of 2.15% for pieces (about 3 ton for the annual production).

Table 10.4 Benefit derived from distance reduction

Distance reduction	10%	20%	50%
Kg CO_2	1.33	1.27	1.07
Benefit	4.6%	9.2%	23.0%

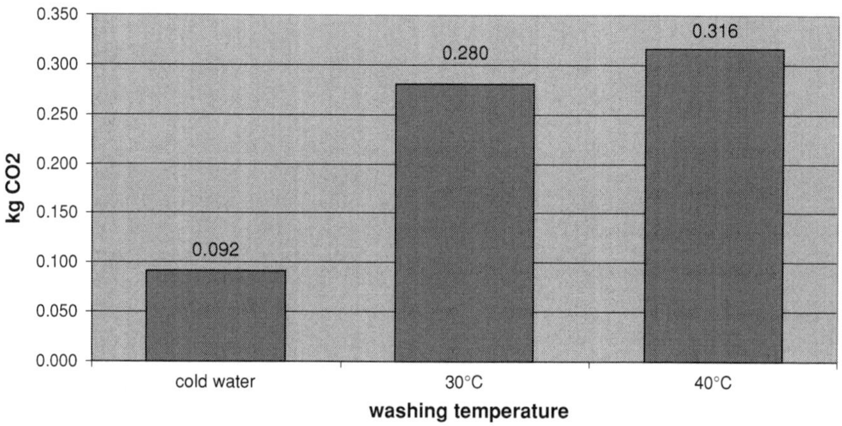

Fig. 10.10 CO_2 production changing the washing temperature

10.6.4 A Change in Consumer Behaviour

The washing phase provides a great Climate Change Impact (0.28 kg CO_2). A change in consumer behavior could decrease considerably this impact. As an example, it should be possible to cut down the number of washing per year to less than the 15 times established in the ENEA study and used in this work to assess the impact of the final life phase. Moreover the consumer could modify the washing temperature (supposed of 30°C in this research) changing considerably the Climate Change Impact (Fig. 10.10).

10.7 Discussion and Conclusions

In the last few years, life cycle thinking has been the focal point in the environmental policy development of the European community standardized by the Integrated Product Policy (IPP). Many other standards are utilized in other countries.

In this scenario, LCA provides the scientific references for all the activities related to the Corporate Sustainability Report that is a tool which more than 500 of the major companies all around the world use to communicate their environmental

Table 10.5 Pedigree matrix

Score:	1	2	3	4	5
U1 Reliability	Verified data based on measurements	Verified data partly based on assumptions OR non-verified data based on measurements	Non-verified data partly based on qualified estimates	Qualified estimate (e.g. by industrial expert); data derived from theoretical information (stoichiometry, enthalpy, etc.)	Non-qualified estimate
	1.00	**1.05**	**1.10**	**1.20**	**1.50**
U2 Completeness	Representative data from all sites relevant for the market considered over an adequate period to even out normal fluctuations	Representative data from >50% of the sites relevant for the market considered over an adequate period to even out normal fluctuations	Representative data from only some sites ($\ll$50%) relevant for the market considered OR >50% of sites but from shorter periods	Representative data from only one site relevant for the market considered OR some sites but from shorter periods	Representativeness unknown or data from a small number of sites AND from shorter periods
	1.00	**1.02**	**1.05**	**1.10**	**1.20**
U3 Temporal correlation	Less than 3 years of difference to our reference year (2000)	Less than 6 years of difference to our reference year (2000)	Less than 10 years of difference to our reference year (2000)	Less than 15 years of difference to our reference year (2000)	Age of data unknown or more than 15 years of difference to our reference year (2000)
	1.00	**1.03**	**1.10**	**1.20**	**1.50**
U4 Geographical correlation	Data from area under study	Average data from larger area in which the area under study is included	Data from smaller area than area under study, or from similar area		Data from unknown OR distinctly different area (north America instead of middle east, OECD-Europe instead of Russia)
	1.00	**1.01**	**1.02**		**1.10**

(continued)

Table 10.5 (continued)

Score:	1	2	3	4	5
U5 Further technological correlation	Data from enterprises, processes and materials under study (i.e. identical technology)		Data on related processes or materials but same technology, OR Data from processes and materials under study but from different technology	Data on related processes or materials but different technology, OR data on laboratory scale processes and same technology	Data on related processes or materials but on laboratory scale of different technology
	1.00		**1.20**	**1.50**	**2.00**
U6 Sample size	>100, continous measurement, balance of purchased products	>20	> 10, aggregated figure in anv. report	>=3	unknown
	1.00	**1.02**	**1.05**	**1.10**	**1.20**

Table 10.6 Basic uncertainty factor

Input/output group	Ub	Input/output group	Ub
Demand of:		*Emission to air of*:	
Thermal energy	1.05	CO_2	1.05
Electricity	1.05	SO_2	1.05
Semi-finished products	1.05	Combustion: NOX, NMVOC total, methane, N2O and NH3	1.50
Working materials	1.05	Combustion: CO	5.00
Transport services	2.00	Combustion: individual hydrocarbons, TSM	1.50
Waste treatment services	1.05	Combustion: PM10	2.00
Infrastructure	3.00	Combustion: PM2.5	3.00
Resources:		combustion: polycyclic aromatic hydrocarbons (PAH)	3.00
Primary energy carriers	1.05	Combustion: heavy metals	5.00
Metals, salts	1.05	Process emissions: individual VOCs	2.00
Land use, occupation	1.50	Process emissions: CO2	1.05
Land use, transformation	2.00	Process emissions: TSM	1.50
Waste heat:		Process emissions: PM10	2.00
Emission to air, water, and soil	1.05	Process emissions: PM2.5	3.00
Emission to water of:		From agriculture: CH4, NH3	1.20
BOD, COD, DOC, TOC	1.50	From agriculture: N2O, NOX	1.40
Inorganic compounds (NH4, PO4, NO3, Cl, Na etc.)	1.50	Radionucleides (e.g., Radon-222)	3.00
Individual hydrocarbons, PAH	3.00	Process emissions: other inorganic emissions	1.50
Heavy metals	5.00	*Emission to soil of*:	
From agriculture: NO3, PO4	1.50	Oil, hydrocarbon total	1.50
From agriculture: heavy metals	1.80	Pesticides	1.20
From agriculture: pesticides	1.50	Heavy metals	1.50
Radionucleides	3.00	Radionucleides	3.00

policies to the market . The use of LCA methodology allows the companies to evaluate and communicate the environmental impact of their processes and products.

With the introduction of more restrictive environmental standards, such as PAS 2050, the carbon footprint analysis could be one of the criteria to evaluate the suppliers from an environmental point of view and to improve the entire supply chain and to support the realization of green products by the eco-design.

The aim of this paper was to assess the carbon footprint associated to a particular product of a leader textile company. The selected item has been utilized as a case study: average characteristics were chosen in order to define the input for the LCA model. This decision was made with the objective of performing a pilot project to assess the feasibility of the LCA methodology to be applied to the entire company production.

A strong support of a leader company was essential to overcome reluctance of some stakeholders in providing data. According to Seuring and Goldbach (2004) inhumane working conditions or contaminations of the (local) environment could be frequently a problem in order to obtain information from suppliers.

With the use of LCA it is possible to face the environmental analysis at different levels. In the proposed case study the main aim was the carbon footprint evaluation but along the study many other aspects were considered.

The LCA method could be a powerful tool to address the ecoefficiency promotion and to provide several benefits to the company. Particularly the work developed in this case study allowed to make the customers awake that the product is environmentally sound; in fact the results obtained were inserted in the marketing campaign. The study also allowed the company to improve production performances in terms of efficient managing and use of resources.

Moreover the collaboration between several stakeholders made this work to be a useful way to increase the environmental consciousness of the employees and of the supply chain operator involved along the production process.

10.8 Appendix

The following tables represent the Pedigree Matrix (Table 10.5) and the basic uncertainty factor Ub values (Table 10.6) as defined in the SimaPrò sotware (2006)

References

Barber A, Pellow G (2006) Life cycle assessment: New Zealand merino industry, merino wool total energy use, carbon dioxide emissions. The Agri Business Group, Auckland

BREF (2003) Reference document on best available techniques for the textile industry, integrated prevention and pollution control (IPPC), July 2003. European Commission, Joint Research Centre, Institute for prospective technological studies

Brent AC (2004) A life cycle impact assessment procedure with resource groups as areas of protection. Int J Life Cycle Assess 9(3):172–179

Byoungho J (2004) Achieving an optimal global versus domestic sourcing balance under demand uncertainty". Int J Oper Prod Manag 24(12):1292–1305

ENEA (2003) Risparmio energetico con la lavatrice, opuscolo 11, Engardio P (2007, January 29th). Beyond the green corporation. BusinessWeek, 50−64.

Frischknecht R, Rebitzer G (2005) The ecoinvent database system: a comprehensive web-based LCA database. J Clean Prod 13:1337–1343

Gereffi G (2001) Beyond the producer-driven/buyer-driven dichotomy. The evolution of global value chains in the Internet era. IDS Bull 32(3):30–40

Halldorsson A, Kotzab H, Skjøtt-Larsen T (2009) Supply chain management on the crossroad to sustainability: A blessing or a curse? Logist Res 1:83–94

Herva M, Franco A, Ferreiro S, Álvarez A, Roca E (2008) An approach for the application of the Ecological Footprint as environmental indicator in the textile sector. J Hazard Mater 156(1–3):478–487

IPCC (2007) Summary for policymakers. In: Solomon S, Qin D, Manning M, Chen Z, Marquis M, Averyt KB, Tignor M, Miller HL (eds), Climate change 2007: The physical science basis. Contribution of working group I to the fourth assessment report of the Intergovernmental Panel on Climate Change. Cambridge: Cambridge University Press

ISO 14040:2006 Environmental management-life cycle assessment-principles and framework, International Organization for Standardization, TC207.

Jimenez-Gonzalez C, Overcash M (2000) Life cycle assessment of refinery products: review and comparison of commercially available databases. Environ Sci Technol 34(22):4789–4796

Langevin B, Basset-Mens C, Lardon L (2010) Inclusion of the variability of diffuse pollutions in LCA for agriculture: the case of slurry application techniques. J Clean Prod 18(2010):747–755

Levi Strauss & Co (2009a) A product lifecycle approach to sustainability, published online at http://www.levistrauss.com

Levi Strauss & Co (2009b) Carbon Disclosure Project CDP 2009 (CDP7), published online at http://www.levistrauss.com

Lo SC, Ma H, Lo SL (2005) Quantifying and reducing uncertainty in life cycle assessment using the Bayesian Monte Carlo method. Sci Total Environ 340:23–33

Mc Curry J (2009) Environment to impact on demand. Int Dye 194(2):9

Nieminen E, Linkeb M, Toblerc M, Vander Beked B (2007) EU COST Action 628: life cycle assessment (LCA) of textile products, eco-efficiency and definition of best available technology (BAT) of textile processing. J Clean Prod 15(13–14):1259–1270

PAS 2050:2008, Specification for the assessment of the life cycle greenhouse gas emissions of goods and services, British Standards Institution

Pre Consultants (2006). SimaPro 6 LCA software: the powerful life cycle solution, http://www.pre.nl/; Swiss Centre for Life Cycle Inventory. Ecoinvent. <http://www.ecoinvent.org> (Accessed 15 March 2009).

Seuring S, Goldbach M (2004) Managing sustainability performance in the textile chain. In: Wagner M, Schaltegger S, Wehrmeyer W (eds) Sustainable Performance and Business Competitiveness. Greenleaf, Sheffield

Weidema BP, Wesnaes MS (1996) Data quality management for life cycle inventories -an example of using data quality indicators. J Clean Prod 4(3–4):167–174

Wiendmann T, Minx J (2007) A definition of carbon footprint, IsaUK Research Report

Yarn production bast fibres/IN U, 2007. http://discover.amee.com/categories/Ecoinvent_textiles_production_yarn_production_bast_fibres_LCI_IN_kg

Chapter 11
Optimizing Sustainability in Products and Services

11.1 Sustainability in Products and Services

In order to effectively integrate sustainability in product and service development, Maxwell and van der Vorst (2003) proposed to optimize sustainability with traditional product criteria. An illustration of the proposed criteria to be optimized in developing sustainable products and services is presented in Fig. 11.1. In addition to the traditional product criteria, e.g. economic, quality, market, customer requirements, technical feasibility, and compliance issues illustrated in Fig. 11.1, the following sustainability criteria have been incorporated:

- environmental impacts;
- social impacts.

Moreover Maxwell and van der Vorst (2003) proposed a procedure for developing sustainable products and services. Figure 11.2 illustrates the main Sustainable Product and Service Developments (SPSD) process steps. Starting at the concept stage, one of the initial steps of SPSD is to consider how the functional requirement can be met through a product, a service, or some combination of a Product Service Systems (PSS) (van Weenen 2000; Reiskin et al. 2000) and optimizing the sustainability impacts of these options with traditional criteria. The use of SPSD may result in a product not being produced at all. This is in circumstances where it is more sustainable and feasible to meet the required functionality by the provision of a service. In practice, complete replacement of a product by a service is difficult to achieve. Some combination of PSS is a more likely possibility (van Hemel 1998). Once it has been determined whether a product, service, or PSS is to be developed, the next stage is to identify the life cycle stages and associated supply chain. A key element of SPSD is that it focuses on the supply chain for the product and/or service rather than solely at an individual company level. The entire supply chain and Supply Chain Dynamics (SCD)

M. Bevilacqua et al., *Design for Environment as a Tool for the Development of a Sustainable Supply Chain*, DOI: 10.1007/978-1-4471-2461-0_11,

Fig. 11.1 Criteria for optimizing sustainability in products and services

are assessed to determine the most effective target organization(s) in the chain for SPSD and how the SCM can be effectively utilized. Once this is determined, SPSD implementation can commence at the company level. Typically, the SPSD implementation occurs at Original Equipment Manufacturer (OEM) level and aspects of it are spread through the supply chain, e.g. through changes in the product and/or service specification. Full SPSD implementation is typically not required by all supply chain companies, just the ones with control over the main life cycle phases with key sustainability issues. However, the input of all relevant supply chain companies into the process (even if solely on an information supply basis) is crucial which is the reason for using SCM to ensure this. The next step is to assess the environmental and then social impacts for each product or PSS life cycle stage from raw materials to end of life. The opportunities for elimination or minimization of these are optimized with the remaining traditional product and service criteria. The specific environmental and social issues to be assessed vary depending on the product and/or service. To ensure a comprehensive approach Maxwell and van der Vorst (2003) proposed a checklist of typical environmental and social impacts to be considered per lifecyclestage. Table 11.1 includes an example of this checklist of the basic functionality, environmental, social, and economic issues which can be considered per life cycle stage in the SPSD process. It should be noted that a life cycle 'approach' is used to assess the sustainability impacts for each life cycle phase. This is not Life Cycle Assessment (LCA) as defined in ISO14040 but a simplified more qualitative approach which incorporates all

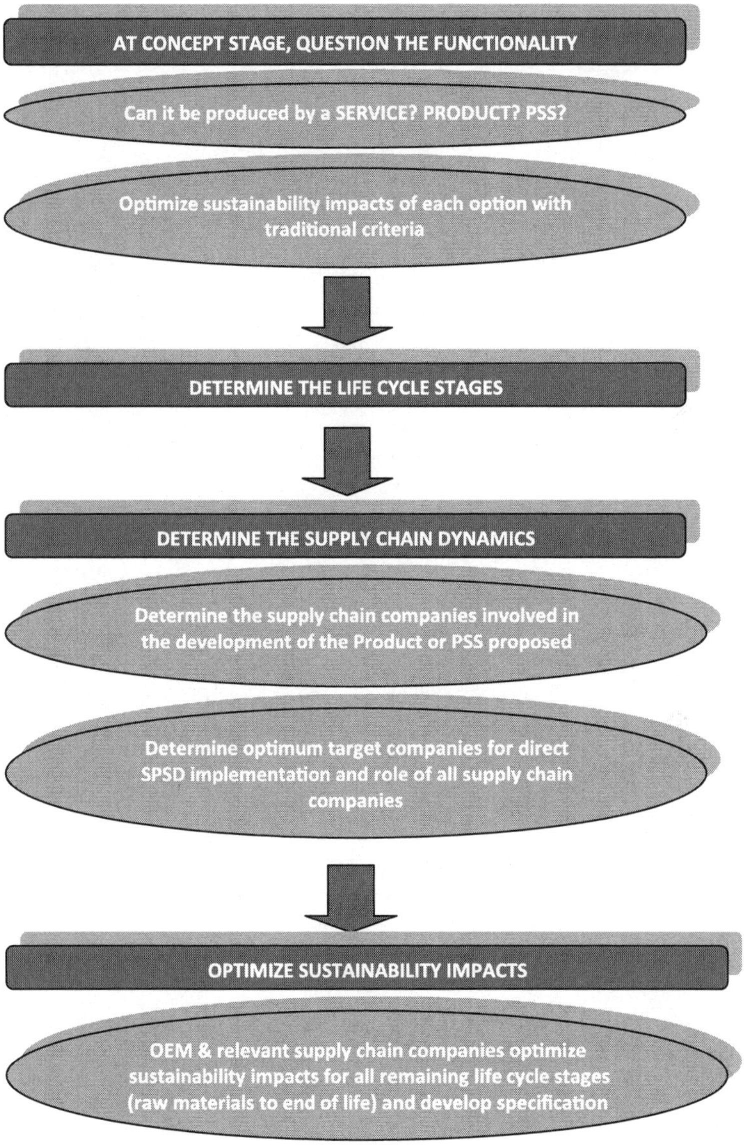

Fig. 11.2 SPSD process summary

Triple Bottom Line (TBL) elements and can be supported by suitable quantitative tools, e.g. abridged LCA if required.

To date, most approaches for reducing the sustainability impacts of products, e.g. eco-design, are aimed directly at individual companies irrespective of their role in the supply chain and the wider product SCD. Further, there is a trend

Table 11.1 SPSD aspects

SPSD criteria	Life cycle	Issues to consider (as relevant)
Optimize functionality	Conception	What is the functionality?
		How can this be achieved?
		Do you need a product?
		Could this be achieved by a service?
		Options for PSS?
Optimize environmental impact	Raw materials	Reduce the volume of materials used (dematerialisation)
		Nature of raw materials
		Eliminate or reduce nonrenewables usage
		Substitution of none/less hazardous raw materials
		Facilitate recovery, reuse, recycling
		Extraction and processing of raw materials
		Transport from supplier
	Production	Optimize production technology
		Eliminate/reduce emissions to air
		Eliminate/reduce effluents
		Eliminate/reduce waste
		Eliminate/reduce energy usage
	Distribution	Is transport necessary
		Volume and nature of transport
		Type of fuel usage
		Eliminate/reduce emissions to air
		Eliminate/reduce waste
	Consumption	Eliminate/reduce waste from product
		Eliminate/reduce waste from packaging
		Eliminate/reduce energy consumption
	End of life	Extend product life
		Design for repair
		Modular design for maximizing upgradability
		Facilitate recovery of components for reuse
		Facilitate recovery of components for recycling and treatment/disposal
Optimize social impacts	Raw materials	Are the raw materials extracted/processed in the developing world?
		Ownership rights
		Are the trading arrangements equitable?
	Production and distribution	Employee conditions of work at company
		Employee conditions of work in subcontract companies
		Impact on local community
		Investment in local community

(continued)

Table 11.1 (continued)

SPSD criteria	Life cycle	Issues to consider (as relevant)
	Consumption	Adverse health/safety impacts for the local community
		Adverse health/safety impacts for the global community
	End of life	Adverse health/safety impacts for the local community
		Adverse health/safety impacts for the global community
Optimize economic aspects	All phases	Is the product and/or service cost effective?
		Does the product and/or service cost the same/less than competing versions?
		Are environmental externality costs (e. g. end-of-life recovery, reuse/treatment/disposal) taken into account?

toward developing ecodesign and Sustainable Product Development tools in line with company size, e.g. large or small- and medium-sized enterprise (SME) in line with traditional environmental improvement approaches.

Product design and supply chain development varies between different products and organizations. Therefore, it is not recommended to integrate environmental issues according to one standard. This process of integration should be elaborated individually within a company. Box Q shows the Alfa Laval attempt to integrate environmental aspects into product design.

Box Q: Integrating Environmental Aspects into Product Design and Development at Alfa Laval

This box describes the main important environmental aspects that have been considered during seven phases of product design and supply chain development at Alfa Laval. As a leading global provider of specialized products and engineering solutions, Alfa Laval is focusing on three key technologies: Centrifugal Separation, Heat Transfer, and Fluid Handling. Jeganova (2004) proposed two ways of integration of environmental aspects into current product design and development process: "high priority" way and "low priority" way. It could be also named as a long-term proposal and a short-term proposal (what could be implemented immediately).

In the phases, the beginning stage and the end stage of product design and development process are the most important. In the beginning stage, designers and decision makers can decide if they choose a "high priority" way or a "low priority" way of consideration of environmental aspects. It should be decided during the step "Needs Analysis", where the needs of Alfa Laval and customers related to Life Cycle Design should be identified. If there is a commercial potential for investing in Product Life Cycle Design, then designers may choose the "high priority" way. If there is no commercial potential and no high demand from customers and other stakeholders for Life Cycle Design, then, designers may choose the "low priority" way.

The final phase of Life Cycle design process is environmental evaluation of product performance. The main outcome of this phase is a document "Environmental Product Profile" which contains environmental information about a product. A standard of this document should be elaborated by Alfa Laval's specialists according to the environmental legal requirements. Usually Environmental Product profiles contain negative and positive environmental impact of a product. The document can have several environmental key areas such as energy consumption, waste and emissions, recyclability, materials, harmful substances, transportation, and packaging. This document can be used as valuable environmental information for Alfa Laval, for the customers, suppliers, government, and other stakeholders. Moreover, Environmental Product profiles can be used in a decision-making process about a new product where designers and other decision makers may look at negative environmental impact of an existing product, and consider this negative impact when they create and develop a new product. This may lead to eco innovation and success!

Coming back to the proposed Life Cycle Design process, each phase of Alfa Laval's product design and development process is analyzed by four main questions: (1) When should it be done? (a name of phase), (2) What should be done? (what kind of aspects and activities), (3) Who should do it (what kind of specialists)? (4) How should it be done? (what kind of tools, approaches, activities, etc.). Moreover, the expected outcome for each gate meeting from each environmental activity will be defined.

The beginning stage of Life Cycle Design

The beginning stage of Life Cycle Design process (see Fig. 11.3) in the proposed version is the Life Cycle Thinking concept. It is proposed to start with integrating this concept to the stakeholders that are related to product design and development process.

Gate (1) The impacts of all life cycle stages need to be considered comprehensively when taking informed decisions on production and consumption patterns, policies, and management strategies. Moreover, Life Cycle thinking should be taken into account during generation if ideas and creation of new products.

It may also lead to the innovation of eco products. If a company considers the product's entire life cycle, it can help to focus not only on the environmental impacts of a product itself, but also on the supply chain in which the product performs.

WHAT: Life Cycle Thinking
WHEN: at the early stage of a development project (during generation of ideas) [before feasibility study]

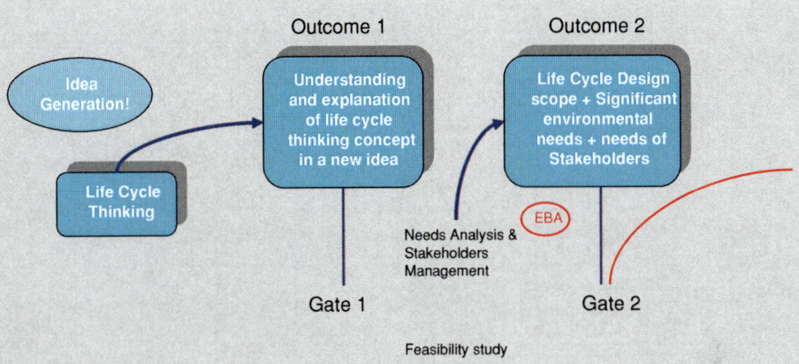

Fig. 11.3 Flow chart: the beginning stage of life cycle design process at Alfa Laval

WHO: Competence Team of environmental specialists or Environmental Coordinator
HOW: via training, seminars, guidelines, principles
OUTCOME/RESULTS for the gate meeting: life cycle thinking should be explained and understood by design team for the gate meeting 1

Gate (2) A new development project should clearly identify Alfa Laval's and its customer's needs. Design is usually focused on meeting these needs. It should be noted that at this step a company should determine a scope of life cycle design project. It is essential to listen and communicate to the stakeholders that can bring the ideas and help to define significant environmental aspects that should be included in the design project.

WHAT: Needs Analysis and Stakeholders Management (Communication)
WHEN: Feasibility study phase
WHO: Environmental Coordinator
HOW: communicate with Alfa Laval's stakeholders and identify significant environmental needs. Here eco benchmarking approach can be applied. Team work
OUTCOME/RESULTS for gate meeting: life cycle design scope, significant environmental needs, needs of stakeholders in terms of environmental issues should be defined for the gate meeting 2.

This stage is the most important stage in Life Cycle Design process. At this stage, designers and decision-makers should decide if they will follow the "high priority" way or the "low priority" way in order to consider environmental aspects in product design and development process. Therefore, in the needs analysis it is important to analyze customer's demands and legislation related to environment. If there is a high demand and commercial potential, then Alfa Laval can follow the "high priority" way. If there is no very high demand and commercial potential, then the company can

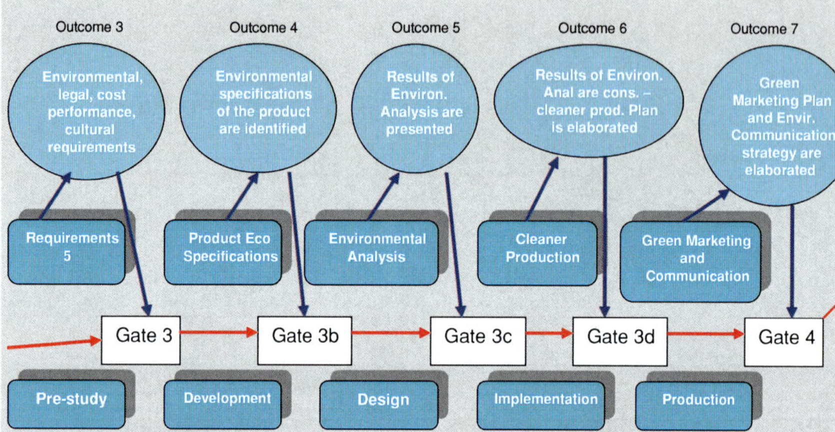

Fig. 11.4 Flow chart: the "high priority" version of life cycle design process

implement the "light" version. Environmental opportunities and related cost-benefit analysis should be conducted and discussed.

The "high priority" way of Life Cycle Design

Five areas of environmental issues have been identified that should be integrated in product development process at Alfa Laval in the "high priority" version: (1) Five Requirements, (2) Identification of Environmental Specifications of a Product, (3) Environmental Analysis, (4) Practices of Cleaner production, (5) Green Marketing and Communication (see Fig. 11.4).

(1) When the company makes a decision to implement the "high priority" version of Life Cycle Design, a formulation of five groups of requirements (environmental, legal, cost, cultural, and performance) related to environmental issues is essential in product development process. To define requirements may be the most critical phase of design. Requirements define the expected outcome. According to Keoleian and Menrey (1993), environmental requirements should minimize raw materials consumption, energy consumption, waste generation, health and safety risks, and ecological degradation. Legal requirements include local, state, and international environmental, health, and safety regulations and mandatory requirements. Cost requirements need to reflect market possibilities, and should help to designers add value to the product system. Cultural requirements define the shape, form, color, texture, and image that a product projects. The choice of them has direct environmental consequences. Successful cultural requirements promote an awareness of how it reduces environmental impacts. Performance requirements are functional requirements. Designers need to offer a high level of performance in order to satisfy the customer's needs. However, not always better performance is good for environment. Innovative

technology may increase performance and decrease environmental impacts; however, it can also increase consumption, which may again generate a negative environmental impact.

WHAT: Identification of five Groups of Requirements
WHEN: Prestudy phase
WHO: Environmental Coordinator
HOW: Building a Matrix of Requirements, data collection, team work
OUTCOME/RESULTS for the gate meeting: five groups of requirements should be identified for the gate meeting 3

(2) After building a matrix of requirements, it is time to identify environmental specifications of a new product via conceptual and detailed design. Design team together with Environmental Coordinator elaborates product design specifications. Environmental coordinator facilitates a process of identification of environmental specifications of a product using data, analysis, and other information that was collected during previous stages. Project-specific priorities are discussed and environmental specifications are chosen.

WHAT: Identification of Environmental Specifications
WHEN: Development and Design phase
WHO: Environmental coordinator with design team members
HOW: team work, brainstorming
OUTCOME/RESULTS for the gate meeting: environmental specifications of a product should be identified for the gate meeting 3b

(3) After identification of environmental specifications, environmental coordinator has to conduct environmental analysis of a product in order to distinguish positive and negative environmental impacts of a product according to selected environmental specifications. Different Life Cycle Design tools, methodologies, databases, and approaches can be applied (for example, eco benchmarking approach). The results of analysis can still influence and/or change the environmental specifications of a product. Therefore, environmental analysis should be conducted before implementation phase.

WHAT: Environmental Analysis
WHEN: Design phase
WHO: Environmental coordinator
HOW: data collection and analysis (depends on the chosen Life Cycle Design tool, methodology, etc. We propose eco benchmarking approach)
OUTCOME/RESULTS for the gate meeting: environmental analysis should be conducted for the gate meeting 3c

(4) In the phase of implementation, the production process is being tested in the 0-series for correct production equipment and correct production material, i.e. drawings, item lists etc. All documents (drawings, specifications etc.) are still preliminary. Therefore, the results of environmental analysis can still be considered.
In this stage, a production unit has to start to integrate cleaner productionpractices (it was failed to find out if cleaner production practices are started at Alfa Laval) or use existing cleaner production methods.

WHAT: Cleaner Production practices
WHEN: Implementation phase and Production phase
WHO: Environmental coordinator, production unit, design team
HOW: team work, brainstorming, seminars,
OUTCOME/RESULTS for the gate meeting: results from environmental analysis should be taken into account and a plant with cleaner production goals should be elaborated for the gate meeting 3d

(5) Environmental Communication Strategy and Green Marketing are needed in order to assist in promotion of the product. Green Marketing should concentrate on the environmental aspects of the product. Marketing specialists together with Environmental Coordinator should create "green" image of a product by using the results of environmental analysis and experience, and knowledge of Environmental Coordinator who followed the whole life cycle design process from the beginning to the end. Environmental Communication Strategy should be elaborated in order to have a continuous communication about environmental aspects of the product with Alfa Laval's stakeholders, particularly with the customers. It should be noted that collection of environmental data and information is crucial because it might be needed to provide the information to different stakeholders.

WHAT: Environmental Communication Strategy and Green Marketing
WHEN: Production and Launching phase
WHO: Environmental Coordinator, Sales and Marketing specialists, Communication specialists
HOW: by creating a green image for the product, promoting the product in the market, communicating with Alfa Laval's stakeholders about environmental aspects of the product, collecting environmental information
OUTCOME/RESULTS for the gate meeting: environmental Communication Strategy and Green Marketing plan should be elaborated and in force for the gate meeting 4

The "low priority" way of Life Cycle Design

If the company, after "Needs Analysis", makes a decision that there is no demand and commercial potential for life cycle design, then "low priority"

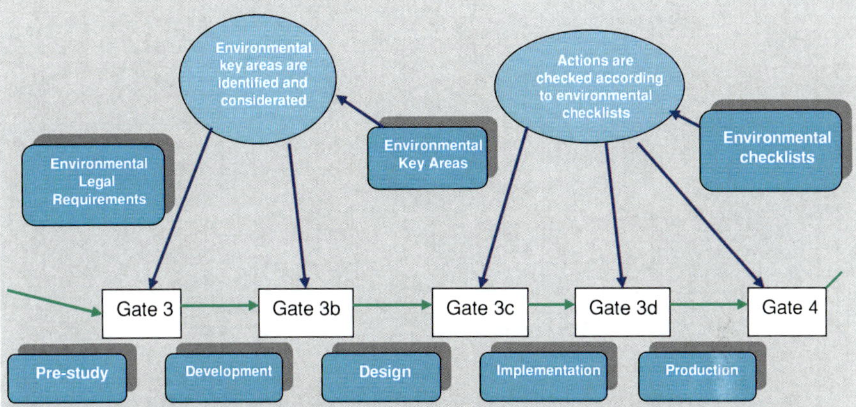

Fig. 11.5 Flow chart: the "low priority" version of life cycle design process at Alfa Laval

way of life cycle design can be implemented (see Fig. 11.5). The "low priority" way differs from the "high priority" way in terms of that there will not be large environmental investments and detailed environmental analysis of a product. Only legal environmental requirements will be considered. For

collecting environmental information, environmental key areas will be identified. For instance, it can be seven focal areas: energy, materials, packaging, harmful substances, transportation, waste, and recyclability. Eco benchmarking approach can be used for identifying environmental key areas. Design, implementation, and production activities can be checked according to environmental checklists. Environmental checklists can be used as an aid to memory in order to remember different environmental criteria during the whole life cycle design process. For example, a checklist about solid waste and toxic emissions may consist of the following questions (Bakker 1995): (1) Is the use of priority substances avoided? (2) Are there options for the product's disposal other than landfill or incineration? (3) Are the toxic substances that are emitted during the product's life cycle minimized in comparison to the products with the same functionality? Environmental checklists should be developed individually by a company according to the needs, goals, and interests of a company and a particular goal of life cycle design process.

(1) When the company makes a decision to implement the "low priority" version of Life Cycle Design, legal environmental requirements about life cycle design practices and a particular product should be considered.

WHAT: Identification of Legal Environmental Requirements about the product and life cycle design process
WHEN: Prestudy phase

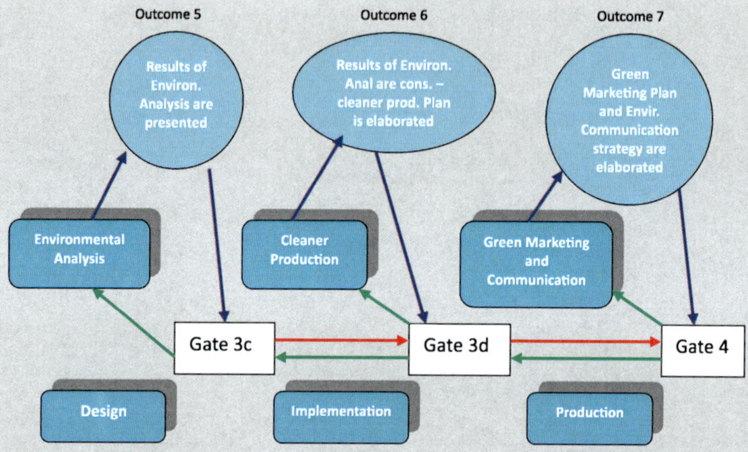

Fig. 11.6 Flow chart: adaptive feedback in life cycle design process

WHO: Environmental Coordinator
HOW: checking appropriate legal environmental regulations and creating a checklist
OUTCOME/RESULTS for the gate meeting: a checklist with legal environmental aspects that should be developed for the gate meeting 3.

(2) After identification of legal environmental requirements, environmental key areas in life cycle design process should be identified. Eco benchmarking approach can be used as one of options in order to collect environmental information about a product. Eco checklists should be created according to environmental key areas.

WHAT: identification of environmental key areas and collection of environmental information
WHEN: development phase
WHO: Environmental Coordinator together with development project members
HOW: for example, eco benchmarking approach
OUTCOME/RESULTS for the gate meeting: environmental checklists are developed for the gate meeting 3c

(3) In the phase of Design, Implementation, and Production, elaborated environmental checklists with environmental key areas should be used.

WHAT: use of environmental checklists
WHEN: Design, Implementation and Production phase

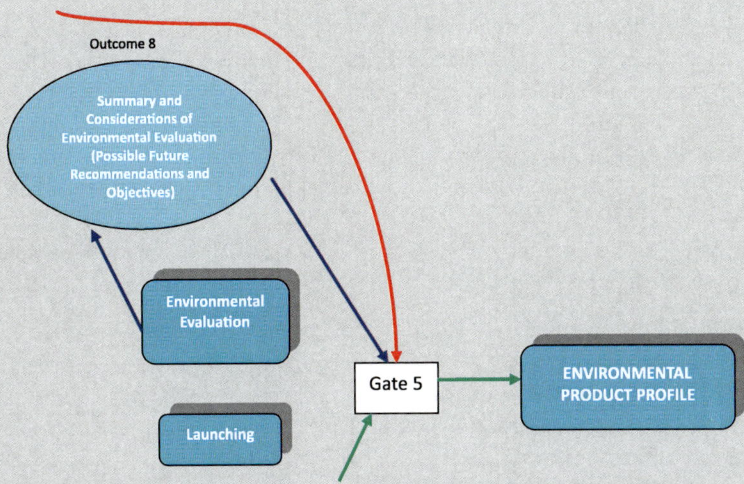

Fig. 11.7 Flow chart: the final stage of life cycle design process at Alfa Laval

WHO: Environmental Coordinator with a project manager controls the process
HOW: according to elaborated environmental checklists
OUTCOME/RESULTS for the gate meeting: environmental information about how life cycle design process has been conducted and what kind of environmental key areas have been considered.

Adaptive Feedback in Life Cycle Design

One of the most important issues in proposed Life Cycle Design process is the adaptive feedback between the meeting gates (see Fig. 11.6). If we look at the meeting gates, it is clear that certain requirements should be met. But what happens if these particular requirements are not met or are not completely fulfilled?

Therefore, the adaptive feedback should be explained. Figure 11.7 shows a fragment from "high priority" version of Life Cycle Design Process where the adaptive feedback is shown with "green" arrows. In the gate meeting 3d, Cleaner production Plan should be elaborated. If the plan is not satisfactory elaborated, then the process comes back to the stage before ("Cleaner production") in order to improve failed Cleaner production Plan. Moreover, there is also the adaptive feedback from one gate to another gate, if designers need to reconsider previous decisions.

The final stage of Life Cycle Design

When the product is produced and launched in the market, Environmental Coordinator can conduct an overall environmental evaluation of the product and its performance in order to identify positive and negative environmental impacts of the product. This information can be used internally for Alfa Laval's employees in order to redesign the product or improve its performance.

Moreover, the company can also set the targets in order to reduce negative environmental impact of the product and its performance for the future.

WHAT: Environmental Evaluation of the product and its performance
WHEN: Launching phase and after the launching phase
WHO: Environmental Coordinator
HOW: Life Cycle Design Evaluation Tools, Methodologies, Approaches, Databases, Software tools, etc.

OUTCOME/RESULTS for the gate meeting: Environmental Evaluation should be conducted and results should be summarized in an Environmental Product Profile for the gate meeting 5.

11.2 Three-Dimensional Concurrent Engineering (3DCE) to Integrate New Product Development (NPD) and Environmentally Responsible Manufacturing (ERM)

The concept of concomitantly designing products, processes, and supply chains is referred to by Fine (1998) as three-dimensional concurrent engineering (3DCE).

Many organizations today view new product development as a source of competitive advantage. Firms that are among the first-to-market commonly generate higher revenue streams initially and preempt later entrants. However, the first-mover advantage is rarely sustainable because competitors soon enter the market with lower-priced, lower-cost alternatives. In the 1980s firms began adopting concurrent engineering, or the simultaneous design of products and processes to accelerate new product introductions (Smith 1997). More recently Fine (1998) introduced the concept of three-dimensional concurrent engineering (3DCE)—concomitantly designing the supply chain with an organizations' products and processes. He argued that supply chain design issues should be addressed at the same time or operational problems will typically occur late in the product development cycle related to logistical support, quality control, and production costs. 3DCE will also help firms respond more quickly to changes in their environment.

3DCE is credited with many potential benefits, including reduced costs, reduced time to market, improved supplier integration, and improved quality (Fine et al.

1998; Balasubramanian 2001; Klassen and Angell 1998) which are generally NPD goals. Simultaneously, it appears that some of the basic ideas of integrating product, process, and supply chain are also being espoused for ERM (Sarkis 2003). The fundamental question is: will using 3DCE as a framework for integrating NPD and ERM yield greater benefits than the separate and uncoordinated application of ERM principles and NPD initiatives? This is a critical question for organizations that wish to pursue ERM, but believe that ERM will hurt their competitiveness by increasing costs, increasing new product introduction times, and generally conflicting with traditional new product development goals.

While traditional NPD focuses specifically on the product, concurrent engineering (CE) represents a revolution of new product development thought by simultaneously focusing on product and process using cross-functional teams (Koufteros et al. 2002). However, due to its popularity, CE by itself no longer provides a source of competitive advantage (Fine et al. 1998). Further, ''in companies that now practice two dimensional CE (product and process only), supply chain development tends to be haphazard'' (Fine et al. 1998).

3DCE represents an important step beyond these approaches, simultaneously focusing on product, process, and supply chain to improve NPD. While NPD and ERM can be integrated without 3DCE, this will likely result in neglect of supply chain design. Supply chain design is critical to both ERM and NPD. Supply chain design specifically considers, "whether to make or buy a component, sourcing decisions, and contracting decisions (such as structuring the relationships among supply chain members). Logistical and coordination decisions include inventory, delivery, and information systems to support on-going operation of the supply chain," (Fine et al. 1998). Thus, the level of supply integration in 3DCE is well beyond that required in CE. If supply chain design is not explicitly integrated as part of NPD and ERM, it is likely that higher costs and reduced performance will ensue. For example, when moving its excellent products and processes to North America for production, it was the supply chain design that caused problems for Toyota. Its sourcing decision to use North American suppliers to make parts wherever possible, the way it structured relationships among suppliers, and its coordination of information within the supply chain were all problematic, delaying product launch by 10 months, and raising development costs by 40%. Toyota quickly redesigned its supply chain to utilize back-up sources from Japan, at a cost of around $1,000,000 per month in premium airfreight charges (Judson 1998). These problems all reflect issues with integrating supply chain design into new product development. Supply chain design in 3DCE goes beyond simply integrating a supplier's technological capability, to include that supplier's information processing and communication capability, inventory management and even the supplier's relationships with its suppliers. 3DCE provides the next level of breakthrough in improving performance. This is documented by Petersen et al. (2005), who demonstrated that early supplier involvement in supply chain design supports improved product and process design. The relationship among and between each of the elements of 3DCE: product, process, and supply chain is depicted in Fig. 11.8. For example, the key issue in simultaneously considering

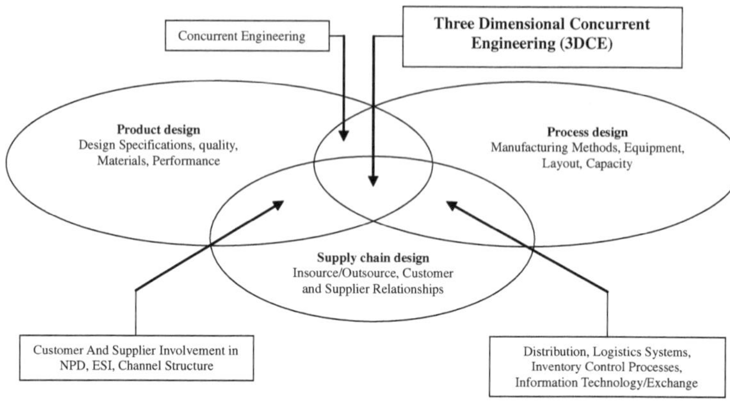

Fig. 11.8 3-D engineering-integrating new product development and environmentally responsible manufacturing. (modified by Ellram e altri 2008)

product and process design is CE. In supply chain and process design, key issues include linking distribution and logistics systems with processes. The relationships among the elements of 3DCE are discussed in more depth in the next section, where the link between 3DCE and ERM is presented.

By incorporating ERM goals at every step of the way in 3DCE, organizations should be able to enjoy all of the benefits of traditional NPD and ERM.

- In ERM, product design applies this focus in making a product that uses environmentally friendly materials, uses fewer materials, and mixes fewer materials together, so that the materials are easier to separate at the end of the product's life (Sarkis and Rasheed 1995; De Ron 1998). For example, environmentally friendly packaging may result in reduced cost, lower weight and associated logistics costs, and reduction in materials sent to landfills (Maxwell et al. 1997).
- Process-related ERM initiatives also include a focus on processes that reduce the source of waste. Waste is any activity that creates additional costs for the organization or consumes any type of resources without an offsetting benefit (Carter and Ellram 2003). Reduction of waste can be accomplished through production process changes, operational improvements that reduce waste, and improved inventory management (Angell and Klassen 1994).
- Essentially, ERM supply chain initiatives focus on the impact of the firm's activities outside of the firm's boundaries, such as the nature of supplier or customer relationships or delivery mechanisms. Supply chain-oriented ERM initiatives may directly involve external stakeholders in order to gain the perspective of those outside of the firm's boundaries. Players external to the firm are engaged for specific reasons, such as supplier involvement to reduce waste or to utilize more environmentally friendly materials in their activities (Min and Galle 1997). For example, perhaps in part because of the negative image that

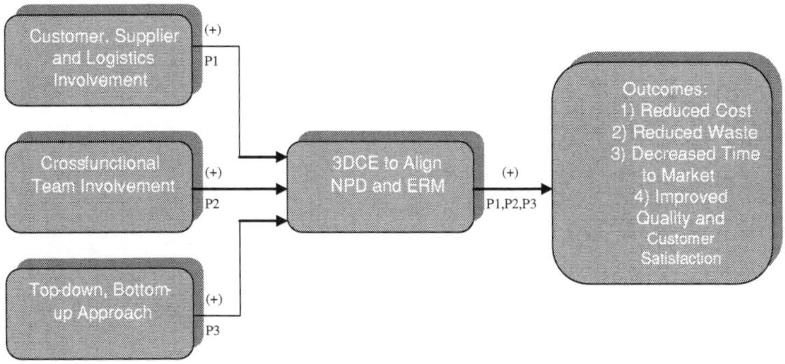

Fig. 11.9 A 3DCE framework for NPD-ERM (modified by Ellram et al. 2008)

WalMart has in some circles, it is responding to consumer's desires for environmentally sound practices by working with its suppliers to reduce waste in the supply chain. This is illustrated by Unilever's redesign of its liquid. All Laundry (product) detergent into "Small and Mighty". It went from a bulky 100 oz bottle to a slim 32 oz. bottle with the same number of loads washed, but Unilever was worried that customers would not notice it next to the large size competitors. This product meets many ERM supply chain initiatives: less packaging, less energy cost, less transportation, less to dispose of or recycle, less shelf space and more. WalMart supported this, heavily promoting the product. Small and Mighty have been very successful not just at WalMart, but in other stores as well (Gunther 2006).

Ellram et al. (2008) demonstrate the strong relationship between 3DCE NPD and ERM (Box R).

Box R: Relationship Between 3DCE NPD and ERM

A key purpose of Ellram et al. (2008) research was to demonstrate the strong relationship between 3DCE NPD and ERM. This was done through the development of four research propositions, and the theoretical model in Fig. 11.9. In practice, many organizations treat NPD and ERM as separate processes, while both independently use many of the 3DCE concepts. The traditional NPD business objectives are outcomes such as reduced time to market, low cost, and improved manufacturability.

The goals of ERM are oriented toward reducing the environmental impact of products throughout the life of the product, including disposal. The approach that is effective in achieving the goals of both NPD and ERM is to simultaneously design the product, the process and the supply chain: in other words, to use 3DCE. This research contends that 3DCE approaches can be effectively used to integrate the goals of ERM efforts with those of NPD efforts. Such synergy provided by 3DCE will increase the acceptance of ERM in NPD. Companies can meet the apparently conflicting goals of sustaining the environment, while satisfying corporate profitability objectives and providing excellent new product performance.

Proposition 1 (P1): companies utilizing 3DCE to integrate NPD-ERM will outperform companies that separately manage NPD and ERM, in areas such as:

(a) reducing environmental waste,
(b) reducing new product time to market,
(c) lowering cost, and
(d) improving product quality.

The benefits of CE in NPD have been widely embraced; however, the outcomes of ERM have been more widely contested within both the academic and practitioner communities. For example, researchers have found a positive (e.g., Klassen and McLaughlin 1996), negative (e.g., Bragdon and Marlin 1972), and no relationship (e.g., Alexander and Buchholz 1978) between environmental and firm performance.

Thus, while this proposition may seem obvious, both research findings and managerial perceptions indicate otherwise.

The literature on traditional NPD and the literature related to NPD with an ERM focus, clearly support that NPD with an ERM focus takes a much more holistic approach, at least in theory (Sidkar 2003). As suggested by Fine (1998), the design of products, processes, and supply chains are inexorably intertwined. 3DCE makes the consideration of the customers, suppliers, and supply chain design explicit rather than an afterthought. A case in point is Chrysler, who was working to understand and improve its Jeep V8 engines, looking at the critical parts along the chain. However, it was completely unaware that one of the critical suppliers to its key supplier for this new concept was planning to get out of what it viewed as an unprofitable business. Without a broader supply chain perspective, Chrysler had become heavily dependent on a supplier who did not wish to continue in this line of business (Fine et al. 1998). While traditional product design is often still performed along functional lines, or at most with an organization-wide focus rather than a supply chain focus, organizations cannot afford to perform sustainable new product development in that manner.

Proposition 2 (P2): companies that utilize the 3DCE approach of integrating key supply chain members early in NPD–ERM will outperform companies that do not have this early integration.

(a) Companies that utilize the 3DCE approach of integrating logistics providers early in NPD–ERM will outperform companies that do not have this early integration, in areas such as (i) lower cost and (ii.) reduced environmental waste.
(b) Companies that utilize the 3DCE approach of integrating key suppliers early in NPD–ERM will outperform companies that do not have this early integration, in areas such as: (i) lower cost and (ii) reduced environmental waste.
(c) Companies that utilize the 3DCE approach of integrating key customers early in NPD–ERM will outperform companies that do not have this early integration, in areas such as: (i) lower cost, (ii) reduced environmental waste, and (iii) improved customer satisfaction.

On the surface, this proposition may seem self-evident. However, all indications from the practitioner and scholarly literature support that such early involvement is not commonplace. In addition, some firms will just involve one external supply chain member, such as key customers, and ignore the potential benefits of involving suppliers and logistics providers. This was the case with Motorola, which developed an excellent new product in cell phones by understanding customer needs and concerns, and even considering end-of-life product returns. However, Motorola did not actively involve its suppliers and logistics providers in NPD and ERM concerns associated with the supply chain for inbound parts. As a result, inbound packaging and labeling problems, pallet problems, and resultant excess injury and waste disposal problems due to poor labeling and improper pallet configuration and materials cost Motorola millions each year. Thus, it seems that the results of testing this proposition could lie on a continuum, from companies which do not involve logistics providers, suppliers or customers in any aspect of NPD–ERM, to those that involve all three in product, process, and inbound and outbound supply chain design (Fiksel et al. 2004).

Proposition 3 (P3): Firms that include the environmental health and safety function in cross-functional coordination during, or prior to, the design phase of new product development within the 3DCE framework will achieve superior environmental outcomes versus firms which do not.

Proposition 4 (P4): a combined, top-down, bottom-up managerial approach is essential for effective NPD–ERM utilizing 3DCE.

From an organization standpoint, there appears to be a gap in the execution of NPD–ERM. The strategic choice literature indicates that top management may initiate changes to organizational structure and systems; however, actual changes in product and process frequently occur from the bottom-up (Daft 1978).

This suggests that a combined, top-down, bottom-up approach might be needed to effectively align the 3DCE integration in the NPD–ERM model. Thus it appears that an alignment of product, process, and supply chain will require the strong support of both top and middle management. Much of the change might be driven initially at the functional level through the new product development process, for example, by marketing and R&D. However, top management support and coordination will likely be necessary in order to orchestrate the coordination of the production process and configuration of the supply chain. These findings suggest the need for the involvement and direction of both top and middle management in effecting 3DCE integration into NPD-ERM, and lead to the next proposition:

Theoretical and managerial implications

Proposition 1 posits that by integrating ERM with NPD processes through 3DCE, organizations can deliver measurable benefits above what they would be able to achieve if ERM and NPD were managed separately. This is illustrated in the middle and right box of Fig. 11.9. These benefits include, but are not limited to: environmental waste reduction, reduced time to market, lower costs, and quality improvements. These benefits should be of great interest to managers of ERM. The best way to glean support for their efforts within an organization is to show that there is a net benefit beyond helping the environment. Too often, sustainable initiatives are viewed as increasing the firm's cost, although this is changing rapidly (Thomson 2006). To be able to demonstrate a net benefit in terms of cost, speed, or other traditional performance measures could greatly increase the acceptance of ERM efforts in the organization. Demonstrating the value of ERM to the organization's success is a major objective of environmental industry groups such as GEMI (2004).

Proposition 2 focuses on the natural synergy that results from key supply chain stakeholder involvement in NPD–ERM efforts. This is shown in the top left box in Fig. 11.9. The inputs and outputs of ERM extend beyond the borders of the organization into the supply chain. Thus, following a 3DCE approach that incorporates logistics issues along with the voice of the supplier and the voice of the customer will result in more successful outcomes in terms of both ERM objectives and traditional NPD objectives reflected in 3DCE. Fine (1998) emphasizes the criticality of choosing the right suppliers and forming the right types of relationships with those suppliers. Prior research suggests that the same holds true when working with customers. Thus, the benefits suggested by proposition 2: lower cost, reduced environmental waste, and improved customer satisfaction, could go a long way in gaining support for NPD–ERM initiatives simultaneously utilizing 3DCE.

Like proposition 2, propositions 3 and 4 focus on the importance of getting the right people involved in NPD–ERM. Proposition 3 (middle left

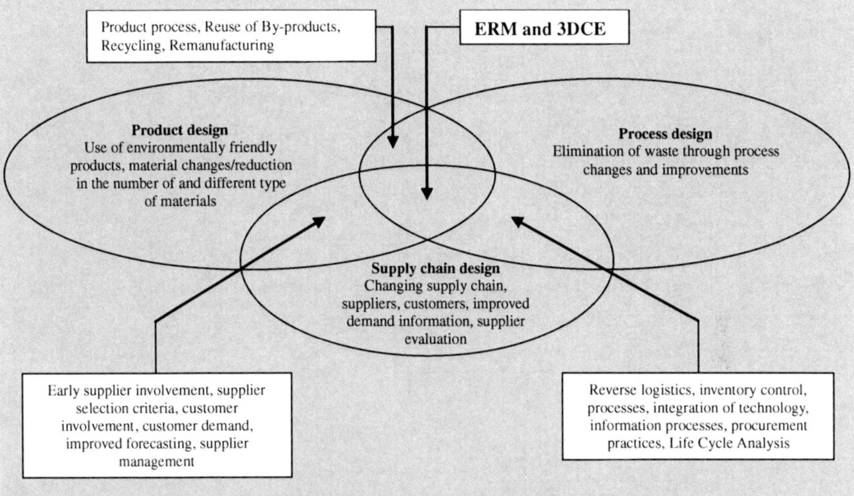

Fig. 11.10 3-D engineering-integrating new product development and environmentally responsible manufacturing. (modified by Ellram et al. 2008)

box in Fig. 11.9) emphasizes the importance of an internal team at the earlier stages of NPD for successful 3DCE. The team should include key internal players, including EHS. The team should play a significant role as an interface with key supply chain members.

Proposition 4 (the bottom left box in Fig. 11.10) implies the need for top management to better understand the potential advantages to appropriately aligning product, process, and supply chain, rather than creating a configuration of the 3DCE variables that they simply feel most ''comfortable'' with. Obviously, this is easier said than done. Miller (1996) suggests that managers can align such a configuration in incremental steps rather than ''giant leaps''. As emphasized in proposition 4, effective NPDVERM utilizing 3DCE requires involvement and support at all levels of the organization. It cannot be a ''top management'' solution. Understanding this is particularly critical as more companies jump on the ''green'' bandwagon. For example, 3M's top management provides high level support and resources for ERM. It relies upon the business units and various levels and functions to identify and implement sustainability initiatives that are in line with 3M's Environmental, Health and Safety (EHS) Management System, and support a life cycle perspective. Cross-functional, new product introduction teams use a Life Cycle Management matrix to address the EHS opportunities and issues at each stage of the product's life. Figure 11.9 shows the incorporation of the four propositions as a basic framework for integrating traditional NPD with ERM. First, support from external supply chain members (proposition 2), internal team members (proposition 3), and both top management and the functional workforce (proposition 4) is needed to create alignment of NPD and ERM goals. With this alignment of goals, positive results will be forthcoming from NPD–ERM utilizing 3DCE, such as environmental waste reduction, reduced time to market, and cost improvements. These benefits will be greater than what could have been achieved had NPD and ERM been pursued separately. Taken together, Fig. 11.9 provides a preliminary roadmap for considering the key people and types of involvement in using 3DCE to link new product development and environmentally responsible manufacturing to yield the desired outcomes.

References

Alexander G, Buchholz R (1978) Research notes corporate social responsibility and stock market performance. Acad Manag J 21(3):479–486

Angell L, Klassen R (1994) Integrating environmental issues into the mainstream: an agenda for research in operations management. J Oper Manag 17:575–598

Bakker C (1995) Environmental information for industrial designers. Technische Universiteit Delft

Balasubramanian R (2001) Concurrent engineeringda powerful enabler of supply chain management. Qual Prog 34(6):47–53

Bragdon J, Marlin J (1972) Is pollution profitable? Risk Manag 19(4):9–18

Carter CR, Ellram LM (2003) Thirty-five years of the journal of supply chain management: where have we been, where are we going? J Suppl Chain Manag 39(2):27–39

Daft R (1978) A dual-core model of organizational innovation. Acad Manag J 21:193–210

De Ron A (1998) Sustainable production: the ultimate result of continuous improvement. Int J Prod Econ 56–57(20):1–19

Ellram LM, Tate W, Carter CR (2008) Applying 3DCE to environmentally responsible manufacturing practices. J Clean Prod 16:1620–1631

Fiksel J, Lambert D, Artman L, Harris J, Share H (2004) Environmental excellence: the new supply chain edge. Supply Chain Manag Rev 8(5):50–58

Fine S, Singer Y, Tishby N (1998) The hierarchical hidden Markov model: analysis and applications. Machine Learning 32:41-62

GEMI (2004) Forging new links. Available at: http://www.gemi.org/GEMI-ForgingNewLinks-June04.PDF. Accessed 11 Jan 2004

Gunther M (2006) The green machine. Fortune. Available at: http://walmartstores.com/GlobalWMStoresWeb/navigate.do?catg. ¼ 651 6 Aug 2006 (Accessed 9 Mar 2007)

Jeganova J (2004) Product life cycle design: integrating environmental aspects into product design and development process at Alfa Laval. LUMES Thesis, Lund University International Masters Program in Environmental Science

Judson J (1998) Integrating supplier designed components into a semiautomatic product development process. In: MIT, LFM Master's thesis, 1998, Idem assessing a new product development process using 3-dimensional concurrent engineering and term paper for course 15.769. MIT, Cambridge, in Charles Fine Clockspeed, pp 152–153

Keoleian GA, Menrey D (1993) Life cycle design guidance manual: environmental requirements and the product system, US environmental protection agency, EPA600/R-92/226, January

Klassen R, Angell L (1998) An international comparison of environmental management in operations; the impact of manufacturing flexibility in the U. S. and Germany. J Oper Manag 16(2–3):177–194

Klassen R, McLaughlin C (1996) The impact of environmental management on firm performance. Manag Sci 42(8):1199–1214

Koufteros X, Vonderembse M, Doll W (2002) Integrated product development practices and competitive capabilities: the effects of uncertainty, equivocality, and platform strategy. J Oper Manag 20(4):331–355

Maxwell D, van der Vorst R (2003) Develop sustainable products and services. J Clean Prod 11:883–895

Maxwell J, Rothenberg S, Briscoe F, Marcus A (1997) Green schemes: corporate environmental strategies and their implementation. Calif Manag Rev 39(3):118–134

Miller D (1996) Configurations revisited. Strateg Manag J 17(7):505–512

Min H, Galle W (1997) Green purchasing strategies: trends and implications. Int J Purch Mater Manag 33(3):10–17

Petersen K, Handfield R, Ragatz G (2005) Supplier integration into new product development: coordinating product, process and supply chain design. J Oper Manag 23(3/4):371–388

Reiskin E, White A, Kauffman J, Votta J (2000) Servicizing the chemical supply chain. J Ind Ecol 3(2–3):19–31

Sarkis J, Rasheed A (1995) Greening the manufacturing function. Bus Horiz 38(5):17–28

Sarkis J (2003) A strategic decision framework for green supply chain management. J Clean Prod 11:159–174

Sidkar S (2003) Sustainable development and sustainability metrics. AIChE J 49(8):1928–32

Smith RP (1997) The historical roots of concurrent engineering fundamentals. IEEE Trans Eng Manag 44(1):67–78

Thomson TS (2006) It's good to be green. Business week. Available at: http://www.businessweek.com/magazine/content/06_19/b3983116.htm?chan ¼ search 8 May 2006 (Accessed 10 Mar 2007)

van Hemel C (1998) Ecodesign empirically explored. In: Design for environment in Dutch SMEs. Boekhandel Milieu-Boek, Amsterdam. ISBN 90-9011667-2

van Weenen H (2000) Product design–practical examples of SMEs. In: Proceedings in sustainable development, SMEs and new enterprises, European Foundation, Ireland, 2–13 Oct 2000

Chapter 12
DfE Procedures in the Development of a More Sustainable Supply Chain

12.1 Introduction

To achieve sustainable supply chain, environmentally conscious design (eco-design) or design for environment (DfE) is becoming an increasingly important topic (Brezet et al. 1994; Den Hagg Brezet and Van Hemel 1997). In many cases, the product development process can be modeled as a product evolution process based on the representation and understanding of a current product. Hence, systematic methodologies for product evolution are necessary from a practical viewpoint (Otto and Wood 1998; Martin and Ishii 2000). In the same way, it is important to systematize a design methodology for evolution of ecoproducts based on modeling and analyzing current products (Coulter and Bras 1997, 1999).

In order to effectively reduce the sustainability impacts of products, the supply chain aspect of product manufacture needs to be incorporated. With the exception of products manufactured in a direct business-to-consumer relationship, most products with significant sustainability impacts, e.g. cars or electronics goods, are manufactured using a number of companies involved in a supply chain. Typically, this will involve an Original Equipment Manufacturer (OEM) and a range of component suppliers, subcomponent suppliers (to many tiers potentially), and assemblers. Typically the OEM has control over the product and/or service design and specification. The other companies, in the supply chain, supply and/or assemble their elements in line with this. In light of this, it is proposed that the supply chain for a proposed product and/or service must be considered in order to determine which company (ies) within the supply chain will be the most effective targets for SPSD as well as its different aspects. Further, it is necessary to target the organizations with control over the product life cycle stages where most sustainability issues can be most effectively tackled typically the product conception and design phases Maxwell and van der Vorst (2003). It is at these stages that most (up to 80%) of the environmental, social, and cost factors for a product are determined (Charter and Tischner 2001). Hence, it is at this stage that truly

M. Bevilacqua et al., *Design for Environment as a Tool for the Development of a Sustainable Supply Chain*, DOI: 10.1007/978-1-4471-2461-0_12,

effective improvements can be made not only in terms of sustainability issues, but also potentially in terms of cost.

There is a trend toward targeting environmental performance improvement methods at organizations based on their size. For example much work has been done to formulate environmental performance improvement approaches, to include eco-design, specifically to engage and meet the needs of SMEs. It is reasonable to apply environmental performance improvement systems, e.g. EMS directly to SMEs as SMEs have control over activities at their production facilities and can manage their environmental impacts. However, the same rationale is not necessarily relevant for products as many companies, especially SMEs, may not have this control and need to work in conjunction with the OEM to change the product specification Maxwell and van der Vorst (2003).

The Hitachi case, in the box S, shows how the Hitachi Group is working on energy-efficient products with lower greenhouse gas emissions.

Box S: Hitachi Case

The Hitachi Group is working on energy-efficient products with lower greenhouse gas emissions. The phase of product use is the part of the product life cycle responsible for the greatest amount of emissions.

Figure 12.1 shows the Percentage of Greenhouse Gas Emissions in the Product Life Cycle while Fig. 12.2 shows the recycling phase of Hitachi firms.

As a measure of how effectively greenhouse gas emissions are being reduced during a product's life cycle, Hitachi Group introduced global *efficiency of warming prevention.* Rating a product's value for function, and during its life span, this measure shows the ratio of this value to the amount of greenhouse gas emissions over the product life cycle. In addition, *factor of global warming prevention* is used to indicate the amount of improvement in *efficiency of global warming prevention* compared with a reference product (Fig. 12.3).

The Hitachi Group uses *resource efficiency* to evaluate how effectively resources are being used during the product life cycle, such as making products lighter and more compact. Rating a product's value for function and during its life span, this measure shows the ratio of this value to the amount of resource used over the product life cycle. In addition, a *resource factor* is used to indicate the amount of improvement in resource efficiency compared with a reference product (Fig. 12.4).

Covering the Entire Supply Chain

Working closely with suppliers and customers, Hitachi Group improved their information management system for chemical substances across the supply chain, from procurement to sales (Fig. 12.5). In November 2008, the

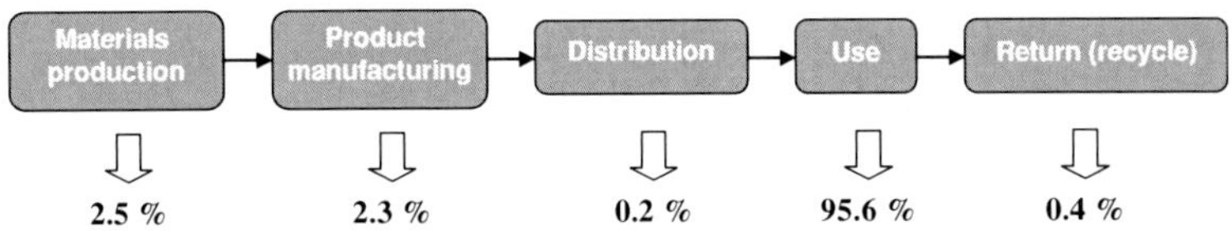

Fig. 12.1 Greenhouse gas emissions

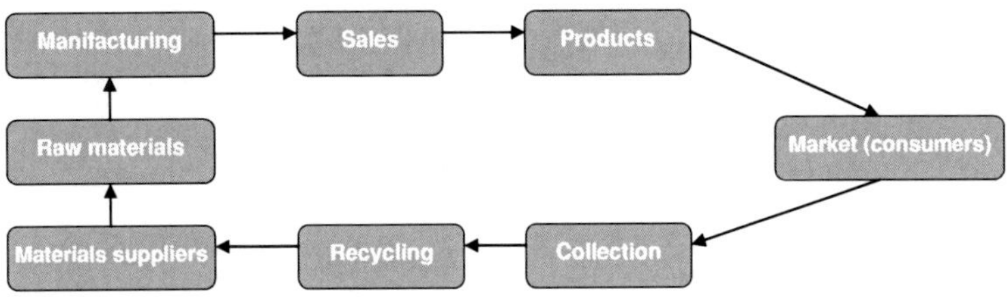

Fig. 12.2 Recycling phase

Definition of efficiency of global warming

$$\text{Efficiency of global warming prevention} = \frac{\text{Product function x product life span}}{\text{Volume of greenhouse gas emissions throughout the product life span}}$$

Definition of factor of global warming prevention

$$\text{Factor of global warming prevention} = \frac{\text{Efficiency of global warming prevention of evaluated product}}{\text{Efficiency of global warming prevention of reference product}}$$

Fig. 12.3 Global warming prevention factor calculation

$$\text{Resource efficiency} = \frac{\text{Product function x product life span}}{\text{Life cycle resource use x value coefficient of each resource}}$$

Definition of resource factor

$$\text{Resource factor} = \frac{\text{Resource efficiency of evaluated product}}{\text{Resource efficiency of reference product}}$$

Fig. 12.4 Resource factor calculation

integrated management system for chemical substances included in products, an IT system which had been used for the RoHS directive, was applied to the REACH regulation. To collect data from thousands of suppliers, we use JAMP information sheets. JAMP is a consortium for managing and disclosing chemical content in products. In addition, an information flow

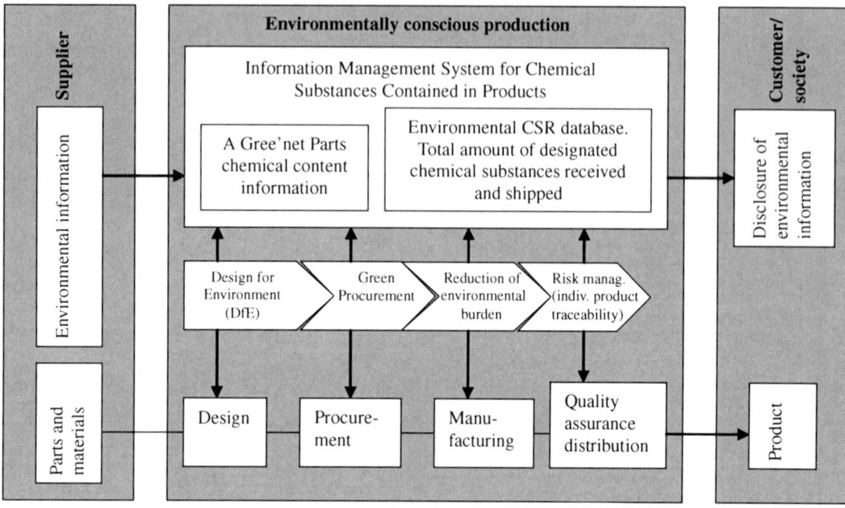

Fig. 12.5 Integrated management system for chemical substances contained in products

trial was started in January 2009 to test the information exchange on chemical substances in the entire supply chain.

For the Hitachi Group's voluntarily controlled chemical substances, Hitachi Group defined 13 prohibited substances (Level 1) and 12 controlled substances (Level 2), and took steps to comply with the RoHS directive and other global regulations for chemical substances in products. We achieved RoHS compliance in July 2006 (Table 12.1).

To comply with REACH, a chemical substance regulation in Europe, in fiscal 2008, Hitachi Group made an all-out effort, including drawing up action guidelines, revising our information infrastructure, and improving our management structure. Hitachi Group now tracks all chemical substances used in the supply chain, from raw materials to end products. To do this, a Group-wide committee compiled a list of controlled substances covered by the REACH regulation (by adding REACH Substances of Very High Concern (SVHC) to the existing list of controlled substances). The committee also drew up a set of procedures, and created an information infrastructure for compliance with the REACH regulation across the entire supply chain: procurement, production, and shipping.

The Hitachi Group has drawn up the Green Procurement Guidelines to gain cooperation from suppliers in the development of environmentally conscious products. Encouraging environmental protection activities by suppliers, Hitachi Group is asking for their understanding and cooperation to develop and supply products with low environmental burdens.

After identifying green suppliers—those who understand the importance of environmental protection and who voluntarily acquire environmental

Table 12.1 Hitachi group's voluntarily controlled chemical substances

Classification	Application	Substance (group) names
Level 1 Prohibited substances	Chemical substances that the Hitachi Group prohibits from being included in procured products. (Chemical substances banned or restricted for use in products (including packing materials) by domestic or foreign regulations and potentially used for procured products for the Hitachi Group)	Cadmium and its compounds, hexavalent chromium compounds, lead and its compounds, mercury and its compounds, bis (tributyltin) oxide (TBTO), polybrominated biphenyls (PBB), polybrominated diphenyl ethers (PBDE), polychlorinated biphenyls (PCB), polychlorinated naphthalene (with 3 or more chlorines), short-chain chlorinated paraffin, asbestos, azo dyes/pigments, and ozone layer depleting substances
Level 2 Controlled substances	Substances that are not restricted for inclusion in procured products but for which monitoring and control are required by domestic or foreign regulations, or for which special consideration for recycling or appropriate disposal is required	Antimony and its compounds, arsenic and its compounds, beryllium and its compounds, bismuth and its compounds, nickel and its compounds (excl. alloys), selenium and its compounds, brominated flame retardants, polyvinyl chloride (PVC), phthalate esters, tributyltins (TBT) and triphenyltins (TPT), ozone layer depleting substances (HCFC), radioactive materials, and potential REACH SVHC

certification—Hitachi Group looks for ways of working with them to further improve their operations from an environmental standpoint. These efforts include mutual exchange of proposals aimed at benefiting from energy and resource efficiency for lower costs, improved quality, and faster delivery.

For developing and making products that result in a lower environmental burden, Hitachi Group asks suppliers to (1) conserve resources (through miniaturization, standardization, etc.), (2) conserve energy, (3) aggressively pursue the 3 Rs. (4) reduce packaging materials, (5) properly manage chemical substances used in products, and (6) provide clear information. Of these, (1) to (4) can make suppliers more competitive by lowering costs and improving product functionality. To help drive home these advantages, Hitachi Group presents case studies and encouragement.

In the last decades many scientific contributions in the management field have been focusing on New Product Development (NPD) and on Supply Chain Management (SCM). However, these two research streams have seldom been considered together and the relation between NPD and SCM has not been deeply explored yet (Crippa et al. 2009).

SCM models and methods to support supply chain design and operations management are developed assuming that product design has been already defined (Simchi-Levi et al. 2003) and cannot be changed. Product design can be seldom altered significantly once the product has entered the marketplace. However, in some cases a redesign of the product could enhance supply chain performances, as well as a redesign of the supply chain for the new product. Moreover, in recent years, globalization, market volatility, product lifecycle contraction, product variety explosion, growing pressure on times, and massive use of outsourcing (van Hoek and Chapman 2006; Prahalad and Krishnan 2008) are some of the most important factors that are pushing companies to rethink the alignment between NPD and SCM.

Box T: New Product Development and Supply Chain Management Alignment

The term "alignment of two processes" indicates that the process owners share a common objective. In the context of SCM–NPD alignment, the common objective is to satisfy the client needs.

To reach the goal to align the two processes, two issues should be considered.

(i) Firstly, it is important to take into account the alignment of the main outcomes of the two processes: product design features and supply chain features. Product features are determined by the NPD process, whereas supply chain characteristics are within SCM decisions. Fisher (1997) claims that alignment between product features and supply chain strategy leads to operational performances maximization. The same goal is pursued by rightly matching product structure and supply chain configuration (Fine 1995).

(ii) Latterly, best practice initiatives carried out by companies such as Dell, Hp, and Zara (Feitzinger and Lee 1997; Ghemawat and Nuevo 2003) demonstrate that a deeper integration between NPD and SCM processes can be the source of a sustainable competitive advantage. Therefore, the issue of NPD–SCM alignment has two equally important dimensions to be explored:

- The alignment between the features of new products and the design of the supply chain;
- The integration between NPD process and SCM processes.

It should be noted that the first dimension is static and structural, since it defines the right matching among supply chain features and product features. The latter is more related to processes and management

New Products features and Supply Chain alignment

The basic assumption of this research is that the success of a new product is determined not only by its features and its quality, but also by the efficiency and the effectiveness of the supply chain in sourcing, making, and delivering the product to customers.

Literature review

A literature review shows that three are the main product features affecting NPD–SCM alignment (Pero and Sianesi 2009): variety, modularity, and innovativeness. Variety affects direct manufacturing costs, manufacturing overhead, delivery times, and inventory levels (Fisher et al. 1999; Miller and Vollmann 1985), or behavioral costs, i.e. "those costs which arise because of the reaction of people to "excessive" variety" (Brun et al. 2006). The right choice of the supply chain or operations management tools or strategy, e.g. flexible automated systems (Akturk and Yayla 2006), tools to enhance supply chain flexibility (Ramdas 2003), application of Variety Reduction Program (VRP) (Koudate and Suzue 1990), can mitigate this effect. Ramdas (2003) distinguishes between variety creation and variety transportation decisions. The first are referring to the definition of the structure of the new product lines, the latter are referring to how the supply chain for the new lines will be structured. Therefore, we can distinguish between the number of different products with their innovation content a company is planning to deliver (Created Variety) that results from variety creation decisions, and the Transported variety that refers to the number of different products a company actually delivers to the market. When the Created Variety equals Transported Variety and Customer Requirements, supply chain performances are maximized. This is reached when the variety creation decisions, i.e. basically NPD-related decisions, variety transportation decisions, i.e. basically SCM-related decisions, are aligned.

Modularity is widely recognized to be a strategy to increase variety while containing costs. To exploit the benefits of modularity, supply chains must be restructured. In the automotive industry, modularization of the product was followed by the "re-definition" of the role of the first tier suppliers (Doran et al. 2007; Doran 2004) that should have the capability to produce the entire system and also to coordinate the work of the other components' suppliers. Indeed, the modularity of the product facilitates outsourcing of value-added activities to suppliers (Doran 2004). Product modularity and interfaces standardization facilitate firms to separate manufacturing from product design activities, i.e. they are respectively performed by the contract manufacturer and the brand name firm (Sturgeon 2002). Finally, alignment between supply chain strategy and product innovativeness should be pursued to enhance supply chain performances. Moreover, Caridi et al. (2008) observed that the degree of innovativeness of new products affects supply chain features—i.e. supply chain structure and intensity of relations among partners.

Empirical evidences

To investigate the relationships among the above-mentioned product features (i.e. variety, modularity, and innovativeness), supply chain features, and supply chain performances, nine firms belonging to the Italian electro-mechanical industry have been studied. One or two innovation projects per firm have been analyzed along with the changes determined by the project in the supply chain. The way each supply chain feature changed after the introduction of the new products has been investigated. To describe the supply chain and measure the changes, Hieber (2002)'s model was chosen. Variety has been measured on a qualitative scale based on the number of new products introduced over the complete catalog. Modularity has been measured on a two-level scale based on the correspondence between functions and module. Finally, Innovativeness has been measured by identifying the position of each new product in the matrix by Weelwright and Clark (1992). Therefore the innovation projects have been divided into Derivative, Platform, and Breakthrough, depending on their innovation content in terms of product and production process. All the interviewed managers claimed that they had to change their supply chains, in response to the introduction of the new product, in order to reach the target supply chain performances. By comparing the case studies, it has been observed that the level of innovativeness affects supply chains more than variety, while modularity mitigates this effect. In particular, highly innovative products call for more complex supply chains, unlike modularity. In other words, a change in the complexity of the basic structure of the product (or line) has a weaker effect in determining the need to change the supply chain than a change in the innovation content of the product (or line). For example, one company, unlike the other cases of low variety products introduction that result in outsourcing decisions, decided to produce the new product internally because of the high innovativeness of the new product. Indeed, this company is planning, in the future, to shift outside the production of the new product when the organization and the supply chain have acquired the knowledge to manage the innovation. We also observed examples of the opposite situation, in which innovations with a low level of innovativeness have been introduced over a modular product: in these cases no changes in the supply chain were observed.

These results suggest that supply chain choices should be done taking into account not only the structural complexity of the product, but also the innovativeness of the product and so the knowledge required to develop and to deliver the innovation. Moreover they support the idea that supply chains must be aligned to new product features in order to reach target performance (Fig. 12.6).

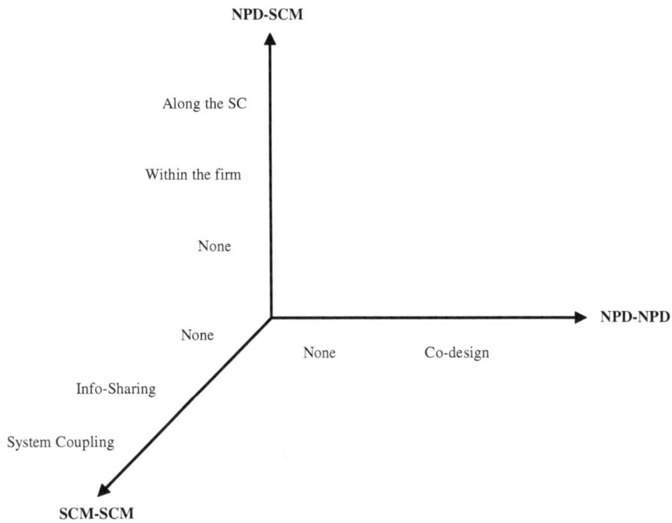

Fig. 12.6 Classification matrix (modified by Crippa et al. 2009)

12.2 Barriers in Implementation of DfE in the Supply Chain Management

Barriers should be taken into account during the implementation of DfE in the SCM. Based on the literature review (Jeganova 2004) it is possible to identify six main implementation barriers:

Insufficient Motivation Barriers

The motivation of senior board to Life Cycle Design issues is important and necessary (Gertsakis et al. 1997). The greater the interest and stronger motivation from managing directors and chief executives, the better and easier is the process of integration of environmental issues into product development. If senior board is aware of the goals of product life cycle design and admits the benefits from integration, then this strong commitment can become the most important driving force for successful implementation of Product Life Cycle Design (Pujari and Wright 1999). For example, senior board can provide enough resources for implementation of Life Cycle Design.

Leadership and Coordination Barriers

Strong leadership and coordination both in environmental issues and Life Cycle Design is the next important issue. Without good managing of coordination of Life Cycle Design Strategy, individuals will not be encouraged to actively participate in the implementation process. Coordination of the process should be established at early stages of implementation.

Top management support and involvement has a crucial impact on any major company initiative, and NPD is no exception. This is perhaps even more vital in environmental new product development (ENPD) since the integration of environmental concerns into the business process can be a major challenge to the existing culture and can require changes that will not occur without the clear leadership and active support of the company's top management. Previous ENPD research has suggested the importance of a senior manager or management committee in coordinating environmental activities (Pujari and Wright 1996) and the need for senior managers to communicate to the corporation that environmental concerns are ''everyone's job'' (Cahan and Schweiger 1993/1994). The NPD process depends upon the willingness of top management to commit resources to new projects. Creating greener products to address increasingly environmentally sensitive markets requires a significant investment in appropriate technologies and capabilities. Successful ENPD is unlikely without top management support, involvement, and resource commitment.

Cost Barriers

Integrating environmental issues into product development and building Life Cycle Design seems to require environmental investments which are usually very expensive. For example, a process of ISO 14001 certification, Life Cycle Assessment studies, external environmental consultancy, environmental investigations, and analysis are considered to be expensive activities. On the other hand, many environmental investments that are resource efficient may bring tangible financial returns in reasonable payback periods (Hawken et al. 1999). However, sometimes it is difficult to measure the benefits from these investments.

Skills and Knowledge Barriers

In many companies there is a lack of knowledge about environmental issues. Therefore, even if environmental objectives are established in the company, a problem could arise from a lack of knowledge and competence about environmental management and life cycle design (Ritzen 2000). A lack of competent human resources is one of the key problems, which leads to the lack of knowledge, information, and communication about Life Cycle Design. This results in a "wrong" attitude toward environmental issues.

Insufficient Internal and External Communication with Stakeholders

Sometimes there is no sufficient communication between the employees related to Life Cycle Design issues. It is necessary to take much time to make clear who is responsible for what and find the 'right' people to get the needed information. Internal communication system in the companies is rather weak in terms that there is no interdisciplinary cooperation between different departments in environmental issues. Information about customers and suppliers could not be summarized or not easily accessible, what does create the barriers in research or business projects, especially in the projects of environmental investigations.

Cooper (1994) argues that since product innovation cuts across traditional functional boundaries and barriers, it requires a cross-functional team approach. Recent research by Song and Parry (1997) has shown that cross-functional integration enhances the diffusion of market and customer knowledge among all members of the project team, not just during development, but also at later stages of test marketing and commercialization. Key elements of such a team would be drawn from marketing and marketing research, engineering, R&D, production, purchasing, and finance (Craig and Hart 1992). Cooper (1994) further concludes that studies of NPD success consistently cite the importance of interfaces between R&D and marketing, coordination among key internal groups, multidisciplinary inputs into the new product project, and the role of teams and team leaders.

For example, Langrish (1972) found poor cooperation or communication among different functional areas as one of the main factors that hampers product innovation. In case of ENPD, cross-functional coordination will be a key element because processes such as LCA, designing-in environmentally benign materials, and designing-out environmentally harmful materials, and environmental impact analysis will inevitably involve several functional areas. These functions are typically represented by design professionals, environmental professionals, and product managers. There is also some empirical work that demonstrates such coordination.

Sullivan and Ehrenfeld's (1992/1993) survey found that marketers were actively involved in physical LCA in environmentally active companies. They suggest that to foster environmental product innovation, life cycle work should be integrated with marketing practices such as market research and other management practices to encourage teamwork and to help identify product characteristics capable of satisfying customers and enhancing the firm's competitiveness.

Insufficient External Pressure

An important barrier is an insufficient external pressure from the customers to start Life Cycle Design. What is the main driving force to invest? The answer is customers' demands. But if customers do not ask for Life Cycle design and environmentally conscious products, will a company start producing 'green' products? Perhaps the answer is "no". This is the reason why some companies are rather skeptical to invest in integration of environmental issues into product design and development process. Moreover if the company's position in a SCM is not so close to the end users as, for example, Tetra Pak's position, the customers do not directly demand environmental information about products. However, because of an overall environmental concern, a concept of sustainable development, and a concept of SCM, the customers, retailers, suppliers, shareholders, and other stakeholders can start demanding environmental certification, information about environmental performance of its products, etc.

12.3 DfE and Supply Chain Stakeholders

According to Bakker (1995), the manufacturer can have strategic and reactive reasons to invest in the ecodesign of its products. Strategic reasons could be the company's corporate image. Reactive reasons can be competition or pressure from the stakeholders.

The complexities of environmental issues require that developers and designers of new 'green' products should research, involve, and learn from stakeholders

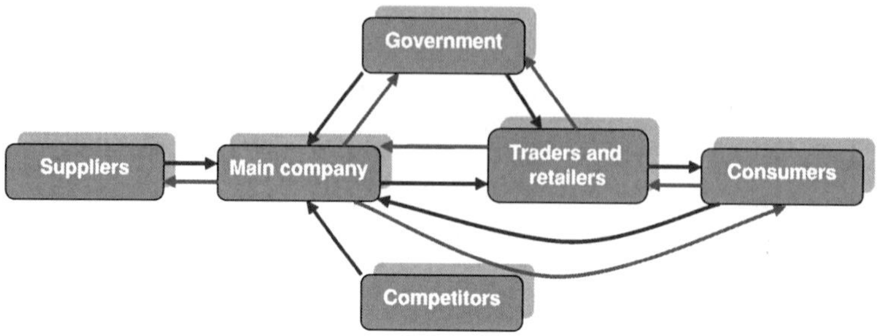

Fig. 12.7 Flow chart: a model of network of actors around product development process (modified by Jeganova 2004)

(Jeganova 2004). The stakeholders often have information that can really help in product life cycle design process in order to develop less environmentally harmful products (Polonsky et al. 1998).

In Fig. 12.7, a simple model of network of actors around the Product Development in a supply chain is shown. It is essential to distinguish 'players' (those actively involved in product development process) and 'stakeholders' (those that are not actively involved in product development process) (Bakker 1995). In many supply chain the players that are involved in product development are the main company itself. Other actors are stakeholders that are not involved in product development process directly: suppliers, trade and retail, customers, competitors, and government.

The blue arrows show the communication flow about environmental issues at main company. Red arrows show the communication flow that is missing at main company and is very essential in product development process (Ulrich and Eppinger 2000).

Customers

The results of product development process should satisfy the customer's needs. Then the product will be successful in the market. Therefore, the customer is the most important stakeholder for the main company who has a direct influence on the product development process. Usually the product is developed or redesigned according to the demands of the customers [personal interview]. Identifying customer's needs is an integral phase in product development process and is very closely related to concept generation, concept selection, competitive benchmarking, and the establishment of product specifications (Ulrich and Eppinger 2000). More and more customers started to demand information about environmental performance of products: harmful substances, energy consumption in usage phase, and other environmental impact. For example, food and beverage producers are very interested in environmental impact of heat exchangers when they use them. It became common that customers demand that their suppliers should have ISO 14000 certificate or participate in EMAS (Barthel 1999). Therefore, it is necessary to present environmental information about the products. Communication with different stakeholders is an effective way to elaborate or enlarge environmental criteria for the products (Colman 2004). Construction companies in Sweden that started to build "environmentally friendly" green houses demand that all the materials, equipment, and appliances they use for building a house and putting to the house should have an ecolabel (Jeganova 2004). Higher life style standards create more demand for ecoproducts. The positive attitude toward environment of main company can play a very important role and influence customers' attentions toward the issue, and vice verse (Henriques and Sadorsky 1999). Market demand for environmentally responsible products has forced companies to consider the environment as a core business issues. Product design strategies that reduce environmental impacts and costs will provide the greatest potential for producers to meet rising consumer expectations (Keoleian et al. 1995).

Suppliers

The environmental performance of the product is closely connected to the materials and energy flows that characterize its life cycle. Automobiles and computers are assembled from many different elements and usually are manufactured by one set of suppliers and recovered or recycled by other suppliers (Fiksel 1996). Therefore, it is essential in life cycle design of the product to set supplier selection criteria and analyze the role of different suppliers. In order to fulfill the principles of life cycle design, it is necessary to have environmental information about raw materials the company receive and the way the suppliers extract these materials for the company. Now the current trend is that suppliers are also becoming co-makers. They supply the manufacturer new technological possibility or develop products or components themselves (Bakker 1995). Nowadays major multinational companies, for example Renault, include not only product developers in ecodesign process but also all component and material suppliers. Renault's strategy vis-à-vis its suppliers is based on long-term relationships. The strategy involves the suppliers in the project at a very early stage of development (Renault's Performance 2002). SCM is really a powerful instrument to influence the companies to improve their environmental performance. Effective and open communication with suppliers or substantial influence on their activities can reduce the environmental burden of many product systems (Keoleian et al. 1995).

Traders and Retailers

The intermediate trade can demand information about the products they buy from manufacturers. Moreover, the intermediate trade can make a choice between different alternatives of the products in order to build up an assortment (Bakker 1995). In the case of Alfa Laval, for example, Tetra Pak AB is considered as the intermediate trade (Jeganova 2004). For instance, Tetra Pak AB provides "integrated processing, packaging, and distribution line and plant solutions for food manufacturing. Tetra Pak's customers get multi-product solutions from a single source, with matching equipment at every stage". It means that Tetra Pak buys products from Alfa Laval in order to integrate it to the system and provide to its customers. Therefore, it is very important for Tetra Pak to have environmental information about products because Tetra Pak works closer to the end users of the product who started to demand environmental information about all products from Tetra Pak.

Poor internal and external communication within a company is the biggest barriers to "selling 'eco-design' (internally) or 'eco-products' (externally)" (Charter and Belmane, JSPD, 1999). Generally, product ecodesign has been an isolated activity within a company. Mostly companies try to elaborate the environmental design for production phase: reduce energy consumption in production, reduce amount of emissions, etc. However, the companies should also consider use phase, distribution phase, and disposal phase.

Environmental and Consumer Organizations, Government

Environmental, consumer organizations, and government can influence product development and ecodesign of the product directly and indirectly (Bakker 1995). Government can influence directly by legislation. In the United States, Europe, or Japan, producers are aware of increasing regulations toward developing cleaner processes. For example, German packaging law excludes from the German market the products which packaging cannot be eliminated, reused, or recycled. In the United States, pollution prevention plans is a requirement for many medium-sized and large enterprises (Cattanach et al. 1995). Environmental and consumer organizations usually demand environmental information more than the intermediate trade and customers.

In the case of chemicals in the United States, Environmental Protection Agency wanted to increase information to users of Toxics Release Inventory (TRI) chemicals[1] by expanding Material Safety Data Sheets to include environmental hazards,[2] or demanding manufacturers to provide "product stewardship" information to their customers (Hanson 1990). In Europe, requirements about providing a "product environmental profile"[3] to the customers are being explored (U.S. Congress, Office of Technology Assessment 1992). Additionally, consumer organizations are interested in product testing whose results are published. Product test results can influence consumer behavior, manufacturer motivation to improve environmental performance of the products, and retailer's strategy to promote the product (Bakker 1995). Cooperation with environmental organizations may result not only in partnership but also in technology development (Fineman and Clarke 1996; Hartman and Stafford 1997). For example, the Danish Railway cooperated with O_2 organization[4] to design more environmentally friendly and cost-effective S-trains. Another example is Foron, German company that cooperated with Greenpeace to produce the Greenfreeze line of refrigerators (Polonsky et al. 1998).

[1] The Toxics Release Inventory (TRI) is a publicly available Environmental protection Agency database that contains information on toxic chemical releases and other waste management activities reported annually by certain covered industry groups as well as federal facilities.

[2] Required by US Occupational Safety and Health Administration.

[3] A product profile is a qualitative description of the life cycle environmental impacts of a product, intended to use by professional buyers, rather than individual consumers (U.S. Congress, Office of Technology Assessment 1992).

[4] A non-profit international network of ecological design professionals.

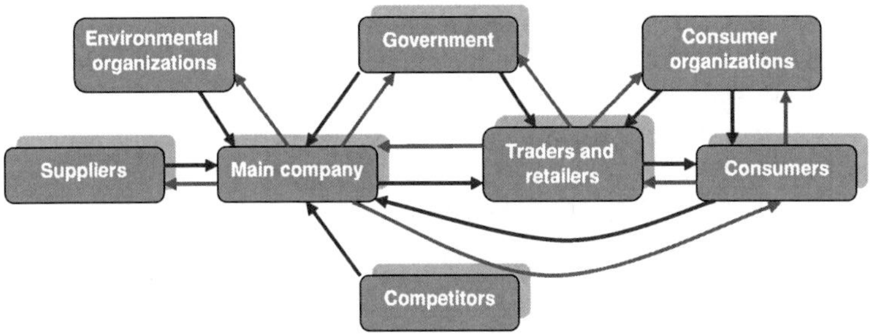

Fig. 12.8 Flow chart: an extended model of network of actors including environmental and consumer organizations around product development process

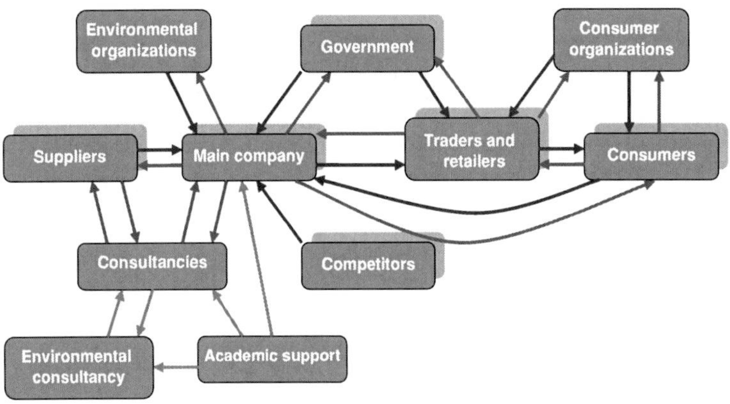

Fig. 12.9 Flow chart: an extended model of network of actors including consultancy and academic support in product development and life cycle design process

Competitors

Competition is an important motivator for change in the company (see Fig. 12.8). If competitors take environmental issues seriously, this can influence the company as well to produce greener products. At the same time, if competitors do not consider environmental issues, then the company has a comparative advantage in greening its products and having a reputation of environmentally friendly company with sustainable products.

Consultancy and Academic Support

In Fig. 12.9, an extended model is shown. In this model all the necessary stakeholders for life cycle design process are involved. A company can use both external and internal environmental consultants in ecodesign process. External environmental consultant works on client contracts in the following areas: water pollution, land and air contamination, environmental impact assessment, environmental audit, waste management, environmental policy, ecological/land management, noise and vibration management, and environmental management (AGCAS 2004).

Design process decisions should not be taken in isolation but in cooperation with all ecodesign consultants (external and internal). Industrial designers after the training in environmental design usually become internal environmental consultants in product ecodesign process (Gangemi et al. 2000). A company can also establish a Competence Team of environmental experts that will develop and facilitate product ecodesign process from the beginning till the end. This Competence Team should be interdisciplinary which means that the Team should involve different specialists.

For example, Environmentally Conscious Products Program at IBM is supported by its Engineering Center for Environmentally Conscious Products which is a Center of Competence for DfE activities. The Centere has a huge division of environmental specialists, product development and procurement engineers, suppliers, and product recycling centres (IBM Environment 2004). The Center works with Product Environmental profiles processes, monitors and documents environmental characteristics of the products. Additionally, The Competence Team or Centre usually cooperates with academies and research centrers to develop new approaches for improvement of environmental performance of products. However, for small- and medium-sized companies the role of internal industrial designer in ecodesign process is very essential.

References

AGCAS, Environmental consultant (online). http://www.prospects.ac.uk/downloads/occprofiles/profile_pdfs/R4_Environmental_consultant.pdf (Aug 2004)

Akturk M, Yayla H (2006) Management of product variety in cellular manufacturing systems. Int J Flex Manuf Syst 17:93–117

Bakker C (1995) Environmental information for industrial designers. Technische Universiteit Delft, Rotterdam

Barthel M (1999) Greening the supply chain. British Standards Institution, London

Brezet JC, Horst Tvd, Riele HRM, Duijf GAP, Haffmans SP, Böttcher HE, Hoo SC, Zweers A, Verkooyen H (1994) PROMISE Handleiding voor Milieugerichte Produkt Ontwikkeling (PROMISE Manual for ecodesign), SdU Uitgeverij, Den Haag, The Netherlands

Brun A, Capra E, Miragliotta G (2006) Behavioural costs in manufacturing: how to balance standardization and variety costs. In: Proceedings of XIV international working seminar on production economics, Innsbruck, Austria, 20–24 Feb

Cahan J, Schweiger A (1993/1994) Product life cycle—the key to integrating EHS into corporate decision making and operations. Total Qual Environ Manag, pp 141–150 (Winter)

Caridi M, Pero M, Sianesi A (2008) The impact of NPD projects on supply chain complexity: an empirical research. In: Proceedings of the exxpand, Bordeaux (France), 20–21 Mar 2008

Cattanach RE, Holdreith JM, Reinke DP, Sibik LK (1995) The handbook of environmentally conscious manufacturing: from design and production to labeling and recycling. Richard D. IRWIN, INC, Chicago

Charter M, Tischner U (2001) Sustainable product design. In: Sustainable solutions. Greenleaf Publishing Ltd, UK (ISBN 187419365)

Charter M, Belmane I (1999) Integrated product policy (IPP) and eco-product development. J Sustain Prod Design

Colman R (2004) Satisfied stakeholders, CMA management, 78, 1, ABI/INFORM Global p 22, March 2004

Cooper RG (1994) New products: the factors that drive success. Int Mark Rev 11(2):60–76

Coulter S, Bras B (1997) Reducing environmental impact through systematic product evolution. Int J Environ Conscious Des Manuf 6(2):1–10

Coulter S, Bras B (1999). Decision support for systematic product evolution. In: Proceedings of the 1999 ASME design engineering technical conferences, ASME, DTEC99/DTM-8747

Craig A, Hart S (1992) Identifying the major themes in NPD research. Eur J Mark 26(11):13–15

Crippa L, Pero M, Sianesi A (2009) New product development and supply chain management alignment. XIV Summer School 'Francesco Turco'

Den Haag Brezet H, Van Hemel C (1997) Eco-design: a promising approach to sustainable production and consumption. UN Environment Programme (UNEP), Paris

Doran D (2004) Rethinking the supply chain: an automotive perspective. Supply Chain Manag Int J 9(1):102–109

Doran D, Hill A, Hwang K, Jacobs G, Operations research group (2007) Supply chain modularisation: cases from the French automobile industry. Int J Prod Econ 106(1):2–11

Feitzinger E, Lee HL (1997) Mass customization at Hewlett Packard: the power of postponement. Harv Bus Rev 75(1):116–121

Fiksel J (1996) Design for environment: creating eco-efficient products processes. McGraw-Hill, New York

Fine C (1995) Clockspeed: winning industry control in the age of temporary advantage. Perseus Book, New York

Fineman S, Clarke K (1996) Green stakeholders: industry interpretations and response. J Manag Stud 33(6):715–730

Fisher M (1997) What is the right supply chain for your product? Harvard Business Review, vol. March–April

Fisher M, Ramdas K, Ulrich K (1999) Component sharing in the management of product variety: a study of automotive braking system. Manag Sci 45(3):297–315

Gangemi V, Malanga R, Ranzo P (2000) Environmental management of the design process. Managing multidisciplinary design: the role of environmental consultancy. Pergamon Renew Energy 19:277–284

Gertsakis J, Lewis H, Ryan C (1997) A Guide to EcoReDesig: improving the environmental performance of manufactured products. Centre for Design at RMIT, Melbourne

Ghemawat P, Nuevo JL (2003) Zara: Fast Fashion. Harvard Business School Case

Hanson D (1990) EPA develops product stewardship, hazard communication regulations, chemical and engineering news, 19 Nov 1990

Hartman C, Stafford E (1997) Green alliances: building new business with environmental groups. Long Range Plan 30(2):184–196

Hawken P, Lovins A, Lovins H (1999) A road map of green capitalism. Harvard Business Review, May–June

Henriques I, Sadorsky P (1999) The relationship between environmental commitment and managerial perceptions of stakeholder importance. Acad Manag J 42(1):87–99

Hieber R (2002) Supply Chain Management—a collaborative performance measurement approach. VDF Hochshulverlang ag an der eth Hjelm O, Lundgren J, Idegren L. Miljö ledd Produktutveckling. Ett Projekt i Skä rningspunkten mellan Miljö ledning och Miljö anpassad Produktutveckling. Slutrapport till NUTEK, Linkö ping University, IVF; 2001 (in Swedish)

IBM Product Stewardship (2004) (Online). http://www.ibm.com/ibm/environment/annual2000/product.shtml. (Aug 2004)

Jeganova J (2004) Product life cycle design: integrating environmental aspects into product design and development process at Alfa Laval. LUMES thesis, Lund University International Masters Program in Environmental Science

Keoleian GA, Koch JE, Menerey D (1995) Life cycle design framework and demonstration projects. Environmental Protection Agency, EPA/600/R-95/107, July

Koudate A, Suzue T (1990) Variety reduction program: a production strategy for product diversification. MA Productivity, Cambridge

Langrish (1972) quoted in Barclay I, Benson, M. Success in new product development: the lessons from the past. Leadership Organ Dev J 1990; 11(6):4–12

Martin M, Ishii K (2000) Design for variety: a methodology for deveioping product platform architectures. In: Proceedings of the 2000 ASME design engineering technical conferences, ASME, DTEC2000/DFM-14021

Maxwell D, van der Vorst R (2003) Develop sustainable products and services. J Clean Prod 11:883–895

Miller J, Vollmann T (1985) The hidden factory. Harv Bus Rev 63(5):142–150 Sept–Oct

Otto K, Wood K (1998) Product evolution: a reverse engineering and redesign methodology. Res Eng Des 10(4):226–243

Pero M, Sianesi A (2009) Aligning supply chain management and new product development: a theoretical framework. Int J Electr Cust Relatsh Manag 3(3):301–317

Polonsky MJ, Rosenberger III, PJ, Ottman JA (1998) Developing green products: learning from stakeholders. J Sustain Prod Design, pp 7–21, April

Prahalad C, Krishnan MS (2008) The new age of innovation: driving concreated value through global networks. McGraw-Hill, New York

Pujari D, Wright G (1996) Developing environmentally-conscious product strategy (ECPS): a qualitative study of selected companies in Britain and Germany. Mark Intell Plann 14(1):19–28

Pujari D, Wright G (1999) Management of environmental new product development. In: Charter M, Plonsky M (eds) Greener marketing—a global perspective on greener marketing practice. Greenleaf Publishing Ltd., Sheffield

Ramdas K (2003) Managing product variety: an integrative review and research directions. Prod Oper Manag 12(1):79–101

Renault's Performance 3 in 2002 (online). http://www.renault.com/docs/finance_gb/re2002_08_Performance2002.pdf. (Aug 2004)

Ritzen S. (2000), Integrating environmental aspects into product development—proactive measures. Ph.D. thesis, Department of Machine Design, Royal Institute of Technology, Stockholm

Simchi-Levi D, Simchi-Levi E, Kaminsky P (2003) Designing and managing the supply chain, 3rd edn. McGraw-Hill, Boston

Song XM, Parry ME (1997) A cross-national comparative study of new product development processes: Japan and the United States. J Mark 61(2):1–18

Sturgeon T (2002) Modular production networks: a new American model of industrial organization. Ind Corp Chang 11(3):451–496

Sullivan MS, Ehrenfeld JR (1992/1993) Reducing life-cycle environmental impacts: an industry survey of emerging tools and programs. Total Qual Environ Manag, 143–57 (Winter)

U.S. Congress, Office of technology Assessment (1992) Green product by design: choices for a cleaner environmental. U.S. Government Printing Office, OTA-E-541, Washington

Ulrich KT, Eppinger SD (2000) Product design and development. Irwin McGraw/Hill, New York

Van Hoek R, Chapman P (2006) From tinkering around the edge to enhancing revenue growth: supply chain-new product development. Supply Chain Manag An Int J 11(5):385–389

Weelwright S, Clark K (1992) Creating plans to focus product development. Harvard Business Review, Mar–Apr, pp 70–82

Chapter 13
Methods for Weighting DfE Choices in the Development of a More Sustainable Supply Chain

13.1 Introduction

In the practice of ecodesign, life cycle assessment (LCA) provides the basic modeling framework for evaluating the environmental load and impact throughout the entire product life cycle from material acquisition to disposal (Wenzel et al. 1997). LCA tools will be playing an increasingly important role in communication between manufacturing companies and their stakeholders, such as in the case of eco-labeling or green procurement. However, it is difficult to integrate the environmental, quality, and cose aspects of a product simultaneously when only LCA results are applied to the improvement of a product.

On the other hand, the importance of ecodesign in earlier phases has been emphasized because decision made in these phases greatly affect the environmental impact throughout the product life cycle (Frei and Zust 1997). Conventional approaches in the early design phases, such as quality function deployment (QFD) (Clausing 1993), consider the usage phase of the product. In order to consider the entire life cycle and fulfill customer satisfaction, an ecodesign methodology must be incorporated into the conventional methods in the early phases of design.

In QFD, customer requirements are incorporated throughout the product development process (Clausing 1993). In order to reflect the customer requirements in the product specifications, QFD matrices are utilized. QFD clarifies the relationship between customer requirements and quality characteristics, and the relationship between quality characteristics and components is clarified.

QFD belongs to the sphere of quality management methods, offering us a linear and structured guideline for converting the customer's needs into specifications for, and characteristics of new products and services. The method involves developing four matrixes, or 'houses', that we enter by degrees as a project for a given product or production process is developed on increasingly specific levels. In this book, our attention focuses on the *planning matrix*, or (*HOQ*) (Fig. 13.1).

M. Bevilacqua et al., *Design for Environment as a Tool for the Development of a Sustainable Supply Chain*, DOI: 10.1007/978-1-4471-2461-0_13,

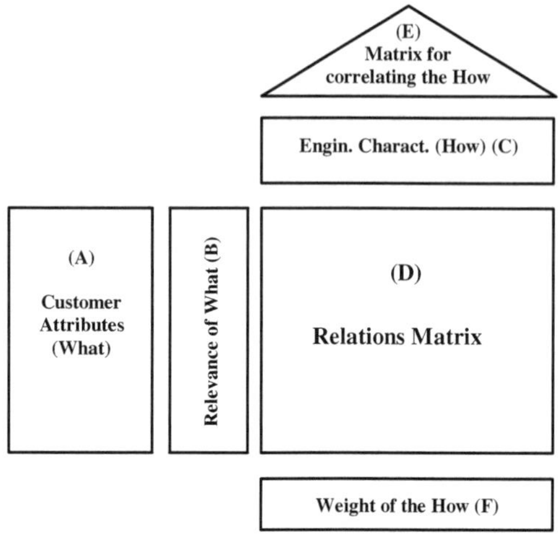

Fig. 13.1 House of quality

The HOQ provides the specifications for product design (or *engineering characteristics*) in terms of their relative importance and of target values that have to be reached in design and production. In a sense, the HOQ is the hub of the whole QFD method: its construction enables us to proceed from the customer's requirements to the design specifications (Smith 1997, 2002). Many authors used QFD technique to integrate DfE and LCA methods. Cristofari et al. (1996) proposed a new methodology using QFD and LCA to document technical requirements. To address this issue, Cristofari et al. (1996) developed the method of Green QFD (GQFD), in which LCA and QFD are combined to evaluate different product concepts.

Hanssen et al. (1996) and Ferde et al. (1995) applied QFD, LCA, and LCC separately for environmentally sound light fittings, but did not form a systematic methodology that could integrate QFD, LCA, and LCC into an efficient tool. Zhang et al. (1999) proposed a new methodology by integrating LCA, LCC, and QFD into an efficient tool that deploys customer, environmental and cost requirements throughout the entire product development process.

Product life cycle simulation (LCS) techniques have been proposed to evaluate the environmental burden and revenue of a company caused by single or multiple product life cycles from a medium-term or long-term viewpoint (Umeda et al. 2000; Murayama et al. 2001). LCS is useful for evaluating the business strategies or modular architecture for an eco-product. However, if the number of components or materials constituting the product is too large, calculations cannot be executed in a practically feasible time because the number of possible combinations of life cycle options increases exponentially.

Zhang et al. (1999) proposed the Green Quality Function Deployment-II (GQFD-II). By integrating life cycle costing (LCC) into QFD matrices and deploying quality, environmental, and cost requirements throughout the entire product development process, GQFD-II elaborates the original GQFD, in which

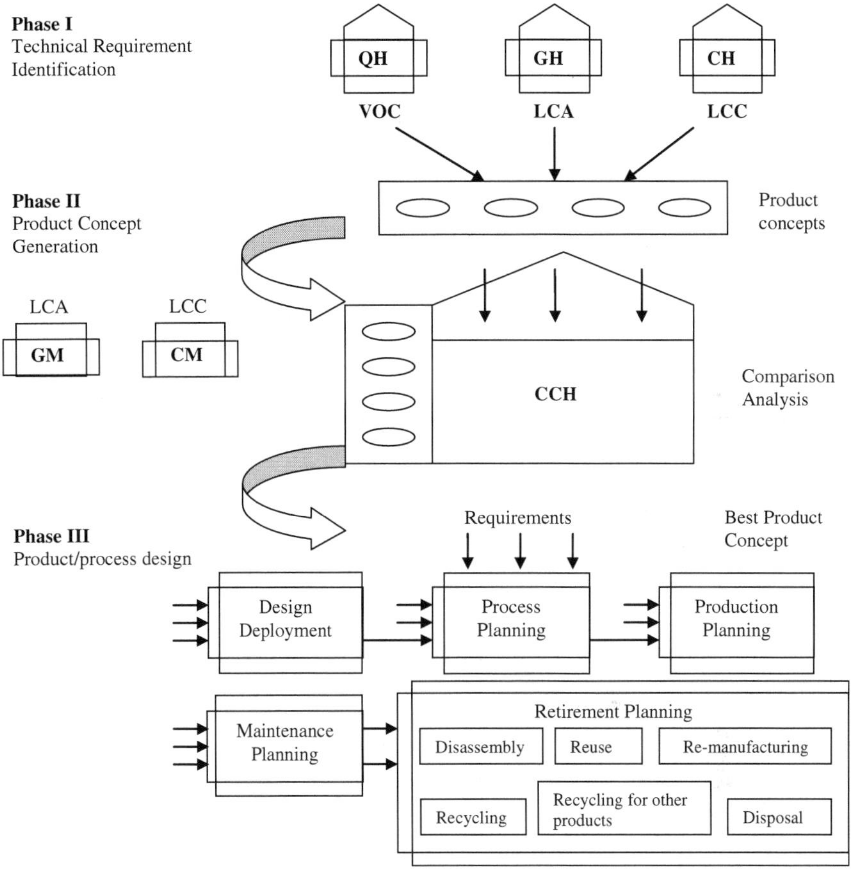

Fig. 13.2 Flowchart of GQFD-II

LCA and QFD are combined to evaluate different product concepts. Figure 13.2 shows the flow chart of GQFD-II method.

GQFD-II includes three major phases. Phase I—Technical Requirement Identification. quality house (QH), green house (GH), and cost house (CH) are established in this phase, where customer, environmental, and cost requirements are established and documented.

Figure 13.3 depicts the structure of GH matrix.

Phase II—product concept generation. A series of product concepts are generated to satisfy the requirements established from Phase I. These concepts can be evaluated with respect to quality, environment and cost. The best product concept is then selected.

Phase III—product/process design. In this phase, the requirements from previous phases are deployed into all product/process design stages, so that a series of

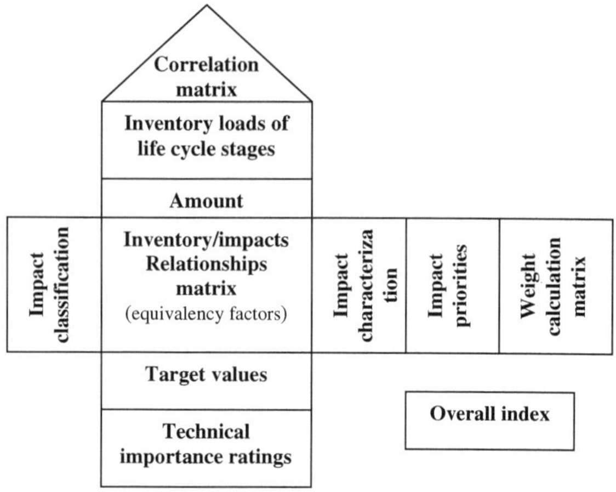

Fig. 13.3 Green house (GH)

matrices can be established, including: design deployment, process planning, production planning, maintenance planning, and retirement planning.

More recently Kobayashi (2005) proposed a method for efficiently planning product life cycles by using quality function deployment and life cycle assessment data. The methodology provides a systematic procedure consisting of the following stages to establish an ecodesign concept (Fig. 13.4). At the first stage, a plan for medium- or long-term production and a collection plan for products are clarified based on the business requirements and product lifetime. The designer creates a rough image of the material flow cycle of the target product in the product family. At the next stage, the specifications of a product and its life cycle are set. The product specifications for fulfilling customer requirements are defined using a QFD-I matrix; the customer requirements are connected to the quality characteristics, including factors related to functions and performance. Reconciliation of the differences between customer requirements and environmental requirements lead to adjustment of the target values for the quality and environmental characteristics. The target price and cost values are given based on the company's methods for marketing and target costing. Then, a new ecodesign concept at the product level is created and the created new product concept is evaluated from the environmental, quality, and cost aspects. Next, solution ideas at the component level are generated which is supported by various analytical methods. For example, the QFD-II matrix and cost-important analysis support customer-oriented solution ideas. Meanwhile, some new analysis methods for considering specific life cycle options, including valve degradation analysis, upgradability analysis, maintainability analysis, reusability analysis, and recyclability analysis, help generate ecosolution ideas. As a result of the idea, generation process mentioned above, an ecodesign concept for the target product is established at component level, and the created concept is evaluated in the same manner at product level.

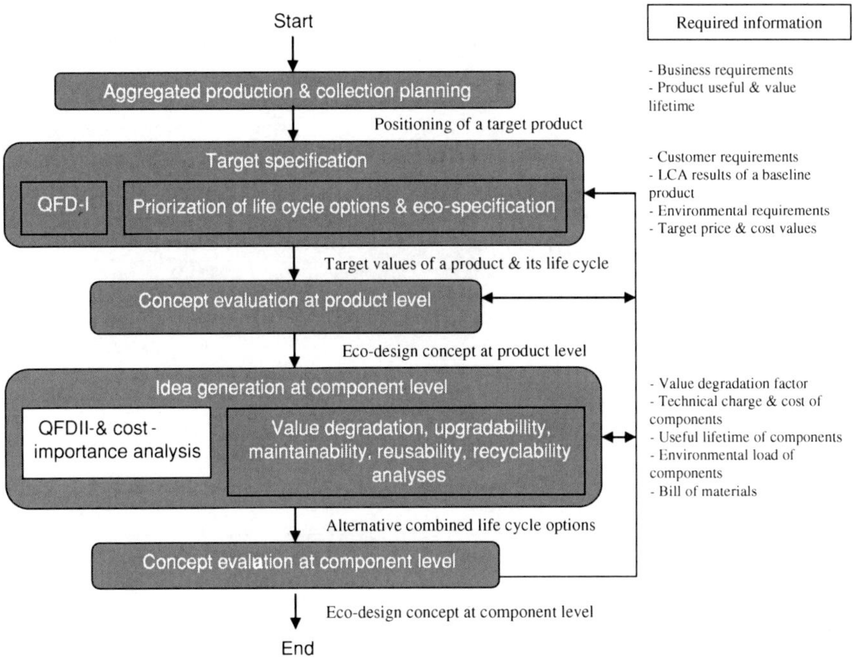

Fig. 13.4 Flowchart of the product LCP process

13.2 Development of a New Method

The process which goes from the design of a machine to its sale can be divided into five main stages: project definition, concept development, prototype development, field test and commercial launch.

For each of these stages there are various activities which involve different aspects of design development (technical, legal, economic, environmental, etc.). It is important to insert environmental considerations in all the design development stages, from the moment the original idea is generated up to the time of launching the product in the market.

In this way the targets that a firm must set for itself are:

- to guarantee environmental performance of the product for all the stakeholders
- to ensure respect of environmental law
- to control that its own environmental performance is coherent with the firm's standards and strategy
- to minimize the environmental impact of the product without compromising other external obligations (cost, safety, functionality)
- to promote the environmental certification of the product

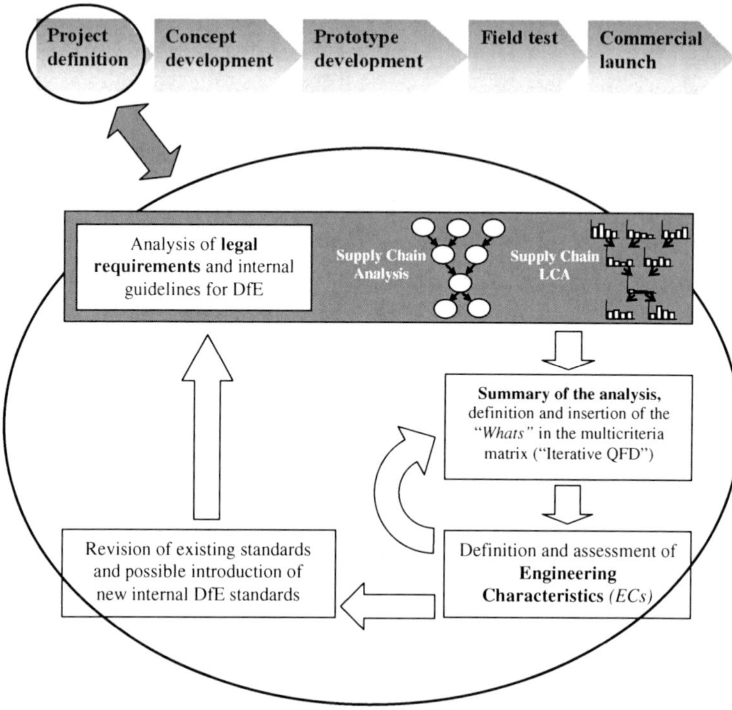

Fig. 13.5 Steps in machine manufacturing design

One critical aspect in machine construction is the understanding of which design choices and parameters mostly influence, from an environmental point of view, the whole supply chain in which the machines are involved. The aim of this study is to propose a solution by redefining the activities which are carried out at the first stage of the machine manufacturing process: project definition (Fig. 13.5).

Starting input at the *project definition* stage can be reformulated as follows:

- an analysis of legal requirements and internal guidelines for design of environmental;
- an analysis of the supply chain in which the machine is involved;
- an LCA study of the whole supply chain in order to understand the contribution of each stage in the production chain to environmental impact. The output of this step is the definition of the most important recommendations (*Whats*) for the supply chain.

The following steps concern the development of several matrices the levels of which are increasingly detailed from the point of view of technical specifications (Fig. 13.6):

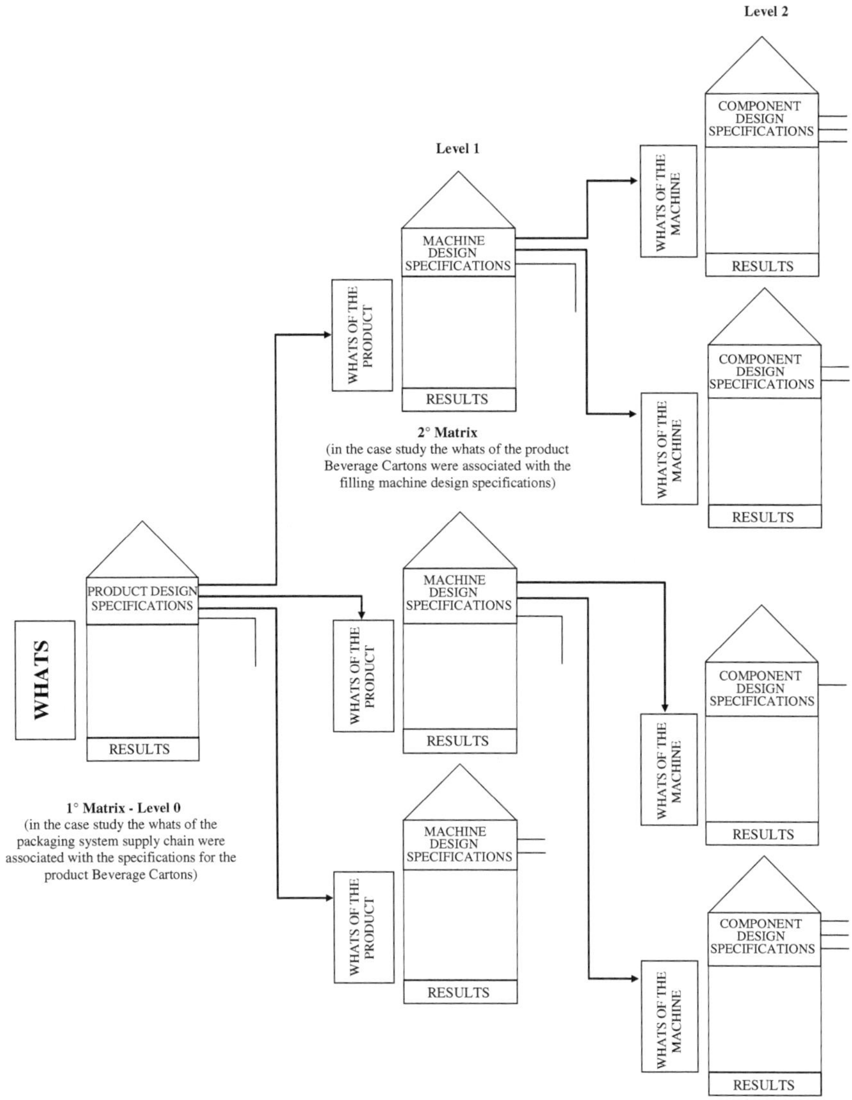

Fig. 13.6 Development of the multicriteria matrix

- the *Whats* are transformed into basic engineering characteristics (ECs) of the product in the supply chain. The ECs are the technical answers to the *Whats* requirements;
- preparation of the relationships matrix "iterative QFD". The team of experts must assess which ECs have an impact on the *Whats* and to what extent;

- the weights of the ECs are fixed at the bottom of the matrix. They represent the key result of the first matrix and the input of the second matrix: the most important requirements of the product "engineering characteristics" in turn will become the recommendations *(Whats)* which by means of suitable matrices will be implemented to the machines used in manufacturing the product itself;
- the use of a cascade of matrices therefore allows the transformation along the whole supply chain of environmental requirements into engineering characteristics which will then be implemented on the machines that are used at the various stages of the supply chain;
- on the basis of the indications provided by the QFD matrices the final stage is that of revising and in some cases of introducing DfE procedures into the machinery;
- the output of the second matrix could in turn become the input of a third matrix which would define the engineering characteristics of the machine components. Suppliers of machine suppliers are involved in this step.

The people involved in the design process of a machine are aware of their role, but they do not have a thorough understanding of the supply chain as a whole, with a consequent weakness in communication between the parties, which makes it impossible to organize environmental improvement plans. To overcome these obstacles and to carry out the project definition activities a panel of experts was formed in order to encourage communication and meetings where the operators and suppliers could contribute their knowledge and information about the processes.

The engineering choices and characteristics, which result from the *"project definition"* stage, are finally reconsidered and verified in the other stages of machine development.

The *"concept development"* stage involves:

- assessing to what extent the environmental requirements fixed by the law and by company policy can really be reached.
- assessing the environmental performance of the product using the tools described by "corporate standards".
- verifying whether it is possible to satisfy environmental targets and to conform with corporate standards and restrictions in the use of dangerous substances.
- documenting the results

The *"prototype development"* stage involves:

- assessing the impact of any changes made.
- optimizing the environmental performance of the product.
- testing and verifying whether the environmental targets and the standards which were considered attainable in the previous stage have been satisfied and whether the risks have been eliminated.

The *"field test"* stage involves:

- optimizing the environmental performance of the final product.
- checking conformity to the laws and to the environmental policy of the firm.
- checking what has been tested and verified during the final steps of the previous stage.
- documenting the results obtained.
- including the environmental information in the technical manuals and sales information.

The *"commercial launch"* stage:

The aim of the design for environment project is to develop the machine using design criteria which respect the environment and to provide clients with information about the environmental performance of the machine. The procedure for certifying this information has been defined on the basis of some reference standards already present on the (European) market, such as the Environmental product declaration (EPD) and ISO 14025.

To assess the validity of the proposed approach a specific supply chain has been considered in the next case study: packaging systems for liquid food substances (beverage cartons).

13.3 Case Study: Packaging Systems for Liquid Food Substances

Two methods have been used in this work to define the most suitable DfE choices for minimizing the environmental impact of a supply chain: QFD, a qualitative method for obtaining the opinion of experts and LCA, a quantitative method for assessing environmental loading.

The quality function deployment (QFD) matrices used in this study have been adapted from the more general QFD matrix, used as a quality control technique for activities and typical features of a product through a series of subsequent steps.

LCA methodology was chosen to identify the stages of the supply chain which are the most critical from the point of view of environmental impact.

The papers available in the literature only take into account a single product rather than all the stakeholders involved in the product life cycle. The procedure proposed in this work, using QFD matrices in series, allows the definition of guidelines for the whole supply chain. This is useful in order to provide "engineering characteristics", information about the manufacturing cycle and about machinery design for decision makers who are involved in the supply chain.

Each stakeholder in the supply chain (clients, manufacturers, suppliers and suppliers of the suppliers) can reduce the environmental impact of a supply chain by making appropriate DfE choices. The contribution of the various stakeholders involved in the production chain must however be guided in order to define the

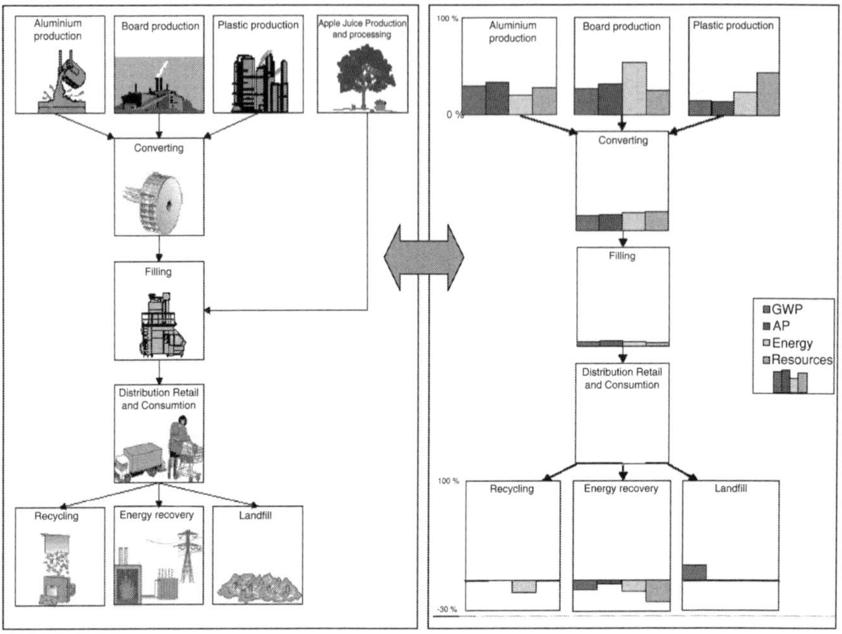

Fig. 13.7 Tree diagram showing the environmental contribution of the various processes

design choices which will have most effect on the overall environmental impact. This method also tries to arrive at an "environmental compromise" which is compatible with the various requirements of the stakeholders involved in the supply chain: suppliers, manufacturers, and clients.

This study focuses on the supply chain for beverage cartons. The beverage carton is the most-used container in Europe for packed juice and milk. In Europe, every day some 100 million packages of liquid, milk, juice, wine, water, soup, etc. are prepared in beverage cartons (Huat 2005).

This type of container protects sensitive products against the effects of oxygen, bacteria and light. Moreover, it reliably protects liquid foods using a minimum of materials.

An LCA study should specify the functions of the system under study. The 'functional unit' measures the performance of the functional output of the product system. The main function of the packaging of beverage cartons is to contain the product; while marking the contents and increasing visual appeal are less important. Therefore, 'g/l' (grams of packaging materials/per litre beverage) was selected as the 'function unit' in this case. The beverage densities are all assumed to be equal to 1 g/ml. A 1 l beverage carton weighs 32 g. The material used for beverage cartons is made up of about 76 wood fibre, 18 polymers and 6% aluminum.

The impact categories used in the study and the "weighting" criteria for the types of impact are referred to the Ecoindicator method. Figure 13.7 shows the tree diagram of the environmental contribution characterizing the various processes which a carton undergoes (the case study refers to packaging for fruit juice) In short the steps involved are:

- *Aluminum production*: aluminum oxide is derived from bauxite extraction and is then transformed into aluminum through an electrolytic process. The aluminum is then laminated and transported to the converting factory.
- *Paper production*: as is well-known the process is based on wood processing.
- *Plastic production*: the plastic material used is LDPE (Low Density Polyethylene) coming from natural gases and petroleum, which when appropriately treated (refining, cracking, polymerization, etc.) make up the LDPE flake.
- *Beverage carton production*: this consists of assembling of aluminum foil, paperboard, and polyethylene through a process of molding and roll forming; the rolls are then delivered directly to the clients using overland transport.
- *The filling process*: this takes place directly on the client's premises. The cartons are made up from the rolls of packaging material and then filled
- *Distribution, sale and consumption*: the cartons are assembled in cardboard packaging then palletized and delivered directly to the retailer.

For the final stage in the life cycle three possible scenarios have been taken into consideration:

- *Recycling*: beverage cartons are recyclable. Carton recycling has been growing across Europe since 1992 and now exceeds 250,000 tonnes annually. This equates to 27% of all cartons placed on the market. Through dedicated paper mills cartons are recycled into other useful products such as tissue and kitchen paper, office stationary, fiberboard, and reel cores for industrial use. The non-fibrous element of cartons is recovered through waste-to-energy plants or extrusion into new products.
- *Energy recovery*: this derives from the incinerating of the beverage cartons taken overland to specific sites. Cartons have high energy content, making them very suitable for incineration with energy recovery. In the European Union, in 2000, 25% of all beverage cartons were used in waste-to-energy plants to produce steam and electricity.
- *Landfill*: here the remains of the cartons decompose with a partial release of methane gas (belonging to the group of greenhouse effect gases).

The results of the LCA study are reported as percentages for the impact categories studied (GWP, AP, Energy and Resource).

Analysis of the results of the LCA study has allowed the relative importance of the various steps in the production line to be estimated, and some useful observations to be made as follows:

- the production of raw materials is responsible for the main contribution in the life cycle of the system studied (aluminum, paperboard, and plastic);

- the coupled carton production (converting) stage is equally relevant from the environmental point of view, in particular because of the waste coming from packaging material processing;
- during the beverage packaging step energy consumption, principally in the form of electricity, together with the AP parameter (acidification potential), prevail over the other types of environmental impact. Moreover neither the waste of packaging material, which accounts for about 25% of the total environmental impact at this stage, nor the loss of the product (in this case fruit juice) should be underestimated;
- the contribution arising from the product distribution, retail, and consumption stage is mainly caused by the transport and double or triple packaging of the product;
- recycling and energy recovery have a "positive" impact on the environment;
- landfill disposal is relevant as far as GWP (Global Warming Potential) is concerned due to the quantity of methane gas released during the decomposition of the packaging material;

13.4 Development of "Iterative QFD"

The methodology used in this study involves the development of two "QFD" matrices, which by degrees go more and more deeply into the design project and into the engineering processes with increasing levels of specificity.

For this study a panel of ten participants was formed, which included two academics, whose research studies are mainly focused on environmental management, three technical operators and three managerial operators involved in the design and production processes of the filling machines and two suppliers of filling machine components. The number of participants, which at first sight may seem rather large, derives from the Delphi technique (1975), adopted for working with panels. The Delphi technique is a structured process which investigates a complex or ill-defined issue by means of a panel of experts. This methodology proves to be an appropriate research design for this type of research and permits individual opinions to be obtained within a structured group and using a communicative process.

The first matrix developed (Fig. 13.8), on the basis of the LCA recommendations for the supply chain of a packaging system, provides the engineering characteristics (ECs) for the product beverage cartons, in terms of their relative importance and the benchmark values that must be reached.

In particular the steps followed for drawing up the matrix were:

1. *Definition of the most relevant environmental aspects.*
   When studying the case the group of experts was asked to define some recommendations (*"Whats"*) for every step of the supply chain of the product

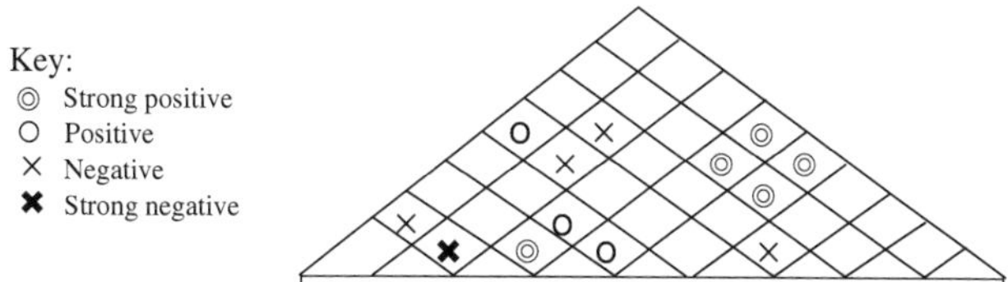

Whats	IMPORTANCE OF THE "Whats"	Produce rectangular beverage cartons	Produce cartons which weigh less	Replace glass fiber	Use easily recyclable materials	Use raw materials coming from renewable resources	Following Design for Disassembly methodology	Produce recloseable beverage cartons	Replacing CFC	Replacing halon
Reduce the quantity of the material required to make the product (aluminum, paper and plastic)	*9*	1	9		1	1		1		
Reduce the waste of packaging material during the production stages (Converting)	*9*	1		1			1			
Improve the energy efficiency of the processes	*3*					1	1			
Optimize the transport and material handling stage	*1*	9	3				3	3		
Favor the recycling of the used cartons	*3*		1	9	9	1	9	1	9	9
Favor the energy recovery of the used cartons	*3*		1	3		1	3	1	3	3
TOTAL SCORE		**27**	**90**	**45**	**36**	**18**	**51**	**18**	**36**	**36**
RELATIVE IMPORTANCE		**7.6%**	**25.2%**	**12.6%**	**10.1%**	**5.0%**	**14.3%**	**5.0%**	**10.1%**	**10.1%**

Engineering Characteristics "ECs"

Fig. 13.8 1° matrix, level 0

Table 13.1 The most important environmental aspects, "Whats"

Supply chain	Customer attributes (*Whats*)
Purchasing of raw materials	Reduce the quantity of the material required to make the product (aluminum, paper and plastic)
Converting	Reduce the waste of packaging material during the production stages
Filling	Improve the energy efficiency of the processes
Distribution retail and consumption	Optimise the transport and material handling stage
Recycling	Favour the recycling of the used cartons
Energy recovery	Favour the energy recovery of the used cartons

beverage cartons. For example, for the first stage of the supply chain, purchasing of raw materials, the "What" is: reduce the quantity of the material required to make the product (aluminum, paper, and plastic). One or more *"Whats"* can be defined for each step of the supply chain (Table 13.1).

Six evaluation factors (*"Whats"*), taken from ISO 14021 (ISO 2000), were considered to assess environmental friendliness.

Table 13.2 Calculation of the importance of "Whats" (1st matrix)

Rules	Importance of the *Whats*
If (GWP and AP and energy and resources) decrease by less than 1%	1
If [(GWP or AP or energy or resources) decrease by between 1 and 2%] and If [(GWP and AP and energy and resources) decrease by less than 2%]	3
If (GWP or AP or energy or resources) decrease by the same as or more than 2%	9

Table 13.3 Product design specifications

Engineering characteristics (Ecs)
• Produce rectangular beverage cartons
• Produce cartons which weigh less
• Replace glass fibre
• Use easily recyclable materials
• Use raw materials coming from renewable resources
• Following design for disassembly methodology
• Produce recloseable beverage cartons
• Replace CFC
• Replace halon

2. *Weighting of the Whats by importance.*
   The importance or priority of the *Whats,* "Imp(What)", was calculated by the team of experts using the LCA analysis. In particular a sensitivity analysis was developed. Using this analysis it was possible to assess as a percentage the impact on the whole supply chain of a 5% reduction in "Whats". Some rules were also created to convert this percentage value into an important judgement (see Table 13.2). The "Imp(What)" was assigned using three levels, "low", "medium" and "high", corresponding to numerical values of 1, 3, and 9, respectively. For example, a 5% reduction in the first *What* ("reduce the quantity of material......") has a relative importance of more than 2% on impact categories along the whole supply chain and it will be associated with an importance value of 9 (see Fig. 13.8). If the total reduction in all the impact categories values had been less than 1% an importance value of 1 would have been associated to this "What", while in all other cases an average importance value of 3 would have been associated. It was hypothesized that the four impact categories all have the same importance.
3. *The Whats are transformed into basic engineering characteristics (ECs).*
   The *ECs* are the technical answers to the *Whats* requirements. Table 13.3 shows the engineering design parameters used for the product. The choice of *ECs* was made on the basis of interviews with experts as well as on the basis of current laws and the design standards.
   By manufacturing rectangular cartons it is possible to optimize space during the distribution stage.
   Plastic materials are used for their light weight, formability, and low cost, but because of their easy degradability over time and in sunlight they need to be

improved using various additives. For example, in order to delay their flammability flame retardants are used which are particularly harmful for the environment.
CFC is used in cooling systems and halon in fire extinguishers. Replacing CFC and halon, the substances having the highest ozone depleting potential, with alternatives having no halon depleting potential is very important for environmental policy.
Another important environmental factor is the glass fiber contained in plastic. This cannot be recycled, since it damages and spoils the mechanical qualities of the reconstituted plastic.

4. *Preparation of the relationships matrix.*
It is necessary to assess which and to what extent the ECs have an effect on the *Whats*. The relationships between *ECs* and *Whats* indicated in the "cells" of the matrix may be positive or negative, of weak (= 1), medium (= 3), or strong (= 9) intensity.
5. *Drawing up the correlations matrix.*
The physical relationships between the specific techniques (ECs) are to be found in the "roof" of the matrix. This step in the construction of the matrix helps in the tracing of pairs of ECs which require parallel improvements and/or those which include ECs in potentially difficult relationships and which therefore involve courses of action which conflict with each other. Therefore this matrix provides positive and negative correlations between pairs of ECs as well as the entity of these correlations.
6. *Other measurements.*
The product development team must estimate the costs, feasibility, and technical difficulty involved in modifying each EC. Objective values must be fixed that is to say measurements which reflect the connection between *Whats* and *ECs*, as well as the clients' requirements.
7. *Action plan.*
The weights of the *ECs* are situated at the base of the matrix. They are the key result of the planning matrix and are calculated using the expression:

$$\text{Weight(EC)}_i = V(\text{EC})_{i1} \times \text{Imp}(\text{What}_1) + V(\text{EC})_{i2} \times \text{Imp}(\text{What}_2) + \ldots + V(\text{EC})_{in} \times \text{Imp}(\text{What}_n),$$

where: V(EC)_{in} is the correlation value of EC_i with What_n and $\text{Imp}(\text{What}_n)$ represents the importance or the priority of What_n.

The construction of the matrix and the results obtained are illustrated in Fig. 13.8.

The results obtained at level 0 are the input of the matrices at detail level 1 (see Fig. 13.6) that is to say the matrices which link the *ECs* of the product with the design characteristics of the machines involved throughout the beverage carton production line. This paper does not consider all the machines involved in the production line, but only those which are used during the filling and packaging

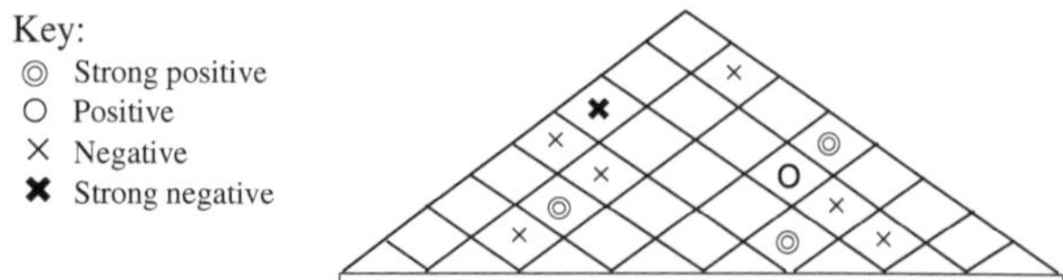

		Engineering Characteristics "ECs"							
Whats	IMPORTANCE OF THE "Whats"	Reduce the volume/weight of the moving parts of the machine	Minimise the need for heating/cooling during product processing	Use high pressure nozzles to reduce water consumption by the machine	Harmless emissions in the air and in water from the production plants	Design the components of the machine according to Design for Rotation Units	Choose machine components with the longest possibile lifetime	Reduce wasting the product (fruit juice) during the filling process	Replace some solvents
Produce rectangular beverage cartons	*3*		1	1		3		1	
Produce cartons which weigh less	*9*		1					1	
Replace glass fiber	*3*	1							
Use easily recyclable materials	*3*								9
Use raw materials coming from renewable resources	*1*		1		1		1		3
Following Design for Disassembly methodology	*3*	1				1	1	1	3
Produce recloseable beverage cartons	*1*			1		1	1	1	
Replacing CFC	*3*		9		3				1
Replacing halon	*3*		9		3				1
TOTAL SCORE		**6**	**67**	**4**	**19**	**13**	**5**	**16**	**45**
RELATIVE IMPORTANCE		**3.4%**	**38.3%**	**2.3%**	**10.9%**	**7.4%**	**2.9%**	**9.1%**	**25.7%**

Fig. 13.9 2° matrix, level 1

stages. The choice of the raw materials used in the beverage cartons, the types of distribution, the possibility of recycling and reutilizing materials are in fact connected with filling and packaging methods. The machines which make beverage cartons and which fill them (filling machines), starting from a roll of prefolded coupled material, composed of wood-fiber, polymers and aluminum have as their output the packaging filled with the product.

Passing from level 0 to level 1 (see Fig. 13.6) the *ECs* of the first matrix become the *Whats* of the second matrix (Fig. 13.9). The relative importance of the $(\text{ECs})_i$ has been used to determine the new values of importance for the *Whats* as shown in Table 13.4.

The values that define the various classes of importance were calculated taking into consideration the ECs number (9 in the case studied) and the number of classes that are required (3 in the case studied). The limit of the first class was calculated as $\frac{100\%}{(\text{ECs number})} - \frac{100\%}{(\text{ECs number})\times(\text{number classes} - 1)}$,

Table 13.4 Importance classes for "Whats" (2nd matrix)

Relative importance *ECs* (%)	Importance of the *Whats* (2° *matrix*)
≤5.5	1
5.5–16.5	3
≥16.5	9

Table 13.5 Machine design specifications

Engineering characteristics (Ecs)
Reduce the volume/weight of the moving parts of the machine
Minimize the need for heating/cooling during product processing
Use high pressure nozzles to reduce water consumption by the machine
Harmless emissions in the air and in water from the production plants
Design the components of the machine according to
Design for rotation units
Choose machine components with the longest possibile lifetime
Reduce wasting the product (fruit juice) during the filling process
Replace some solvents

while the limit of the second class as $\frac{100\%}{(\text{ECs number})} + \frac{100\%}{(\text{ECs number})\times(\text{number classes - 1})}$, (see note[1]).

In order to build the second matrix steps 1–7, as previously explained for the formation of the first matrix, were repeated. Table 13.5 shows the engineering parameters "ECs" for the design of the filling machines used. Even in this case the choice of the EC parameters was arrived at by a team of experts not only on the basis of their knowledge of both present and future legal requirements and design standards which filling machines must satisfy, but above all on the basis of DfE principles.

13.5 Discussion

The environmental sustainability of the manufacturing, packaging, and distribution process for liquid food substances is influenced by many factors. In recent years the areas that many firms in the liquid packaging sector have focused on in order to attain important environmental targets are:

- partnerships with suppliers of aluminum in order to produce aseptic packages with less aluminum content but having the same product protection features.

[1] In the case studied the limit of the first class is 100/9–100/(9·2) ≈ 5.5; while the limit of the second class is 100/9 + 100/(9·2) ≈ 16.5.

- partnerships with suppliers of plastic materials in order to reduce the emissions deriving from packaging material production processes.
- the introduction of a new package format which is lighter and cheaper, designed for the protection and the aseptic packaging of beverages, in particular of milk, to be delivered in less time to those areas of the world which are the most difficult to reach.

From the analysis carried out in this paper using the first matrix it can be seen that the three most important ECs to implement for beverage cartons are:

1. manufacture lighter beverage cartons so as to reduce the consumption of raw materials
2. use materials which can be easily recycled
3. choose design for disassembly methodology

The drawing up of the matrix of correlations between the various ECs allows us to have an overall view of the integration between the different phases in the life cycle of the product and the environmental aspects, thereby ensuring an adequate solution to the tradeoff associated with most decisions which characterize a design. The various EC modification proposals, with their relative importance values, arising from the two matrices must be assessed by estimating from an environmental and economic point of view the effects of the change to a set of product design parameters. Using the roof of the first matrix it is possible to identify three types of tradeoff.

- *Concerning environmental aspects:* minimizing the weight of a product sometimes threatens the possibility of recycling. In the last 20 years, for example, improvements to this aspect of the product have resulted in the average weight of a beverage carton being reduced by 20%. Nevertheless, designing for environmental benefit must always be balanced with the need to avoid compromising consumer safety by undermining the true function of the packaging product—its ability to protect and preserve its contents.
- *More or less tangible and more or less emotional environmental, economi, and social benefits*: investments in order to reduce energy consumption during the use of the product could lead to an environmental impact which is lower than that which would be obtained by choosing other materials to optimize the features of the product in the final stage of its life cycle.
- *Environment, technical aspects, and quality*: the use of some materials which can lead to benefits for the environment may undermine the reliability and the durability of the product.

It is important to identify additional possibilities for recycling the repulping residuals (the non-fiber materials—polyethylene and aluminum—in the package that remain after repulping). An important multinational firm, operating in the packaging field, has developed a new recycling process based on plasma technology, which separates these residuals into pure aluminum, which can be reused to produce new packages and paraffin, which can be used in the plastics industry (Huat 2005). In recent years there has also been growing demand for cartons that can be reclosed—the product lasts longer and less is wasted. However, more materials are required to make such packaging. Many firms have designed the innovative Flexi-Cap that can be reclosed and, being made of plastic, is extremely lightweight and can be flat-packed when empty, allowing for extremely efficient distribution to customers' filling plants.

The second matrix developed shows that the three most important engineering characteristics (ECs) to consider in a DfE for filling machines are:

1. Replace some solvents. Emissions of VOC (volatile organic compounds) mainly arise from solvents used in printing inks and to some extent from printing plate production. There has been a downward trend in VOC emissions from firms over the last three years, which is partially due to a decrease in the production of packaging material requiring solvent-based printing inks and to the installation of cleaning equipment.
2. Minimize the need for heating and cooling the machines. Several factories have established energy teams to ensure a focused and structured approach to energy saving. Submonitoring systems are increasingly applied in order to facilitate energy management.
3. Reduce wasting the product (fruit juice in this case study) during the filling process.

The design specifications chosen must be reviewed and analyzed according to an environmental checklist which covers the entire life cycle of the product and considers: the choice of materials, technical optimization of the production line, optimization of the distribution system and optimization of the product so as to attain better management in the final stage of the lifecycle.

It is necessary to highlight some aspects and problems that came out during this study:

- the panel of experts created for this work worked for a period of about 2 weeks, and the sessions were planned on a three-round Delphi process.
- one of the most important problems in the procedure proposed is to create a large enough panel of experts (ten people) willing to invest time in this study.
- moreover, initially it was not easy to find a common communication language since the group members came from various production sectors and also had different functional specializations and skills.

13.6 Conclusions

Owing to the speed at which technology and knowledge in general evolve and because of the complexity and the variety of the products, it is not easy to define a general strategy for integrating environmental features into the design processes.

This study investigated a new model for integrating environmental aspects into machine and machine component design. The first part of this paper showed how DfE principles can be introduced into all stages of machine development. In particular as far as the first stage, "project definition", is concerned a virtuous circle has been set up based on two types of input:

- LCA of the entire supply chain
- analysis of legal requirements and guidelines for a DfE

which by using QFD matrices lead to the definition of the main ECs, to be considered by designers in the subsequent stages of product development. These matrices also allow the machinery involved in manufacturing the product as well as the machine components to be defined.

The most important design specifications must be integrated according to the internal design standards of the firms. Current laws and regulations do not specify levels of environmental performance which must be reached. Therefore on one hand it is more difficult for the firms to make comparisons with their main competitors, while on the other the firms involved have more freedom to act whatever their initial level of "environmental maturity" may be.

It is important to underline that many firms design and assemble components and modules for filling machines, and therefore most of the machine production and manufacturing process takes place outside the firm itself. Being aware of the importance of the environmental impact of the whole life cycle of the product means that it is necessary to try to control more carefully the stages at the start of the chain of values and then insert more levels of detail (level 2, level 3 etc.). It becomes important to ask for the collaboration of the suppliers and to carry out more and more frequent inspections to check up on their activities and their internal procedures.

This study, which allowed the definition of engineering characteristics connected with the most relevant environmental aspects of filling machines, is of general value and can be applied to all production systems involved in the beverage carton production line; from the machines used for the production of the raw materials to the machines which manufacture the roll of coupled material made up of wood-fiber, polymers and aluminum.

References

Clausing D (1993) Total quality development. ASME Press, New York

Cristofari M, Deshmukh A, Wang B (1996) Green quality function deployment. In: Proceedings of the 4th international conference on environmentally conscious design and manufacturing, Cleveland, 23–25 July 1996, pp 297–304

Fariborz YP, Rafael AC (2002) Quality function deployment for the good of soccer. Eur J Operat Res 137:642–656

Ferde J, Rekdal E, Karvag SE, Ronning A, Hanssen OJ (1995) Environmentally sound product development of light fittings: case-project report from the NEP-project at Glamox AS in Molde, Norway, NEP-project Report No. 10/95, November, Nordic project on environmentally sound product development

Frei M, Züst R (1997) The eco-effective product design—the systematic inclusion of environmental aspects in defining requirements. In: Krause F-L, Seliger G (eds) Life cycle networks. Chapman & Hall, London, pp 163–173

Hanssen O, Rydberg T, Rønning A (1996) Integrated life-cycle assessment in product development and management. In: Curran M (ed) Environmental life-cycle assessment. McGraw-Hill, New York, pp 1–4

Huat LO (2005) Implementing the global vision at a local level. Tetra Pak publication

ISO 14021 (2000) Environmental labels and declarations—Self declared environmental claims, International Organization for Standardization, TC207

ISO 14025 (2006) Environmental labels and declarations—Type III environmental declarations—Principles and procedures, International Organization for Standardization, TC207

Kobayashi H (2005) Strategic evolution of eco-products: a product life cycle planning methodology. Res Eng Des 16:1–16

Linstone HA, Turoff M (1975) The Delphi method techniques and application. Addison-Wesley, London

Murayama T, Hatakeyama S, Narihiko N, Oba F (2001) Life cycle profitability analysis and LCA by simulating material and money flows. In: Proceedings of the 2001 IEEE international symposium on electronics and the environment, pp 139–144

Smith RP (1997) The historical roots of concurrent engineering fundamentals. IEEE Trans Eng Manag 44(1):67–78

Umeda Y, Nonomura A, Tomiyama T (2000) Study on life-cycle design for the post mass production paradigm. Artif Intell Eng Des Anal Manuf 14(2):149–161

Wenzel H, Hauschild M (1997) Environmental assessment of products, methodology, vol 1. Chapman & Hall, UK

Wenzel H, Hauschild M, Alting L (1997) Environmental assessment of products, vol 1. Chapman & Hall, London

Zhang Y, Wang HP, Zhang C (1999) Green QFD-II: a life cycle approach for environmentally conscious manufacturing by integrating LCA and LCC into QFD. Int J Prod Res 37(5): 1075–1091

Chapter 14
Enviromental Marketing and New Product Development

14.1 Introduction

Marketing has been called "the interface between consumption and production" (Charter et al. 2002). Marketing activities are involved with both production and consumption: they influence the product portfolio and the communication efforts of the producer. Accordingly, marketing has a key role to play in the incipience of a product or service development process as well as at its 'end'. Production and consumption patterns in society are increasingly seen as crucial to our route to a sustainable future. Hence, as stated by Charter et al. (2002): "Marketers have an involvement in the sustainability debate both at macro and micro level, even though they may not know it."

Another often used definition of marketing is that of the Chartered Institute of Marketing, which states that marketing is: "the management process responsible for identifying, anticipating and satisfying customer requirements profitably" (Peattie 1995). Based on this definition, Peattie defined green marketing in 1995 as: "the holistic management process responsible for identifying, anticipating and satisfying the requirements of customers and society, in a profitable and sustainable way".

The challenge of responding appropriately to concern about the natural environment has changed many aspects of the way businesses operate and has become an integral part of purchasing, marketing, and corporate strategy (Pujari et al. 2003). Where once environmental responsiveness was viewed as involving compliance, expense, and tradeoffs with other corporate goals, increasingly, it is being portrayed as an opportunity. Although the "win–win" logic of being "green and competitive" (Porter and van der Linde 1995) is disputed by others (e.g., Walley and Whitehead 1994), the literature (USAEP 2001) points to external benefits that arise from environmental improvement, including:

- **Economic benefits from increased efficiency**. By reducing wastes, companies decrease handling expenses, fines, and even costly inputs. Supplier's savings may be passed along to buyer companies.

M. Bevilacqua et al., *Design for Environment as a Tool for the Development of a Sustainable Supply Chain*, DOI: 10.1007/978-1-4471-2461-0_14,

- **Competitive advantage through innovation**. Efficient production is enhanced through the use of cleaner technologies, process innovation, and waste reduction. Reduction in wastes equals dollars earned.
- **Improved product quality**. Supply chain partnerships help maintain relationships between buyers and suppliers leading to increased control over product quality.
- **Consistent corporate environmental goals**. In an era of multi-faceted, non-vertical manufacturing, companies include supplier outreach to address corporate environmental goals.
- **Improved public image**. Consumers, investors, and employees respond positively to companies with a reputation for good environmental performance.

Menon and Menon (1997) argue that there is an emerging consensus among business leaders that the goals of social good and business success are no longer an either/or proposition but are being increasingly interwoven into an "ecopreneuring" paradigm. Porter and van der Linde (1995) also point towards an underlying logic linking the environment, resource productivity, innovation, and competitiveness. This means tackling the socioenvironmental impacts of marketing strategies, both in terms of nonmarket outputs (through pollution prevention and more sustainable sourcing) and market outputs (through innovative products and a product stewardship orientation; Ottman 1994). Product stewardship efforts encompass all aspects of managing products and their performance and impacts, through both the products economic and physical (cradle to grave) life cycles.

The effective development of new, environmentally improved (or greener) products will clearly be crucial in creating successful environmental strategies, and in helping to move companies and economies towards environmental sustainability.

Lampe and Gazdat (1995) highlighted the most important catalysts and pressures that have resulted in green marketing. Table 14.1 summarizes these factors comparing the European and US situation.

Another area that attracted a great deal of interest early on was the characteristics of the green consumers. Numerous such studies were made. Diamantopoulos et al. (2003) summarized the socio-demographic surveys made in 1966–1994 to describe the green consumer and found, among others, 39 studies on education, 31 studies on sex, 35 about age, and 21 surveys dealing with social class. Table 14.2 shows the distribution developed by Peattie (1992). A more recent analysis has been carried out by Charter et al. (2002) (Table 14.3).

Many of the surveys aimed at identifying typical demographic qualities of the green consumer. For example, females, young people, and people with a relatively high education and income were identified as most likely to engage in green consumer behavior (Straughan and Roberts 1999). However, in the plethora of surveys made, many contradictory views of the probable green consumer were put forward (Peattie 2001).

Table 14.1 Pressure in green marketing

Catalysts and pressures	Description
Environmental damage and the media	A major catalyst of public opinion to preserve the environment has been environmental damage or the threat of environmental harm. These events are usually carried to the public by the media. Europeans are particularly sensitive to these issues "by virtue of the size of the region (only one-fourth that of the United States) and the [population] density" which is much greater than in the US (Flattau 1990). A relative scarcity of natural resources is another important factor in Europe (Blumenfeld and Gilbert 1990). Examples of disasters impacting Europeans include chemical spills on the Rhine River; the poisoning of North Sea seals; contamination of Mediterranean beaches; the Chernobyl nuclear power meltdown; the revelations regarding the shocking mess in Eastern Europe; and the large oil spill off the Shetland Islands. Americans have been influenced by events such as Love Canal; Times Beach; the barge of New York garbage which could not find a dump; and the Valdez oil spill. Both Europeans and Americans have become particularly concerned about damage to the Earth's ozone layer; global warming; reaching the limits of landfills; and acid rain.
Public opinion and social concern for the environment	Public opinion in both Europe and the US, as influenced by environmental damage, media coverage, and other factors has strongly favored action to protect the environment. "Due to varying perceptions and cultural backgrounds, the North American consumer responds to these issues differently than does the European consumer" (Blumenfeld and Gilbert 1990). Economic problems in the US have led to a weakening of environmental concerns. This trend is subject to reversal by economic recovery or by environmental disaster. European public concern for the environment seems to be more stable even in an economic recession.
Social forces and the greening of business	Public concerns have translated into potent forces for the environment including green political power and green consumerism. These forces, and institutional pressures from investors and employees (including management), have been major catalysts for the greening of business, which, in turn, has given rise to the concept of green marketing. Germany provides an example of the greening process. In the early 1970s the media publicized environmental calamities including the death of Bavarian forests, toxic waste spills in the Rhine and chemical dumping in the North Sea which devastated its seal population. Citizen protests gave rise to Europe's first significant Green Party. By 1983 the party had gained enough political power to force the creation of a powerful German environmental ministry and the passage of tough environmental legislation. A poll taken in the late 1980s showed that 82% of those surveyed in West Germany said that they take environmental considerations into account when shopping at the supermarket (by comparison in the

(continued)

Table 14.1 (continued)

Catalysts and pressures	Description
	Netherlands 67% say the same, 55% in the UK, and 50% in France). Since the early 1970s the German industry has spent a tremendous amount of money for a cleaner environment (82.3 billion ECUs). Meanwhile, the country's economy has prospered and the environment is improving. Early development of green technologies and products has provided German companies with international business opportunities as other countries go green (Evans 1990).
Green political power	"Once thought of as little more than business-bashing cranks, the Greens have moved from the political fringe to the center" (Tully 1989). In 1981, Belgium became the first country to elect Greens to Parliament. In addition to Belgium and Germany, Green parties were represented in the 1990 national legislatures of Italy, Portugal, the Netherlands, Luxembourg, Austria, Switzerland, Sweden, and Finland. Green Party representatives in the 1990 European Parliament were elected from Belgium, Germany, Netherlands, France, Italy, Spain, and Portugal. The British Green Party won strong support (15% of votes cast in 1989), but gained no seats in the European Parliament because of the electoral system (Flattau 1990). American public opinion has had an important influence on environmental public policy-making. As previously discussed, several factors led to early environmental reforms by US lawmakers. Unlike Europe, the Green Party only recently began to organize in the US. Politicians are directly responsive to the voters when opinion polls indicate increasing support for environmental issues. Pressure groups such as the Sierra Club, National Resources Defense Council and Environmental Defense Fund play a key role influencing government.
Consumer attitudes and green purchasing	Europe's green consumerism has strongly entwined with green politics and the Green Party. Environmental groups in both Europe and the US have educated and pressured consumers through boycotts and other campaigns. A 1990 report by the Arthur D. Little Center for Environmental Assurance compared the strength of green consumption in the US and Europe: "In Northern and Central Europe, environmental concerns have already reached levels that significantly affect consumers' purchasing and political voting decisions. Although comparable growth in demand for 'green' products… has yet to be seen in the United States, initial signs of such a trend can be observed in areas of traditionally high environmental awareness such as the West Coast and New England" (Blumenfeld and Gilbert 1990). However, reports confirmed that green products have not gained mass acceptance in the US market. For many consumers pro-environmental sentiments do not lead to green purchases. In Europe (and Japan) the environmental marketing movement continues to have much more vitality

(continued)

Table 14.1 (continued)

Catalysts and pressures	Description
Environmental law	Green political power has resulted in the proliferation of environmental laws. The EC has been setting green rules, yet its members increasingly disregard them (The Economist 1991a). US laws such as the Comprehensive Environmental Response, Compensation and Liability Act (CERCLA, also commonly known as 'Superfund') are making it difficult and expensive for some businesses to get financing or insurance. This is because of the potential liability or penalties for environmental accidents or violations. By contrast, European liability standards are not as onerous as those in the US. EC legislation tends to require certain results while allowing member nations' discretion in implementation. In most European countries regulations are not as extensive, specific, or rigid as in the US, and as a result there is less litigation over liability (Bloom and Morton 1991, p. 75). European governments increasingly favor economic incentives to encourage consumers and industry to behave in ways that do little harm to the environment. Examples include taxes in several countries on ozone depleting chemicals, and tax concessions in Germany to stimulate manufacture and sales of less polluting cars. Since most of these laws require businesses to protect the environment, they also provide a green marketing opportunity. Some companies join green lobbyists to support stricter environmental regulations. This is particularly true where a company already meets the proposed standards that its competitors do not meet. Self-regulation, to avoid further government regulation, also provides a green marketing opportunity.
Institutional pressures—investors and employees	In addition to the forces of politics, law and consumers, pressures on business to protect the environment emanate from investors and employees including all levels of management. In the US, socially responsible investing is a growing trend. An increasing number of mutual funds offer investments screened on environmental criteria. Individual and institutional investors may avoid companies with a poor environmental record, or seek those with a positive environmental performance. Such choices are made for both ethical and financial reasons. Stockholders are offering an increasing number of environmental resolutions at annual meetings, such as initiatives for adoption of the *Valdez Principles* (Sheets 1990).

(continued)

Table 14.1 (continued)

Catalysts and pressures	Description
The Greening of business	Business has been profoundly influenced by the social, political and legal pressures for environmental protection. The "Business Council for a Sustainable Environment," a group of 48 chief executives from companies such as Minnesota Mining and Manufacturing Co., Nippon Steel Corp. and the Royal Dutch/Shell Group, launched a 'green business' manifesto. They called upon their colleagues to take the initiative on saving the environment by promoting a brand of market led environmentalism. London-based Greenpeace criticized the effort as a "veiled attempt to minimize environmental controls on big business". This skepticism is shared by the American public. There is a growing realization that, in addition to green marketing, the environment also offers an opportunity for the development and sale of green technology to other businesses and government. As previously mentioned, German companies, with their nation's head start, have already benefited from the sale of 'green' technology. For an increasing number of companies the environment is becoming an integral part of business strategy, and green marketing is a key element of that strategy.

Table 14.2 Share of consumers according to typology by Ogilvy and Mather (Peattie 1992)

Consumer typology by Ogilvy and Mather		
16%	Activists	Likely to buy green products and services
34%	Realists	Are worried about the environment but sceptical about the green bandwagon
28%	Complacents	See the solution as somebody else's problem
22%	Alienated	Are unaware of green issues or see them as transient

Table 14.3 Share of consumers according to typology by US Roper Starch Worldwide 2000 (Charter et al. 2002)

Consumer typology by US Roper starch worldwide 2000		
11%	True blue greens	Major green purchasers and recyclers
5%	Greenback greens	Will buy or give green but will not make lifestyle changes
33%	Sprouts	Care but would only spend a little more to buy green
18%	Grousers	Environment is somebody else's problem
31%	Basic browns	Essentially do not/will not care

Figure 14.1 summarizes the key influences on the performance of new products in the market, and in environmental terms, as determined by the following review of the environmental management and NPD literatures.

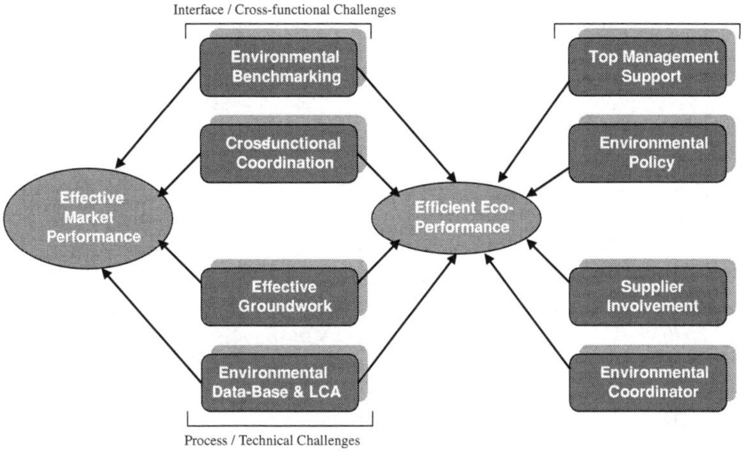

Fig. 14.1 Influences on the performance of environmental new product

Pujari et al. (2003) carried out a study (see Box U) in order to define which factors positively influence market performance and eco-performance of environmental new product development (ENPD). This study reports the findings of a large-scale research project on environmental new product development (ENPD) within British manufacturers.

BOX U: Market Performance and Eco-Performance of ENPD

Two factors were analyzed in Pujari et al. (2003). Factor 1 is named as "Effective Market Performance (MARPERF)," which include variables relating to market, competitiveness, and financial measures. The variables, which load heavily on this factor, are increased market share, creation of new domestic and international markets, achieving competitive advantage, and good return on investment.

The second factor shows variables having a common dimension of nonfinancial performance, primarily relating to environmental aspects of product performance. This factor is named as "Efficient Eco-Performance" (ECOPERF). Variables loading heavily on this factor are enhanced environmental image of the firm and overall reduction in environmental impact of product.

In this study, for both regression equations, MARPERF and ECOPERF, hierarchical regression method (Cohen and Cohen 1983) is applied for specifying regression models (see Table 14.4). The issue of multicollinearity in regression analysis was resolved because factor analyses were run with varimax method.

Analyzing the coefficients in Table 14.4, Hypothesis 1, stating, "An explicit environmental policy positively influences the level of eco-performance of ENPD," could not be supported in this study. This was not an expected result, given the emphasis that many companies place on environmental policy development and the role that such policies could be expected to play in developing an organizational climate in which activities such as ENPD can flourish. The result may reflect the lack of a perceived direct connection between such a policy and the relatively operational issue of a specific product's eco-performance; and it may also reflect the tendency of many such statements to be perceived as "little more than a bit of judicious PR" (Shimnel 1991).

Hypothesis 2, stating, "A higher degree of top management support for ENPD will positively influence the level of eco-performance of ENPD," is supported. This confirms that the important role of top management in the market performance of new products (Maidique and Zirger 1984) is duplicated in relation to the eco-performance of environmental products. This reflects Johne and Snelson's (1988) view that clear messages from the top about the importance of ENPD can become a critical success factor.

Hypothesis 3, stating, "A higher degree of integration of environmental coordinator in ENPD process positively influences the level of eco-performance of ENPD," is also supported. This result further emphasizes the need for a more integrated role for the environmental coordinator in the ENPD process. This role may involve issuing environmental guidelines to ENPD teams and contributing in the implementation process and conducting environmental assessment at every stage gate of the ENPD process. Environmental coordinators can be agents of change in the firms and can certainly facilitate environmental product stewardship programs (Barrett 1993).

Hypothesis 4, stating, "A higher degree of supplier involvement positively influences eco-performance of ENPD," is also supported. This further supports the augmented supply chain view of the ENPD wherein adverse environmental attributes can be removed by monitoring/auditing or even partnering with suppliers. It is now increasingly being accepted that it would require the consideration of a whole supply chain of materials and inputs required to make an environmental new product.

Hypothesis 5a, stating, "A higher degree of effective groundwork positively influences market performance and eco-performance of ENPD," is also supported. This result confirms Cooper's (1988) findings on the general importance of groundwork for market performance and shows that it also affects the eco-performance of ENPD. However,

Hypothesis 5b, stating, "A higher degree of product testing and experimentation for environmental impact positively influences market performance

Table 14.4 ENPD performance: results from regression analysis

Factors	MARPERF		ECOPERF		
Effective groundwork	0.255 (0.003)	0.257 (0.001)	0.207 (0.013)	0.204 (0.011)	0.204 (0.004)
Product experiment	0.053 (0.526)	0.052 (0.481)	0.019 (0.816)	0.018 (0.817)	0.018 (0.794)
Environmental database for LCA	0.171 (0.042)	0.172 (0.022)	0.286 (0.001)	0.285 (0.000)	0.285 (0.000)
Environmental benchmarching and measurement		0.332 (0.000)		0.103 (0.196)	0.103 (0.143)
Cross-functional coordinator		0.297 (0.000)		0.242 (0.003)	0.242 (0.001)
Top management support		–			0.354 (0.000)
Environmental product policy		–			0.051 (0.472)
Supplier involvement		–			0.136 (0.050)
Environmental coordinator		–			0.227 (0.002)
R^2	0.097	0.297	0.125	0.194	0.392
Adjusted R^2	0.076	0.269	0.105	0.162	0.348
F	4.648 (0.004)	10.795 (0.000)	6.178 (0.001)	6.148 (0.000)	8.876 (0.000)
n	134	134	134	134	134

and eco-performance of ENPD," could not be supported. This may indicate that companies have not yet developed testing procedures orientated towards measuring eco-performance.

Hypothesis 6, stating, "A high degree of integration of environmental impact databases in existing information systems positively influence the market performance and eco-performance of ENPD," is also accepted. Effective management of environmental information should be fed into ENPD projects by assessing life cycle impacts of materials used. The environmental database ideally should become a part of the existing information system and be accessible to the ENPD team. Moreover, this information can eventually lead to developing accurate environmental claims, thereby earning credibility and gaining success in the marketplace (Wallace and Suh 1993; Orlin et al. 1993/1994.

Hypothesis 7, stating, "A higher level of environmental benchmarking positively influences market performance and eco-performance of ENPD," is accepted for MARPERF but could not be supported for ECOPERF. This only partially supports those authors who have argued that environmental benchmarking will lead to effective market and efficient eco-performance (Peacock 1993; Azzone and Manzini 1994). Systematic and detailed environmental benchmarking and performance measurement processes are important, not just

for ENPD, but also to create an organization behind the products, which pursue continuous environmental improvement and progress towards sustainability.

Hypothesis 8, stating, "A higher degree of cross-functional coordination positively influences market performance and eco-performance of ENPD," is supported. This suggests a need to integrate LCA work conducted by environmental professionals with other functional activities within the ENPD process (Vigon and Curran 1993; Sullivan and Ehrenfeld 1992).

14.2 Green Marketing

Green marketing is defined as the marketing response to the environmental effects of the design, production, packaging, labeling, use, and disposal of goods or services.

Technology push and market pull factors are relevant drivers for technological innovations in general (Pavitt 1984) but also for environmental innovations (Rennings 2000). In an empirical study on the differences in environmental process and product innovations, Cleff and Rennings (1999) find that market considerations are especially important for environmental product innovations. Firms may use environmental improvements to differentiate their products from others and thus gain a competitive advantage (Reinhardt 1998). However, many consumers are reluctant to pay premium prices or trade off other product qualities solely for a product's green attributes (Peattie 2001). Additionally, consumers' claims of prioritizing green attributes have mostly not matched their actual purchasing behavior (Prakash 2002). The eco-marketing literature suggests that green products which besides their public benefits also have private (environmental) benefits for the customer will generate stronger consumer demand (Reinhardt 1998; Belz 2001; Belz and Bilharz 2005). Such customer benefits can have different sources, e.g. cost and energy savings through more efficient appliances, improved product quality and durability, better repair, upgrade, and disposal possibilities, as well as reduced health impacts.

These customer benefits help firms to overcome the second externality of environmental innovations: by shifting some portion of the environmental benefit from the public to the customers, firms can deliver an added value. Thus they are able to increase the demand for their environmentally improved products and can thereby monetize on their environmental investments.

Green marketing in Europe and the US has primarily centered on the areas of product (including packaging and labeling) and promotional strategies. There is considerably less information and emphasis on the other major areas of marketing–

Table 14.5 Areas and activities in green marketing

Areas of green marketing	*Description*
Product variety	Although most of the green products are in the household cleaning and paper products areas, the variety of green products (goods or services) that are already being marketed is abundant. Following is a brief listing of some of these products sold with environmental themes or benefits: -Eco-warriors (ecological soldiers for children) -Dating services (to help find an environmentally friendly mate) -Ice cream (containing nuts from rain forests) -Batteries (mercury free) -Golf tees (biodegradable) -Toilet paper (from recycled materials) -Hollywood films (with ecological themes) -Stock mutual funds (investing in green oriented companies) Given this list one wonders whether just any product can be a green product. Simon Williams, the CEO of The Michael Peters Group, indicates that there are four areas that help to define a product as being green: (1) content (2) structure and packaging (3) message and (4) positioning (Mathews 1990). It is precisely in these areas where much of the controversy about green products and marketing has taken place.
Labeling	A very important area of consideration has been the area of eco-labels, labeling which certifies that a given product is environmentally safe or friendly. A key consideration is whether this labeling should be done by a private, independent, and presumably impartial organization, or whether it should be done by a government entity. Another consideration is whether there should be a single label for the whole country or several labeling organizations. The approach varies from country to country. Germany, Canada, and Japan have a single primary label which is government backed. EC ministers have adopted a common green labeling system for its members. This EC eco-label is a voluntary program. In June of 1993 the first two European eco-label products were introduced—a washing machine and a dishwasher. The US has several private labels (The Economist 1991b). An important issue is whether these labels will be helpful to consumers or prove confusing as the number and variety of such labels increases, especially when the manufacturers' own green labels are included.

(continued)

Table 14.5 (continued)

Areas of green marketing	*Description*
Packaging	An important part of the total product for many companies is the packaging which not only provides information but also serves as a type of promotion for the product. The packaging also is a major source of environmental waste. Much of the green marketing discussion, both in Europe and the US, is over excessive packaging and the material that the packaging is made of. The demand for recycled or recyclable materials shows it is important to consumers. Many companies in the US have instituted green packaging campaigns mainly because of consumer and consumer group pressure. For instance, Coca-Cola, in addition to using recycled aluminum and glass, has started to market the first soft drink bottles made of recycled plastic. The packaging used in Le Menu frozen food dinners was reduced by 25%; and 75% of its packaging comes from recycled materials (Voss 1991). McDonald's switched from plastic foam containers to paper wrappers on its hamburgers. Factors such as this have prompted worldwide consideration of what is being called source reduction—building less packaging into the design of the product. Examples of this in Europe are Proctor and Gamble's refills for detergents and other household products. In the US the cartons on Secret and Sure deodorants were eliminated resulting in the saving of 6 million pounds of paper pulp and keeping 80 million deodorant cartons from going to the landfill (Reitman 1992). The source reduction resulted in higher marketing costs because of sturdier shipping containers, a tighter fitting lid, and different promotion on the container itself, but the environmental savings were tremendous. Source reduction has become a major trend worldwide.
Pricing	The pricing of green-oriented products has typically been higher in both Europe and the US to reflect the added costs of modifying the production process, the packaging, or the disposal process. An additional reason for higher prices was the perception that consumers would pay more for green products. Surveys indicate that consumers say that they will pay from 7 to 20% more for environmentally friendly products (Reitman 1992). What consumers say they will do and what they actually do, in terms of purchasing green products, may be quite different. A Simmons market research study in 1991 is believed to be the first to link actual buying behavior with consumer attitudes on

(continued)

Table 14.5 (continued)

Areas of green marketing	*Description*
	the environment. In a survey of 23,500 people conducted in 1990 and 1991 in the United States, 60% of women aged 18 and older agreed that people should not use aerosol sprays, but 49% of these same women said they bought aerosol hair sprays "most often", as compared to alternatives such as pump sprays. It was agreed by 58% of men 18 and older that aerosol sprays should not be used, but 87% of those same men purchased aerosol shaving cream "most often" over green alternatives such as tube shaving gels (Mandese 1991).
Promotion	It has been suggested that "catering to environmental worries might be the hottest sales strategy since advertising agencies discovered sex in the 1950s" (The Economist 1990b). A major problem for marketers, and the consumer, has been the confusion with many of the environmental terms used in promoting products. Terms such as biodegradable, recyclable, and environmentally friendly, have come under harsh criticism and now, in many cases, are being avoided by companies because of the difficulty in defining and documenting them. What began as positive promotion has turned into negative publicity in some cases. Several US companies stopped using the terms in their promotions. Mobil Corporation stopped promoting its Hefty trash bags as photo-degradable because of consumer and governmental complaints that the bags would not degrade in landfills for a long time. McDonald's stopped using the "clam shell" plastic foam containers, which it said were recyclable, after criticism that very few facilities could recycle them. Willamette Industries eliminated the words biodegradable and recyclable from the bottom of its paper bags distributed at supermarket checkout counters. This pinpoints a real problem for marketers—their ability to properly promote their green actions and products. A survey of 80 manufacturers in the US found that corporate environmental marketing programs were 'embryonic' compared to the actual level of corporate action. Easily understandable and accurate environmental terminology is a key issue that needs to be addressed.

(continued)

Table 14.5 (continued)

Areas of green marketing	*Description*
	The problem facing companies attempting green marketing is that consumers are increasingly not believing environmental claims. Perhaps part of the answer lies in the fact that when looking at information about the environment, consumers are least likely to believe information that corporations supply them in advertisements, on product labels, or printed on packaging. While companies have been backing away from green claims, they have been attempting to better target green consumers. A process called 'environmental targeting' attempts to identify consumers who are most likely to buy green products. A company called TDI has developed a computer-based marketing program to identify and track these consumers. They then sell advertisements on buses and commuter railroads in specific areas in major US cities to target environmentally oriented consumers. Their program is appropriately called Green Targeting.
Distribution	An important part of the distribution process for green products is through retailers who sell the goods to the end consumer. In many instances they share the responsibility of the claims made by the manufacturers of green products. For this reason it has been suggested that retailers help verify the claims as an aid to the consumer. Wal-Mart stores was one of the first retailers to develop their own green line of products and require that manufacturers meet certain criteria. Fred Meyer, a grocery chain in Portland, Oregon, stopped buying house brand tuna from Taiwan, Korea, and Japan, since it was caught by drift-net fleets. They have endorsed and use the Green Cross program to certify environmentally friendly products and packaging. While these efforts continued in some stores, the recession and other factors made it harder for US manufacturers to get other retailers to stock green products. Part of the problem is the higher prices that typically are charged for green products.

pricing and distribution. Miles and Russell (1997) carried out an analysis in each of these four areas. Table 14.5 provides to explain these areas of marketing with a perspective of green marketing activities.

Kammerer (2009) carried out a research focused on the electrical and electronic appliances (EEA) industry in Germany. The aim of this paper is to analyze the influence of customer benefit and regulation on environmental product innovation.

He focused on the following four environmental issues in this study: energy efficiency, toxic substances, material efficiency, and electromagnetic fields. Three different sectors within the EEA industry have been selected for the sample:

Tab. 14.6 Number and share of firms with environmental product innovations (EPI_ANY) by sector and environmental issue (overall 2% of the cases have missing data for this variable)

"Has your company implemented any environmental improvements in your products in the past 3 years?"	Household appliances (HA)	Information and communication technology (IT)	Medical appliances (ME)	Total
Energy efficiency (EFF)	21 (81%)	23 (85%)	17 (46%)	61 (68%)
Toxic substances (TOX)	22 (85%)	26 (96%)	37 (97%)	85 (93%)
Material efficiency (MAT)	19 (73%)	19 (73%)	27 (73%)	65 (73%)
Electromagnetic fields (EMF)	15 (60%)	22 (88%)	28 (80%)	65 (77%)
Total	77 (75%)	90 (86%)	109 (74%)	276 (78%)

information and communication technology (IT); household appliances including lamps and lighting fixtures (HA); and medical appliances (ME).

As can be seen in Table 14.6, in 78% of the cases, an Environmental Product (EP)—Innovation has been implemented. Broken down by sector, IT (86%) clearly exceeds the other two sectors (74 and 75%, respectively).

Looking at the issue level, almost all companies have implemented EP-innovations regarding toxic substances (93%). For each sector it is the issue with the most EP-innovations, ranging from 85 (HA) to 97% (ME). The second issue is electromagnetic fields for which 77% have implemented EP-innovations. For this issue, considerably fewer HA companies (60%) have EP-innovations than ME (80%) and IT (88%) ones. In contrast, EP-innovations regarding material efficiency have been implemented evenly over the sectors by around 73% of companies. Energy efficiency is the issue for which the least companies (68%) have implemented EP-innovations. This relatively low rate is mainly caused by the ME sector where only 46% of the companies have been innovative in this area, compared to 81% for HA and 85% for IT.

The potential for customer benefit has been measured at the environmental issue level. The 4-point ordinal scale ranges from no benefit to large benefit. Looking only at the issues, companies rated customer benefit most frequently moderate for the issues EFF and EMF, making it the median category for these issues (see Table 14.7). For TOX and MAT, customer benefit has been rated lower with little benefit being the median and most frequent answer category.

Depending on the sector, companies gave different ratings for customer benefit from increased energy efficiency: while moderate is the median category for HA and IT companies, ME companies consider the EFF issue less beneficial for their customers with the median lying between little benefit and moderate benefit. In contrast to the major study variables, firms' green capabilities have been measured at firm level. Firms may allocate resources and develop specific knowledge for certain environmental issues, however the underlying green capabilities are the same. Therefore these factors were surveyed at firm level. Green capabilities have

Tab. 14.7 Customer benefit (CUST_BEN) by environmental issue and sector (overall 1% of the cases have missing data for this variable)

"How do you rate the direct benefit to your customers from product improvements?"	No benefit (%)	Little benefit (%)	Moderate benefit (%)	Large benefit (%)	Total (%)
Energy efficiency (EFF)	11	29	35	25	100
HA sector	8	19	46	27	100
IT sector	11	26	30	33	100
ME sector	13	37	32	18	100
Toxic substances (TOX)	18	43	26	13	100
HA sector	31	38	15	15	100
IT sector	15	48	30	7	100
ME sector	11	42	32	16	100
Material efficiency (MAT)	19	37	29	15	100
HA sector	31	42	15	12	100
IT sector	19	44	22	15	100
ME sector	11	29	42	18	100
Electromagnetic fields (EMF)	18	25	39	18	100
HA sector	28	26	20	16	100
IT sector	8	16	60	16	100
ME sector	18	24	37	21	100

Tab. 14.8 Green capabilities of firms by sector

Does your company...	... use the env. attributes of your products in marketing? (%)	...have set up voluntary env. targets for products? (%)	...conduct systematic env. analyses of your products? (%)	...train its product developer in env. issues? (%)	...have a certified env. management system (e. g. ISO 14001)? (%)
HA sector	54	46	31	23	23
IT sector	52	48	26	26	19
ME sector	34	34	21	16	13
Total	45	42	25	21	18

been measured with 5 indicators (see Table 14.8). Overall, the most prevalent measures are the use of products' environmental attributes in marketing (45%) and voluntary environmental targets for products (42%). Few companies have systematic environmental analyses of products (25%) and environmental trainings for product developers (21%).

The statistical analyses clearly show that customer benefit plays a key role for environmental product innovations. Firms that attribute a large potential for customer benefit to an environmental issue are significantly more likely to implement EP-innovations for this issue. Furthermore, they implement their EP-innovations for more products and their EP-innovations are more often market novelties. In short, customer benefit fosters the implementation of EP-innovations, their broad application, and their level of novelty.

However, as shown in the descriptive statistics section, customer benefit is not constant within an environmental issue and/or industry sector.

14.3 Comparison Between Traditional and Green Marketing

"Marketing activities that recognize environmental stewardship as a business development responsibility and business growth opportunity" is how Coddington (1993) defines the emerging practice of environmental marketing. Ottman (1993) defines environmental marketing as having two objectives: (1) developing and offering environmentally compatible products and (2) creating a quality-based image that suggests to all stakeholders that the firm is environmentally sensitive.

Miles and Munilla (1993) in defining the eco-marketing concept suggest that "consumers are… concerned with a holistic view of corporate image, particularly with regard to social concern and (environmental) responsibility." Sheth and Parvatiyar (1995) suggest that corporations' internal environmental marketing efforts should be augmented by government policy, resulting in the two-dimensional construct of "sustainable marketing" which includes: (1) "proactive corporate strategies that would benefit both corporations and society;" and (2) "government intervention for sustainable development." Environmental marketing, green marketing, eco- marketing, and sustainable marketing are all different perspectives of the attempt by businesses to adapt to the growing environmental concerns of various stakeholders.

The US Federal Trade Commission's 1992 guidelines on environmental marketing focus on advertising and define the parameters that are acceptable when making environmental claims (Coddington 1993). Table 14.9 summarizes the differences between the sustainable/green/environmental marketing perspective and traditional marketing. Table 14.10 provides an adaptation of Van Waterschoot and Van den Bulte (1992) revised 4P marketing mix classification framework.

The aim of green marketing is to include environmental issues in the marketing efforts. The idea is that if we provide consumers with better information about the green properties of the products offered, for example by the use of eco labels, they can (and will) include this information in their purchasing decisions. This, in turn, will push companies to produce products that are better from an environmental point of view.

Rex and Baumann (2007) find that although a great deal of effort has been invested in making them more effective and efficient, the market share of eco-labeled products is still low, partly because they have been addressed mainly to 'green' consumers.

It has been argued that green products have certain specific inherent negative characteristics that have to be overcome in order to market them successfully. One that has gained special focus is how to demonstrate the environmental qualities of the products (Meyer 2001). Governmental and non-governmental organizations have recognized this information problem and made great efforts to facilitate the

Tab. 14.9 Environmental/sustainable marketing perspectives compared to traditional marketing

Objective/ perspective	*Environmental/green/sustainable marketing*	*Traditional marketing*
Objective	Satisfy customer needs in an environmentally sustainable way, while earning a profit	Satisfy customer needs at profit
Perspective of customer	The buyer of the product, and the victim of all externalities; or all stakeholders	The reason for existence
Perspective of government	An ally in the creation of a sustainable economy to work and manage	A regulator and limiter. To be managed
Perspective of demand	The redirection of demand towards products with low levels of externality production	The stimulation of for any and all products. Most efforts placed on highest margin products

process, by introducing and maintaining trustworthy environmental labeling schemes. Several publications have pointed out the lack of data or knowledge about whether or not eco labels improve the overall environmental quality of society (e.g. Refs. Rubik and Frankl 2005a, b). There have even been suggestions that eco labels may be counterproductive, for example by acting as “barriers to environmental innovation” because the criteria are based on current products (Erskine and Collins 1997). However, for specific products or product groups, assessments can be found showing environmental improvements thanks to the introduction and use of eco labels. For example, the food retailer Coop Sweden has stated that, as a result of consumers’ choice of ecological food products during 2004, the amount of pesticides used for food production was reduced by 14,000 kg and the amount of artificial fertilizers by 1,000,000 kg (Coop Sweden 2004). A survey of changes in household detergents in Sweden showed decreased use of chemical products by 15% since the introduction of eco labels for this product group and that the surfactants used had been replaced by more biodegradable ones. One problem with these kinds of isolated examples is that it is hard to relate to ‘ordinary’ material flows such as total product sales and overall material flows in society. Another problem is to determine what improvements can be attributed to the existence of eco labels. In the case of household detergents, the eco-labeling organisation had an expert panel to review the detergents survey and concluded, in a general way, that “the eco label had had significance for the changes” (Rex and Baumann 2007). The expert panel, however, pointed out that the effects of the eco label should be seen in combination with other efforts.

Also for assessments of market penetration for eco labels, information is fragmentary and covers a very limited number of products. It can be concluded that the share of eco-labeled products differs considerably with the type of product and country. Some specific products in certain regions show a remarkable market share of green products: for example, the market share of eco-labeled printing paper is more than 70% in the Nordic countries (Rubik and Frankl 2005a, b), and

Tab. 14.10 Environmental/sustainable marketing adapted to the Van Waterschoot e Van den Bulte improved marketing mix classification framework

Marketing mix function	*Environmental/green/ sustainable marketing*	*Traditional marketing*
Product Need satisfying instruments in an exchange.	Environmental design of product; products designed to facilitate long-term use, energy efficient, efficient recycling, "life cycle recyclability/ responsibility" & consider both the total cost of production and consumption	Planned obsolescence, designing products to have shorter lives, disposable products, no concern about externalities from the production or consumption of product
Mass communication Non-personal message with goals of creating awareness, interest, and desire	Environmental labeling. Move towards rational consumption	Stimulation of both primary and selective demand utilizing mass media. Attempt to create desire for unsought goods. Focus on image and emotion of products
Place Where and how of availability	Life cycle assessment, Total cost assessment of distribution	Distribution based on interrelationship between costs of distribution and strategic objectives
Personal communication Personal messages with goals of maintaining awareness and interest and stimulating desire and sales.	Focus on meeting consumer needs at minimal cost to environment, while achieving long-term profits	Focus on meeting customer needs at a profit
Price The cost and method of payment	Total cost assessment, full cost accounting, or the explicit internalization of all external costs must be considered in setting price in relationship to strategic objectives and demand	A strategic decision based on interrelationship between marketing objectives, financial objectives, and demand

in Sweden, the market share of eco-labeled laundry detergents is around 90%. On a more aggregated level, however, market penetration is much lower.

Rex and Baumann (2007) gave an explanation of this low market share for green products. They found that there are mainly two aspects: the market for green products is either non-existent or saturated and the environmental information given needs to produce more efficient results. From a green marketing perspective, the low share of green products sold is explained by the modest size of the green market segment.

As mentioned in Table 14.3, the 'true blue greens' are only about 10–15% of consumers. The motivation for more effective and efficient eco-labeling schemes has generally been that consumers need environmental information.

Fig. 14.2 The marketing strategy (modified by Kotler et al. (2001)

In a theoretical exposition of marketing theory, Rex and Baumann (2007) find that green marketing could learn from conventional marketing in discovering other means than labeling to promote green products. Examples include addressing a wider range of consumers, working with the positioning strategies of price, place, and promotion and actively engaging in market creation.

The marketing strategy consists of a continuous loop where the sequence of demand measurement, segmentation, targeting, and positioning result in competitive advantages (see Fig. 14.2).

Using the framework given by the marketing strategy and the marketing mix, the propositions and the findings in green marketing research can be described as follows:

1. *Demand measurement*—The rise of green issues in the early 1990s led to numerous polls showing an increasing number of green consumers. These surveys measured market demand by identifying the size of the green market. Demand was mainly presented in terms of percent of the consumers who reported themselves as being green.
2. *Segmentation and targeting*—Segmentation and targeting are closely related concepts. Market segmentation is the process of dividing a market into groups of consumers with different needs, characteristics, or behaviors. In the targeting process these segments are evaluated and the company decides which segment or segments to address (Kotler et al. 2001). The green consumer typologies presented above (see Tables 14.2 and 14.3) are examples of market segmentation on the basis of the green characteristics of the consumers. In line with the general marketing theory, when the segments have been identified, the characteristics of these consumers need to be identified to allow for efficient positioning. As we have seen, a great deal of effort in green marketing has also been put into identifying the characteristics of the green consumer.
3. *Positioning*—Once the green consumer has been identified, the product needs to be positioned on the market to appeal to this group. This is achieved through a careful design of the marketing mix: product, place, price, and promotion. Promotion was recognized early as important in this perspective. However, eco labels came to be the focus of the attempts to position green products.

The most tangible discrepancy between green and conventional marketing perspectives was found regarding positioning. Although intended to be used by the industry and business, it should be remembered that environmental labels (especially type I) have developed into policy instruments rather than marketing means. In the general marketing literature, labels have a subordinate role as positioning tools. A label is considered part of the technical information attached to the product. As such it is classified as 'product' in the marketing mix (Kotler et al.

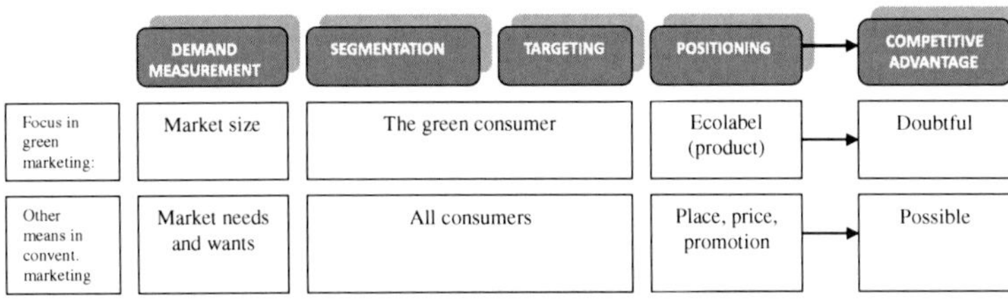

Fig. 14.3 The marketing strategy from two perspectives: current and past focal areas in green marketing and examples of additional marketing means suggested in the conventional marketing literature

2001), and not as a promotion tool. The importance of labels for product positioning is not accentuated in the conventional marketing literature. Instead, the emphasis is on promotion. This stands in stark contrast against the emphasis placed on eco labels in the more recent green literature.

Strategically, market demand measurements have been more focused on the number of existing green consumers than on market needs (and wants) and new market potentials. Tactically, focus has been on informing consumers about technical qualities and not on influencing consumers through promotion. In comparison with green marketing, conventional marketing seems both richer (in terms of means used) and more active (in terms of industry involvement).

Figure 14.3 summarizes the comparison of green and conventional marketing, using the framework of the marketing strategy suggested by Kotler et al. (2001). The figure illustrates that the past and current focal areas in green marketing have been the measurement of market size, identification of the green consumer, and positioning through eco labels.

14.4 Environmental Marketing Claims

Green claims can be found in many advertisements and labels today. They are the marketing response to consumers' increasing interest in protecting the environment.

Institutional consumers also care about buying "green." Indeed, the President of the United States recently issued Executive Orders encouraging federal procurement officers to purchase recycled and environmentally preferable products.

But what do green claims really mean? And when are they considered misleading?

Vague, unsubstantiated, confusing, or misleading information has the potential to reduce consumer confidence in the market, as well as disadvantaging environmental-related business.

As each jurisdiction differs in approach, agencies may like to provide definitions for what constitutes 'environmental marketing claims' in their guidance. These definitions can be specific or generic in nature. Some jurisdictions apply the definitions and classifications as provided in various standards, one example being

Table 14.11 Main aspects of these guides

Aspect	Description	Examples
Distinction between benefits of product, package and service	An environmental marketing claim should be presented in a way that makes clear whether the environmental attribute or benefit being asserted refers to the product, the product's packaging, a service, or to a portion or component of the product, package, or service	**Example 1**: A box of aluminium foil is labeled with the claim "recyclable," without further elaboration. Unless the type of product, surrounding language, or other context of the phrase establishes whether the claim refers to the foil or the box, the claim is deceptive if any part of either the box or the foil, other than minor, incidental components, cannot be recycled. **Example 2**: A soft drink bottle is labeled "recycled." The bottle is made entirely from recycled materials, but the bottle cap is not. Because reasonable consumers are likely to consider the bottle cap to be a minor, incidental component of the package, the claim is not deceptive. Similarly, it would not be deceptive to label a shopping bag "recycled" where the bag is made entirely of recycled material but the easily detachable handle, an incidental component, is not.
Overstatements of environmental attribute	An environmental marketing claim should not be presented in a manner that overstates the environmental attribute or benefit, expressly or by implication. Marketers should avoid implications of significant environmental benefits if the benefit is in fact negligible	**Example 1**: A package is labeled, "50% more recycled content than before." The manufacturer increased the recycled content of its package from 2% recycled material to 3% recycled material. Although the claim is technically true, it is likely to convey the false impression that the advertiser has increased significantly the use of recycled material. **Example 2**: A trash bag is labeled "recyclable" without qualification. Because trash bags will ordinarily not be separated out from other trash at the landfill or incinerator for recycling, they are highly unlikely to be used again for any purpose. Even if the bag is technically capable of being recycled, the claim is deceptive since it asserts an environmental benefit where no significant or meaningful benefit exists.

(continued)

Table 14.11 (continued)

Aspect	Description	Examples
Comparative claims	Environmental marketing claims that include a comparative statement should be presented in a manner that makes the basis for the comparison sufficiently clear to avoid consumer deception. In addition, the advertiser should be able to substantiate the comparison	**Example 1**: An advertiser notes that its shampoo bottle contains "20% more recycled content." The claim in its context is ambiguous. Depending on contextual factors, it could be a comparison either to the advertiser's immediately preceding product or to a competitor's product. The advertiser should clarify the claim to make the basis for comparison clear, for example, by saying "20% more recycled content than our previous package." Otherwise, the advertiser should be prepared to substantiate whatever comparison is conveyed to reasonable consumers. **Example 2**: An advertiser claims that "our plastic diaper liner has the most recycled content." The advertised diaper does have more recycled content, calculated as a percentage of weight, than any other on the market, although it is still well under 100% recycled. Provided the recycled content and the comparative difference between the product and those of competitors are significant and provided the specific comparison can be substantiated, the claim is not deceptive.

Table 14.12 Environmental marketing claims

Aspect	Description	Examples
General environmental benefit claims	It is deceptive to misrepresent, directly or by implication, that a product, package, or service offers a general environmental benefit. Unqualified general claims of environmental benefit are difficult to interpret, and depending on their context, may convey a wide range of meanings to consumers. In many cases, such claims may convey that the product, package or service has specific and far-reaching environmental benefits.	**Example 1**: A brand name like "Eco-Safe" would be deceptive if, in the context of the product so named, it leads consumers to believe that the product has environmental benefits which cannot be substantiated by the manufacturer. The claim would not be deceptive if "Eco-Safe" were followed by clear and prominent qualifying language limiting the safety representation to a particular product attribute for which it could be substantiated, and provided that no other deceptive implications were created by the context. **Example 2**: A product wrapper is printed with the claim "Environmentally Friendly." Textual comments on the wrapper explain that the wrapper is "Environmentally Friendly because it was not chlorine bleached a process that has been shown to create harmful substances." The wrapper was, in fact, not bleached with chlorine. However, the production of the wrapper now creates and releases to the environment significant quantities of other harmful substances. Since consumers are likely to interpret the "Environmentally Friendly" claim, in combination with the textual explanation, to mean that no significant harmful substances are currently released to the environment, the "Environmentally Friendly" claim would be deceptive. **Example 3**: A pump spray product is labeled "environmentally safe." Most of the product's active ingredients consist of volatile organic compounds (VOCs) that may cause smog by contributing to ground-level ozone formation. The claim is deceptive because, absent further qualification, it is likely to convey to consumers that use of the product will not result in air pollution or other harm to the environment.

(continued)

Table 14.12 (continued)

Aspect	Description	Examples
Degradable/ biodegradable/ photodegradable	It is deceptive to misrepresent, directly or by implication, that a product or package is degradable, biodegradable or photodegradable. An unqualified claim that a product or package is degradable, biodegradable or photodegradable should be substantiated by competent and reliable scientific evidence that the entire product or package will completely break down and return to nature, i.e., decompose into elements found in nature within a reasonably short period of time after customary disposal.	**Example 1**: A trash bag is marketed as "degradable," with no qualification or other disclosure. The marketer relies on soil burial tests to show that the product will decompose in the presence of water and oxygen. The trash bags are customarily disposed of in incineration facilities or at sanitary landfills that are managed in a way that inhibits degradation by minimizing moisture and oxygen. Degradation will be irrelevant for those trash bags that are incinerated and, for those disposed of in landfills, the marketer does not possess adequate substantiation that the bags will degrade in a reasonably short period of time in a landfill. The claim is therefore deceptive. **Example 2**: A commercial agricultural plastic mulch film is advertised as "Photodegradable" and qualified with the phrase, "Will break down into small pieces if left uncovered in sunlight." The claim is supported by competent and reliable scientific evidence that the product will break down in a reasonably short period of time after being exposed to sunlight and into sufficiently small pieces to become part of the soil. The qualified claim is not deceptive. Because the claim is qualified to indicate the limited extent of breakdown, the advertiser need not meet the elements for an unqualified photodegradable claim, i.e., that the product will not only break down, but will also decompose into elements found in nature.

(continued)

Table 14.12 (continued)

Aspect	Description	Examples
Compostable	It is deceptive to misrepresent, directly or by implication, that a product or package is compostable. A claim that a product or package is compostable should be substantiated by competent and reliable scientific evidence that all the materials in the product or package will break down into, or otherwise become part of, usable compost (e.g., soil-conditioning material, mulch) in a safe and timely manner in an appropriate composting program or facility, or in a home compost pile or device.	**Example 1**: A manufacturer indicates that its unbleached coffee filter is compostable. The unqualified claim is not deceptive provided the manufacturer can substantiate that the filter can be converted safely to usable compost in a timely manner in a home compost pile or device. If this is the case, it is not relevant that no local municipal or institutional composting facilities exist. **Example 2**: A lawn and leaf bag is labeled as "Compostable in California Municipal Yard Trimmings Composting Facilities." The bag contains toxic ingredients that are released into the compost material as the bag breaks down. The claim is deceptive if the presence of these toxic ingredients prevents the compost from being usable. **Example 3**: A manufacturer makes an unqualified claim that its package is compostable. Although municipal or institutional composting facilities exist where the product is sold, the package will not break down into usable compost in a home compost pile or device. To avoid deception, the manufacturer should disclose that the package is not suitable for home composting.

(continued)

Table 14.12 (continued)

Aspect	Description	Examples
Recyclable	It is deceptive to misrepresent, directly or by implication, that a product or package is recyclable. A product or package should not be marketed as recyclable unless it can be collected, separated or otherwise recovered from the solid waste stream for reuse, or in the manufacture or assembly of another package or product, through an established recycling program. Unqualified claims of recyclability for a product or package may be made if the entire product or package, excluding minor incidental components, is recyclable.	**Example 1**: A packaged product is labeled with an unqualified claim, "recyclable.'' It is unclear from the type of product and other context whether the claim refers to the product or its package. The unqualified claim is likely to convey to reasonable consumers that all of both the product and its packaging that remain after normal use of the product, except for minor, incidental components, can be recycled. Unless each such message can be substantiated, the claim should be qualified to indicate what portions are recyclable. **Example 2**: A nationally marketed 8 oz. plastic cottage-cheese container displays the Society of the Plastics Industry (SPI) code (which consists of a design of arrows in a triangular shape containing a number and abbreviation identifying the component plastic resin) on the front label of the container, in close proximity to the product name and logo. The manufacturer's conspicuous use of the SPI code in this manner constitutes a recyclability claim. Unless recycling facilities for this container are available to a substantial majority of consumers or communities, the claim should be qualified to disclose the limited availability of recycling programs for the container. If the SPI code, without more, had been placed in an inconspicuous location on the container (e.g., embedded in the bottom of the container) it would not constitute a claim of recyclability.

(continued)

Table 14.12 (continued)

Aspect	Description	Examples
Recycled content	A recycled content claim may be made only for materials that have been recovered or otherwise diverted from the solid waste stream, either during the manufacturing process (pre-consumer), or after consumer use (post-consumer). To the extent the source of recycled content includes pre-consumer material, the manufacturer or advertiser must have substantiation for concluding that the pre-consumer material would otherwise have entered the solid waste stream. In asserting a recycled content claim, distinctions may be made between pre-consumer and post-consumer materials. Where such distinctions are asserted, any express or implied claim about the specific pre-consumer or post-consumer content of a product or package must be substantiated.	**Example 1**: A manufacturer routinely collects spilled raw material and scraps left over from the original manufacturing process. After a minimal amount of reprocessing, the manufacturer combines the spills and scraps with virgin material for use in further production of the same product. A claim that the product contains recycled material is deceptive since the spills and scraps to which the claim refers are normally reused by the industry within the original manufacturing process, and would not normally have entered the waste stream. **Example 2**: A manufacturer purchases material from a firm that collects discarded material from other manufacturers and resells it. All of the material was diverted from the solid waste stream and is not normally reused by the industry within the original manufacturing process. The manufacturer includes the weight of this material in its calculations of the recycled content of its products. A claim of recycled content based on this calculation is not deceptive because, absent the purchase and reuse of this material, it would have entered the waste stream. **Example 3**: A dealer of used automotive parts recovers a serviceable engine from a vehicle that has been totaled. Without repairing, rebuilding, remanufacturing, or in any way altering the engine or its components, the dealer attaches a "Recycled" label to the engine, and offers it for resale in its used auto parts store. In this situation, an unqualified recycled content claim is not likely to be deceptive because consumers are likely to understand that the engine is used and has not undergone any rebuilding.

(continued)

Table 14.12 (continued)

Aspect	Description	Examples
Source reduction	It is deceptive to misrepresent, directly or by implication, that a product or package has been reduced or is lower in weight, volume or toxicity. Source reduction claims should be qualified to the extent necessary to avoid consumer deception about the amount of the source reduction and about the basis for any comparison asserted.	**Example 1**: An ad claims that solid waste created by disposal of the advertiser's packaging is "now 10% less than our previous package." The claim is not deceptive if the advertiser has substantiation that shows that disposal of the current package contributes 10% less waste by weight or volume to the solid waste stream when compared with the immediately preceding version of the package. **Example 2**: An advertiser notes that disposal of its product generates "10% less waste." The claim is ambiguous. Depending on contextual factors, it could be a comparison either to the immediately preceding product or to a competitor's product. The "10% less waste" reference is deceptive unless the seller clarifies which comparison is intended and substantiates that comparison, or substantiates both possible interpretations of the claim.
Refillable	It is deceptive to misrepresent, directly or by implication, that a package is refillable. An unqualified refillable claim should not be asserted unless a system is provided for: (1) the collection and return of the package for refill; or (2) the later refill of the package by consumers with product subsequently sold in another package. A package should not be marketed with an unqualified refillable claim, if it is up to the consumer to find new ways to refill the package.	**Example 1**: A container is labeled "refillable x times." The manufacturer has the capability to refill returned containers and can show that the container will withstand being refilled at least x times. The manufacturer, however, has established no collection program. The unqualified claim is deceptive because there is no means for collection and return of the container to the manufacturer for refill. **Example 2**: A bottle of fabric softener states that it is in a "handy refillable container." The manufacturer also sells a large-sized container that indicates that the consumer is expected to use it to refill the smaller container. The manufacturer sells the large-sized container in the same market areas where it sells the small container. The claim is not deceptive because there is a means for consumers to refill the smaller container from larger containers of the same product.

(continued)

Table 14.12 (continued)

Aspect	Description	Examples
Ozone safe and ozone friendly	It is deceptive to misrepresent, directly or by implication, that a product is safe for or "friendly" to the ozone layer or the atmosphere.	**Example 1**: A product is labeled "ozone friendly." The claim is deceptive if the product contains any ozone-depleting substance, including those substances listed as Class I or Class II chemicals in Title VI of the Clean Air Act Amendments of 1990, Pub. L. No. 101-549, and others subsequently designated by EPA as ozone-depleting substances. Chemicals that have been listed or designated as Class I are chlorofluorocarbons (CFCs), halons, carbon tetrachloride, 1,1,1-trichloroethane, methyl bromide and hydrobromofluorocarbons (HBFCs). Chemicals that have been listed as Class II are hydrochlorofluorocarbons (HCFCs). **Example 2**: An aerosol air freshener is labeled "ozone friendly." Some of the product's ingredients are volatile organic compounds (VOCs) that may cause smog by contributing to ground-level ozone formation. The claim is likely to convey to consumers that the product is safe for the atmosphere as a whole, and is therefore, deceptive.

ISO 14021:1999 for self-declared environmental claims. Other jurisdictions provide a generic approach and apply their guides to all advertising and other forms of marketing that apply to any claim about the environmental attributes of a product, package, or service. In addition, guidance may wish to refer to environmental marketing claims as 'green claims' or provide an explanatory note that the guide will utilize both terms throughout the guide.

One aim behind environmental claim guidance is to educate business and consumers that self-declared marketing claims should be made in accordance with good marketing practices in compliance with the domestic legislation. This may be that the claims should not be misleading, fraudulent, deceptive, or unfair depending on legislative provisions.

Environmental marketing claims are difficult by nature. It is hard for a consumer to verify the accuracy of an environmental claim by looking at the product or packaging or without having the relevant scientific knowledge.

Some agencies include some of the following principles in their current guidelines:

- make it clear between benefits of product, package, and service;
- not to overstate or exaggerate a benefit or environmental attribute;
- provide clear, prominent, and understandable qualifications or disclosures (when required) to prevent deception or misunderstanding;
- present comparative claims in a way that avoids consumer deception/misunderstanding and is able to be substantiated/verified; and/or
- provide consideration of the whole product life cycle.

Federal Trade Commission developed guides for environmental advertising and marketing practices. They provide the basis for voluntary compliance with such laws by members of the industry (Table 14.11).

These guides apply to environmental claims included in labeling, advertising, promotional materials, and all other forms of marketing, whether asserted directly or by implication, through words, symbols, emblems, logos, depictions, product brand names, or through any other means, including marketing through digital or electronic means, such as the Internet or electronic mail. The guides apply to any claim about the environmental attributes of a product, package or service in connection with the sale, offering for sale, or marketing of such product, package or service for personal, family or household use, or for commercial, institutional or industrial use.

An important part of these guides is made up of Environmental marketing claims. Table 14.12 shows the main points.

References

Azzone G, Manzini R (1994) Measuring strategic environmental performance. Bus Strategy Environ 3(1):1–14

Barrett J (1993) The future for environmental managers. ENDS—Environ Manag Bus 82–87

Belz F (2001) Integratives Oeko-marketing—Erfolgreiche Vermarktung oekologischer Produkte und Leistungen. DUV, Wiesbaden
Belz F, Bilharz M (2005) Nachhaltigkeits-Marketing in Theorie und Praxis. DUV, Wiesbaden
Bloom GF, Morton MS (1991) Hazardous Waste is Every Manager's Problem. Sloan Manag Rev Summer:75–84
Blumenfeld K, Gilbert J (1990) The Green Consumer. Arthur D. Little Center for Environmental Assurance, Cambridge
Charter M, Peattie K, Ottman J, Polonsky MJ (2002) Marketing and sustainability, association with the centre for sustainability design. Centre for Business Relationships, Accountability, Sustainability and Society (BRASS), UK
Cleff T, Rennings K (1999) Determinants of environmental product and process innovation. European Environ 9(5):191–201
Coddington W (1993) Environmental marketing: positive strategies for researching the green consumer. McGraw-Hill, New York
Cohen J, Cohen P (1983) Applied multiple regression/correlation analysis for the behavioral sciences, 2nd edn. Lawrence Erlbaum Associates, Hillsdale
Coop Sweden (2004) Miljökvitto 1, 2004 [Environmental receipt 1, 2004]. http://www.coop.se/includefiles/MODULER/CCMS/show_page.asp?iMappeID¼246&sSideNavn¼Milj%F6kvittoI%2Cþ2004. Accessed Jan 2006
Cooper RG (1988) Pre-development activities determine new product success. Ind Mark Manag 17:237–247
Diamantopoulos A, Schlegelmilch BB, Sinkovics RR, Bohlen GM (2003) Can socio-demographics still play a role in profiling green consumers? A review of the evidence and an empirical investigation. J Bus Res 56(6):465–480
Erskine CC, Collins L (1997) Eco-labelling: success or failure? The Environmentalist 17:125–133
Evans R (1990) The Earth's new friends? Int Manag 7:26–31
Flattau E (1990) Wiping out Pollution. Europe 13–15
Frechette P (2009) Beyond recycled content. http://www.fivewinds.com/english/news-and-events/. Accessed 21 July 2009
Johne A, Snelson P (1988) Auditing product innovation activities in manufacturing firms. R&D Manag 11(2):227–233
Kammerer D (2009) The effects of customer benefit and regulation on environmental product innovation. Empirical evidence from appliance manufacturers in Germany. Ecol Econ 68(2009):2285–2295
Kotler P, Armstrong G, Saunders J, Wong V (2001) Principles of marketing, 3rd European edn. Prentice Hall, Harlow
Lampe M, Gazdat GM (1995) Green marketing in Europe and the United States: an evolving business and society interface. Int Bus Rev 4(3):295–312
Maidique MA, Zirger BJ (1984) A study of success and failure in product innovation: the case of the US electronics industry. IEEE Trans Eng Manag EM-31:192–203
Mandese J (1991) New study finds green confusion. Advertising Age, 21 October
Mathews R (1990) US headed for total normalcy: environmental conservation is imminent and is a great marketing opportunity. Grocery Marketing, May
Menon A, Menon A (1997) Enviropreneurial marketing strategy: the emergence of corporate environmentalism as market strategy. J Mark 61:51–67
Meyer A (2001) What's in it for the customers? Successfully marketing green clothes. Bus Strategy Environ 10:317–330
Miles MP, Russell GR (1997) ISO 14000 Total quality environmental management: the integration of environmental marketing, total quality management, and corporate environmental policy. J Qual Manag 2(1):151–168
Miles MI, Munilla LS (1993) Eco-orientation: an emerging business philosophy. J Market Theory Pract 1(2):43–51

Orlin J, Swalwell P, FitzGerald C (1993/1994) How to integrate information strategy planning with environmental management information systems—Part 1. Total Qual Environ Manag Winter:193–202
Ottman JA (1994) Green marketing: challenges and opportunities. NTC Business Books, Lincolnwood
Ottman JA (1993) Green marketing. NTC Business Books, Lincolnwood
Pavitt K (1984) Sectoral patterns of technical change: towards a taxonomy and a theory. Res Policy 13(6):343–373
Peacock M (1993) Developing environmental performance measures. Ind Eng 25:20–22
Peattie K (1995) Environmental marketing management: meeting the green challenge. Pitman Publishing, London
Peattie K (1992) Green marketing. Pitman Publishing, London
Peattie K (2001) Golden goose or wild goose? The hunt for the green consumer. Bus Strategy Environ 10(4):187–199
Porter ME, van der Linde C (1995) Green and competitive: ending the stalemate. Harv Bus Rev 73:120–133
Prakash A (2002) Green marketing, public policy and managerial strategies. Bus Strategy Environ 11(5):285–297
Pujari D, Wright G, Peattie K (2003) Green and competitive influences on environmental new product development performance. J Bus Res 56:657–671
Reinhardt FL (1998) Environmental product differentiation: implications for corporate strategy. California Manag Rev 40(4):43–73
Rennings K (2000) Redefining innovation—eco-innovation research and the contribution from ecological economics. Ecol Econ 32(2):319–332
Rex E, Baumann H (2007) Beyond ecolabels: what green marketing can learn from conventional marketing. J Clean Prod 15(2007):567–576
Rubik EF, Frankl EP (eds) (2005a) The future of eco-labelling. Greenleaf Publishing, Sheffield
Rubik EF, Frankl EP (eds) (2005b) The future of eco-labelling. Greenleaf Publishing, Sheffield
Sheets K (1990) Business's Green Revolution. US news and World report, 19 February, pp 4548
Sheth JN, Parvatiyar A (1995) Ecological imperatives and the role of marketing. In: Polonsky MJ, Mintu-Wimsatt AT (eds) Environmental marketing: strategies, practice, theory, and research. Haworth Press, New York, pp 3–20
Shimnel P (1991) Corporate environmental policy in practice. Long Range Plan 24(3):10–17
Straughan RD, Roberts JA (1999) Environmental segmentation alternatives: a look at green consumer behaviour in the new millennium. J Consumer Market 16(6):558–575
Sullivan MS, Ehrenfeld JR (1992/1993) Reducing life-cycle environmental impacts: an industry survey of emerging tools and programs. Total Qual Environ Manag 143–157
The Economist (1990b) Friendly to Whom? 7 April
The Economist (1991a) The Dirty Dozen. 20 July
The Economist (1991b) Eco-babble. 21 September
Tully S (1989) What the "Greens" mean for business. Fortune, pp. 159–164, 23 Oct 1989
USAEP (2001) Greening the supply chain, U.S.—Asia environmental program. http://www.usaep.org/ctem/greening.htm
Van Waterschoot W, Van den Bulte C (1992) The 4P classification of the marketing mix revisited. J Market 56(4):83–93
Vigon BW, Curran MA (1993) Life-cycle improvement analysis: procedure development and demonstration. Proc IEEE Int Symp Electron Environ 151–156
Voss B (1991) The green marketplace. Sales and marketing management, July, pp. 74–76
Reitman V (1992) "Green" product sales seem to be wilting. Wall Str J B1
Wallace DR, Suh NP (1993) Information-based design for environmental problem solving. Ann CIRP 42(1):175–180
Walley N, Whitehead B (1994) It's not easy being green. Harv Bus Rev 72:46–52

Chapter 15
Supply Chain Environmental Policy

15.1 Introduction

The idea of a sustainable link between business and environment, initially proposed about two decades ago, revolves around the central thesis that the goals of environmental conservation and the goals of business need not be disparate and conflicting (Barbier 1987; Hawken et al. 1999; Holliday). Proponents of environmental sustainability have now taken this thesis one step further with the argument that environmentally conscious and ecologically friendly strategies could, in fact, lead to competitive advantages and superior financial performance (Esty and Winston 2006; Hart 2005). While earlier views were dominated by notions that environmental objectives were a constraint to the economic goals of a business or that the economic objectives of a business were a direct threat to environmental conservation, the recent approach treats both economics and ecology as two sides of the same coin. It has been argued by Hart (2005) that when properly focused, the profit motive of business can accelerate the transformation toward global sustainability, with nonprofit, government, and multilateral agencies, all playing crucial roles as collaborators. Savitz and Weber (2006) suggest that a sustainable supply chain is one that creates profit for its shareholders while protecting the environment and improving the lives of those who meet interacts. Recently, Lash and Wellington (2007) suggested that firms involved in a supply chain will be at a competitive disadvantage if they do not pay attention to sustainability issues.

Marketing has been long concerned with understanding environmentally conscious consumers and devising appropriate strategies to target such consumers (Antil 1984; Ellen et al. 1991). In both marketing as well as management strategy, it has been argued that managerial decision-making must incorporate environmental issues, including ideas on resource conservation and environmental sustainability (Drumwright 1994; Hart 1995; Shrivastava 1995). Incorporating consumers' and managerial concerns on the natural and physical environment contributes not only to superior business performance, especially in terms of

M. Bevilacqua et al., *Design for Environment as a Tool for the Development of a Sustainable Supply Chain*, DOI: 10.1007/978-1-4471-2461-0_15,

competitive advantage, but also to enhance corporate reputation (Menon and Menon 1997; Sisodia et al. 2007).

Baylis et al. (1998) argue that larger firms have better opportunities and abilities to reduce environmental impacts due to their higher amount of financial and human resources. Additionally, Greening and Gray (1994) contend that larger firms may be subject to greater public scrutiny. Several empirical studies show that firm size has a positive effect on firms' environmental activities in general (Coglianese and Nash 2001; Melnyk et al. 2003) and on EP-innovation in particular (Cleff and Rennings 1999; Hitchens et al. 2000; Rehfeld et al. 2007). In contrast, Wagner (2008) as well as Seijas-Nogareda (2007)and Engels (2008) do not support this influence of firm size on EP-innovation. R&D expenditure is a common proxy for and closely related to a firm's innovation activity (Acs and Audretsch 1988). Although R&D does not automatically lead to innovations, R&D is still the most widely used strategy aiming at innovation. Rehfeld et al. (2007) found empirical evidence that R&D activities also have a positive influence on EP-innovation.

15.2 Ecological Reputation and Supply Chain Strategies

Within marketing, there are two streams of research that could be used to support the link between sustainability and superior financial performance. First, resource-based theory suggests that better access and utilization of resources will lead to competitive advantage and therefore better performance in terms of profitability (Hunt and Morgan 1995). Second, empirical evidence suggests that ecologically conscious policies lead to better customer retention which again leads to better performance (Sisodia et al. 2007). The means to achieve such environmental sustainability is through the maximization of value addition with the least use of resources, least amount of waste, and least pollution (Lovins et al. 1999).

The latter approach, with its sights focus clearly on profits and competition as well as on environmental sustainability, has had the greatest appeal for scholars within various business disciplines.

There is now a growing recognition that environmentally-friendly product strategies gain better customer endorsements and therefore, contribute to long-term profits. Companies with clear environmental positioning in the market, such as Ben and Jerry, Body Shop, and Patagonia, are often cited as successful societal marketing examples (Kotler 2003).

Since these firms enjoy an ecological reputation, they seek business suppliers that also are ecologically-conscious. Thus, environmentally responsible actions not only target an otherwise ignored subset of environmentally-conscious customers, but also by building a green supply chain, they also enable the firm to develop distinct advantages over their competitors (Winsemius and Guntram 1992).

Sharma et al. (2010) focused on the domain of business-to-business marketing and develop a framework that highlights the role of marketing in environmentally sustainable supply chain strategies. In developing this framework, they integrate prior research from several business disciplines, highlight ignored areas of study, and provide justifications for additional theoretical and empirical research in this area. In addition, they develop propositions that will hopefully guide future research (see Box V).

Box V: Sustainability and Business-to-Business Marketing

Sharma et al. (2010) develop a framework (Fig. 15.1) that is based on two major objectives in sustaining environments. First, when firms do not manufacture more units than are required (over-produce), a reduction in over-supply occurs that leads to lower levels of products needing to be disposed (that may need recycling or remanufacturing), leading to a more sustainable environment. Sharma et al. (2010) label this strategy as reducing surplus supply. Second, firms can reduce the number of products that need recycling. Sharma et al. (2010) label this strategy reducing reverse supply and suggest that firms need to develop repairable products as well as more complete recycling and remanufacturing strategies.

Their focus in this paper is on identifying the important role of marketing in environmentally sustainable strategies. Figure 15.1 highlights the critical role of internal and external marketing in the successful implementation of environmental strategies. As will be observed in subsequent sections, both internal marketing and external marketing are important for a successful approach to sustainability. The recent literature within marketing makes a distinction between external marketing and internal marketing. External marketing refers to marketing strategies and activities outside the firm,

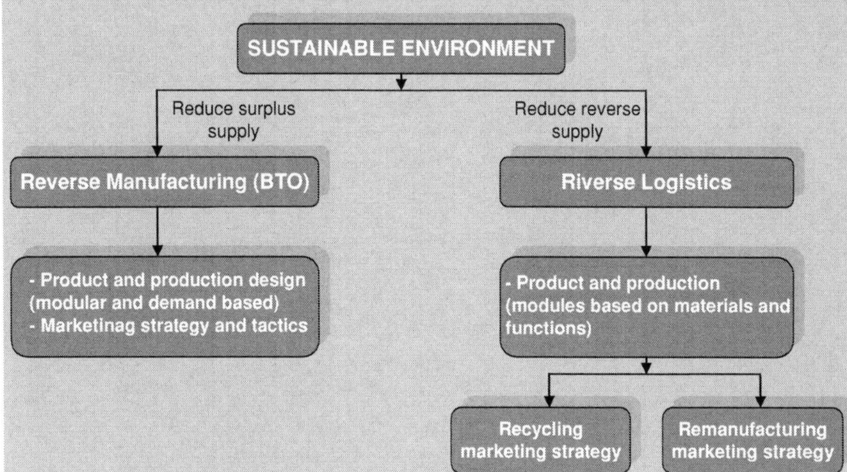

Fig. 15.1 The sustainable market framework

Table 15.1 Three paths to sustainable environments

Strategy	Reduce surplus supply–produce to demand	Reduce reverse supply chain	
		Recycle products	Remanufacture products
Key benefit to sustainable environments	Reduce demand of unneeded functions. Reduce recycling of new products.	Recycling resources. Meet emerging government regulations.	Reduce demand for recycling. Reduce demand for new parts.
Internal process changes/ internal marketing requirements	High Product design changes for minimizing inputs and maximizing output configurations. Manufacturing changes. Supply chain changes. Marketing changes.	Moderate Product design change for recycling.	Low in the short term Moderate in the long term. Outsource initially. Product design changes for remanufacturing. Marketing changes.
Key customer groups	Current customers. Customer who seek customization	Recycling firms	Emergine markets (cannot afford new products now but may in the future). Non-locational developing markets
External marketing focus	Customization/costs	Easy for recycling	Remanufactured products provide similar benefits
Firm's ability to implement	Low	Moderate	High
Set-up cost	High	Low	Low to moderate
Expected ROI	High	Low	Moderate
Ability to match regulations	Low	High	Moderate
Current research	High in manufacturing/ supply chain. None in marketing	Moderate in manufacturing/ supply chain. Dated in marketing	Very high in manufacturing/ supply chain. None in marketing

i.e., those employed to attract or retain customers or build market share. Internal marketing refers to the marketing of process changes within firms, especially the communications necessary to successfully deploy new organizational strategies. While this distinction has been advanced in the area of services marketing (Gronroos 1990), internal marketing is necessary to achieve a greater interdepartmental consensus and coordination for all firms (Narver and Slater 1990). In the context of environmental sustainability as well, internal marketing efforts are especially important, since a coordinated effort is required across all functional departments. In discussing their framework, they provide the implications on both internal and external marketing efforts of the firm. The three major strategies discussed in the framework are also elaborated in Table 15.1.

Surplus supply (over production) typically has two implications for firms—discounting or non-consumption of the product. First, discounting results in business customers purchasing features that they will not utilize during the life of the product. Therefore, the materials that were utilized in providing unnecessary functionality will result in an unnecessary need for recycling. Examples of such features are DVD players in laptops that business customers may not use but are bundled with the system. Second, due to over supply, some products may never be sold and may go to recycling without reaching the consumer. Such recycling may result in materials going back to the same cycle of non-use or may result in harvesting for spare parts. Oversupply thus increases the burden on costly recycling efforts.

One key strategy to reduce surplus supply is to produce only after an order has been placed or build-to-order (BTO).

Due to changes in the internal processes, BTO renders internal marketing quite difficult. Moreover, the focus of the entire marketing function also changes. BTO and lean manufacturing will increasingly make the marketing function responsible for "supply management" (Sharma and LaPlaca 2005). The customer will be the starting point for marketing activities for multiple reasons. The increasing diversity in needs, wants, and resources of businesses and households will make customer behavior inherently less predictable and forecasting less accurate (Sheth et al. 2000). This may lead to a shift to qualitative techniques of forecasting from traditional quantitative techniques. In such an environment, companies that succeed will be those that can rapidly adjust their supply to meet demand; in other words, those that practice demand-driven supply management. For example, airlines use yield management to optimally allocate available capacity across fare classes, and to manage demand to match capacity. Many airlines such as Continental Airlines are now able to dynamically manage capacity by exchanging crew-compatible fleets.

Reduce reverse supply—recycle products

Recycling has been discussed extensively in the marketing context. Recycling enables firms to meet regulations and aids in the creation of sustainable environments. While recycling is costly, it is easy to implement, as third party vendors undertake the actual recycling efforts. For example, Wal-Mart is currently investing about $500 million to develop a system that would reduce the firm's energy consumption, greenhouse gas emissions, and production of solid waste (Zimmerman 2006). One program it is encouraging is that of closed-loop recycling, whereby the firm's waste would be sold to recycling centers for use as raw material. It is selling waste paper to Georgia-Pacific Corp. which converts such waste into paper towels and tissues that can be sold in Wal-Mart stores under its own private label (Zimmerman 2006). This reduces the supply of waste paper that needs to be recycled. In the context of environmental conservation, a key contribution of firms is to engage in modular product designs that allow customers to readily change one part of the product without discarding the entire product. This leads to a reduction in recycling commitments. For example, the average cost of a television or a PC is today less than $1000. However, when a TV stops working, customers cannot diagnose the problems with the television. Customers need to decide whether they want to pay service personnel to diagnose the problem, and then pay additional repair costs. A large proportion of customers do not like the uncertainty of repair costs and throw the television away rather than have it repaired (Lewis 2008).

Reduce reverse supply—remanufacturing products

One of the more attractive strategies for sustainable environments is remanufacturing of products. Typically, in remanufacturing products, a used product is examined and repaired so as to make the product functional and to extend the life of the product. This strategy is becoming more prevalent for several reasons. First, firms are under pressure to contribute to environmental sustainability. This pressure is largely from the government through rules and regulations and to a smaller degree from customers. This focus is more critical in the electronics industry where components are harder to recycle and contain materials that harm the environment. Second, there is a growing segment of marginal customers who cannot afford new products now (but may in the future) but are interested in acquiring the product (c.f., Robotis et al. 2005).

Internal marketing may be less difficult, as firms need not initially change their internal processes. However, as stated earlier, internal marketing is more difficult when product design changes are proposed. Marketing's focus on the customer also changes since there are now two key customer groups—one that buys only new products and another that is content with remanufactured products. Adverse impacts on the firm's image can be avoided by marketing remanufactured products to markets where the firm currently has no presence. Such a strategy is used by Caterpillar to sell its remanufactured products in developing countries.

15.3 Environmental Labels and Declarations

Consumer product manufacturers are beginning to find it necessary to label their products with stickers that claim adherence to certain environmental impact production standards. Labels allow consumers to chose products based on

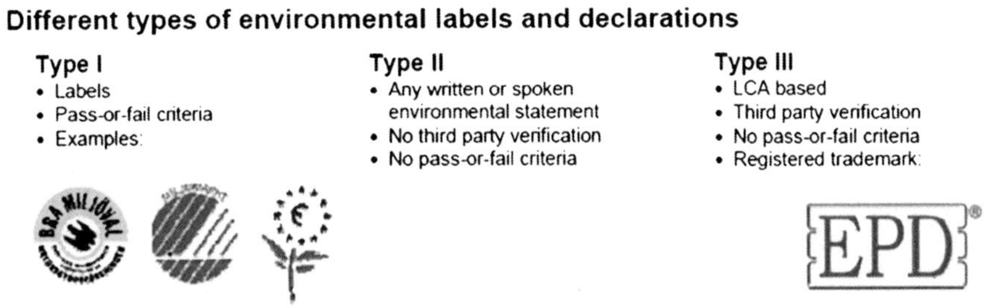

Fig. 15.2 Different types of environmental labels and declarations

environmental characteristics, thus signaling their preferences to the producer through marketplace purchasing.

Internationally, consumers are increasingly demanding quantifiable information on the environmental impacts of the products they buy and governments are increasingly utilizing market-based policies to implement positive environmental change in the marketplace. These trends have led to the development of multiple labeling systems which notify customers of the environmental characteristics of the products they are purchasing (see Box X and W).

Box X: ISO 14000 Environmental Labels

The **ISO 14000 Environmental Labels and Declarations Collection** establishes the guiding principles for the development and use of environmental labels and declarations. It also provides the requirements for self-declared environmental claims, developing environmental labeling programs, and quantifying environmental declarations.

The ISO 14000 Environmental Labels and Declarations Collection

The ISO 14000 Environmental Labels and Declarations Collection comprises:
ISO 14020:2000 Environmental labels and declarations—General principles

ISO 14021:1999 Environmental labels and declarations—Self-declared environmental claims (Type II environmental labeling)

ISO 14024:1999 Environmental labels and declarations—Type I environmental labeling—Principles and procedures

ISO 14025:2006 Environmental labels and declarations—Type III environmental declarations—Principles and procedures (Fig. 15.2)

Box W: Sustainable Products from the Start

An eco-labeling tool helps a leading supplier stay ahead of the curve

Today's shopper is no stranger to eco-labels. ENERGY STAR®, EPEAT®, and Green SealTM are only a few of the innumerable logos certifying that products meet certain environmental requirements. In light of this trend, many consumer-facing companies are considering showcasing eco-labels on their products as an attempt to win the hearts-and wallets-of eco-conscious consumers in a struggling economy.

But what if a company does not sell directly to the consumer, and is in fact only one element of the end product's supply chain? Can these suppliers still take advantage of growing consumer interest in eco-labels? BASF, a leading chemical manufacturer, has demonstrated leadership in its supply chain by proactively measuring its products' environmental impacts against the eco-label requirements faced by its customers with the help of a firm of consulting services versatile eco-labeling tool.

Thinking ahead to the end consumer

As part of a *Product Sustainability Round Table* pilot project in 2007, BASF requested to research the specific requirements associated with the most popular eco-labels on the market. BASF hoped to learn more about the criteria its customers are required to meet when they attempt to achieve certification. "We want to be strategic partners with our customers, understanding their product specifications in order to help them meet their requirements," said Patrick Meyer, Senior Product Stewardship Specialist with BASF.

To begin the project, the consulting firm analyzed the current labels on the market and created a tool that was able to search and sort through a database depending on the user's request. For instance, if the user wished to learn about labels that are used in the electronics industry in Europe, the tool would generate a list of labels pertaining to those specifications. BASF was especially interested in discovering the most influential or popular labels in specific geographic locations or industries. Identifying these eco-labels would help the company in its aim to compare their products' environmental performance against the labels their customers often choose to pursue. To broaden its versatility, the consulting firm enabled the tool to document the number of eco-labels that share similar requirements, such as the percentage of recycled content required. Lastly, BASF requested that the tool include information on when each eco-label is updated with new criteria and how this process occurs. This feature allows the company to consider the possibility for input and participation into the labels' review process. "We are using this tool for internal assessment, but will also work in partnership with our customers and other external stakeholders" said Meyer.

BASF-which has developed the tool even further into the BASF SELECT™ (Sustainability, Eco-Labeling and Environmental Certification Tracking) Tool-is now tracking eco-labeling trends to better understand their customer's drivers of environmental performance. Most importantly, the company can now compare the eco-labels' requirements to the sustainability features of their products. "In the value chain, BASF is upstream from the end product, yet it is demonstrating leadership with its interest in global eco-labels," said Laura Flanigan at Five Winds. "BASF aims to fully understand the environmental aspects associated with the products that use its materials, and it uses that information to improve its environmental profile."

The future is bright for eco-labels

BASF and other companies that consider their customers' drivers for sustainability stand to benefit greatly from the growing use of eco-labels and other factual data on environmental performance. The collective need of consumers for an easy way to identify environmentally preferable products is not bound to fade anytime soon. The European Commission has recently recommitted to fostering green procurement and has focused on scaling up the use of the EU flower—a Europe-wide symbol of environmentally-friendly products. At the same time, several U.S. thought leaders are striving to tackle eco-label confusion and help citizens recognize the information they can trust. Most recently, the World Resources Institute received a grant from Wal-Mart to develop a Green Standards Guide that will rate various eco-labels to help companies decide which ones best fit their needs. These and other efforts across the world will continue to raise the expectations and profiles that are communicated on products for decades to come (extracted from http://www.fivewinds.com/; 2010).

15.3.1 Environmental Product Declaration (EPD)

The majority of labeling programs are limited to products that have published Environmental Product Declaration (EPD) reports that quantify their life cycle impacts on the environment. For the most part, EPD programs have similar characteristics and constraints.

The labeling programs set out to achieve the same end: create a market driven move toward sustainability. In order to accomplish this, labeling programs lay out objectives very similar to those of Sweden's EPD program; credible, neutral, comparable, open to all products and services, open to all interested parties, environmental impact oriented, instructive, and continuously updated. In order to meet these objectives, the programs set up environmental standards for products with the help of multiple stakeholder groups, including industry. For example, in Canada's Environmental Choice Program, criteria for products go through a stringent review process by industry, consumer groups, environmental groups, and government.

In addition, labeling programs require ISO 14000 pre-certification for industry involvement. The majority of EPD systems set environmental criteria for products through the same stringent step-by-step process. Initially, industry must perform a Life Cycle Assessment (LCA) with the boundaries of the assessment agreed upon by all the stakeholder groups. Once all the information are gathered, the stakeholders must agree to a set of cut off criteria over which products can be eligible for the label. For example, for a refrigerator to qualify for an EU ecolabel it must have an ozone depletion potential of zero and a global warming potential of 15 or under. These key criteria normally fall under pre-determined ecological criteria set by the labeling organization for each group of products. Under the EU system, ecological criteria include the reduction of ozone-depleting and global warming creating substances.

In order to notify the customer of product approval, labeling programs brand themselves differently. Ecolabel Denmark has a swan while the EU Ecolabel has a flower. Customers know, when they see a label, it implies that a product surpasses certain standards. However, there are short comings to these labeling programs that need to be highlighted. Labeling programs require industry buy in, limit consumer preferences, and often set differing criteria for products. Because labeling programs require extensive LCAs, it is imperative to have industry involved in every step of the process or it is likely they would chose to disengage from the program all together. However, one must question the setting of criteria and the transparency of the LCA if industry is helping to decide where to draw the boundaries of LCAs and where to set the criteria limits. It is often in their best interest to keep certain environmental risks under wraps.

In addition, labels give a generic stamp of approval for limited and specific criteria, making it impossible for a consumer to signal if he or she has different environmental preferences than those set by the program. If a consumer is more concerned with landfill waste than energy usage, but labeling programs only consider energy usage, that consumer is not able to choose a product according to his or her preferences. Moreover, labels are often chosen for their design appeal and not for their educational value. The lack of information on a label further negates consumer preferences by making it unclear what the environmental criteria really are. This problem is compounded, specifically in the European Union where labeling programs abound and each program sets criteria differently.

The INTEND project, funded by the EU commission's Life and Environment Program until 2005, is trying to rectify this problem through harmonizing schemes across countries but it is currently unclear how successful they will be. Perhaps the biggest set-back to these programs is that they are what the International Standards Organization (ISO) names "Type III" programs. Type III programs label products with a published EPD with quantifiable information based on an LCA (see Box Z). While useful in their own right, Type III programs do not allow direct comparisons or the weighting of products against one another, they only allow for criteria to be set. The customer can then make his or her own comparisons based on the published criteria. This creates an incentive for more industry partners to buy into the programs, knowing that their products will not specifically be chosen against. However, it does not give the consumer the ultimate information that he or she really needs: all else being equal, which product is the best environmental choice?

Box Z: Type III Environmental Declarations

Type III environmental declarations present quantified environmental information on the life cycle of a product to enable comparisons between products fulfilling the same function. Such declarations:

- are provided by one or more organizations
- are based on independently verified life cycle assessment (LCA) data, life cycle inventory analysis (LCI) data or information modules in accordance with the BS ISO 14040 series and where relevant, additional environmental information
- are developed using predetermined parameters
- are subject to the administration of a programe operator, such as a company or a group of companies, industrial sector or trade association, public authorities or agencies, or an independent scientific body or other organization.

It includes information about the environmental impacts associated with a product or service, such as raw material acquisition, energy use and efficiency, content of materials and chemical substances, emissions to air, soil and water, and waste generation. It also includes product and company information.

EPD is voluntarily developed information and the purpose of an EPD is to provide easy accessible, quality-assured and comparable information regarding environmental performance of products.

There are two documents, which control how the calculations and data collection behind an EPD should be done and what information the EPD must contain; Requirements for the EPD system (MSR) and Product specific requirements (PSR). The MSR contains general requirements for all EPDs and a PSR contains more detailed requirements for each product group.

Description of the product and the company

The first part of the EPD is very straight forward with descriptions of the product and the manufacturer.

The functional unit, which is the unit to which all calculations are referred, can be stated here or in the second part. The functional unit reflects the actual function of the product.

Environmental performance

In most EPDs, important air and water emissions are expressed both as inventory data and as potential influence on different environmental impact categories, for example global warming (GWP). In this case, all emissions contributing to global warming are included in the impact category GWP. Resource consumption is divided into non-renewable and renewable resources. All results of calculations are presented per functional unit, which for e.g. chemicals is 1000 kg of the product. EPDs could also include a presentation of environmental impact from a typical transport to customer. An EPD must be certified by a third party and the declaration is valid for three years. If important process changes are made during the period of validity, updates are needed.

Those responsible for developing Type III environmental declarations and programes based on BS ISO 14025 will need to pay attention to the level of awareness of the target audience, be they large, small, medium sized enterprises, public procurement agencies, or consumers.

BS ISO 14025 specifically establishes the use of the BS ISO 14040 Environmental management series in the development of Type III environmental declaration programes and environmental declarations. It also establishes principles for the use of additional environmental information.

This standard is primarily intended for use in business-to-business communications, but can be used in certain conditions when communicating in a business-to-consumer capacity.

The U.S. product market is dominated by ISO Type I and Type II programs. Type I programs are run by independent third party interests like the Forest Stewardship Council, the Marine Stewardship Council, and LEEDS. In these programs, the third party creates the criteria with no or little industry input. In Type II programs, labels are created by industry setting their own standards. For example, Sony set its own standards for an "eco-info" label that means, among other things, that the product was built with lead-free solder and packaging is made from 100% recycled paper4. In addition the EPA runs a Type III program called "energy star" which, with industry input, labels appliances that are energy efficient.

Like Type III programs, Type I and II programs have there own inherent problems. Often Type I programs suffer from limited information availability, specifically when they are trying to get started. Type II programs, completely company drives, have no system for external auditing and the incentives for a company to reveal limited information are high.

Box AA shows the analysis carried out by Zackrisson et al. (2008) about the experiences of ten European SMEs who have tried to use Stepwise EPDs for market Communication

Box AA: Stepwise Environmental Product Declarations

Since EPDs are based on LCA, SMEs must try to understand and be able to follow a LCA-study if they want to get the full benefits out of using the EPDs.

In short, making an EPD involves the following activities:

A. Making a simplified or streamlined life cycle assessment, LCA, to identify the most significant environmental aspects and impacts of the product;
B. Formulation of Product Category Rules together with interested parties;
C. Making a detailed life cycle assessment to validate and supplement the results of the initial assessment;
D. Drafting of EPD; and
E. Independent verification of the life cycle assessment and the EPD.

Stepwise EPD only consists of steps A, D, and a simplified version of step E. Thus, Stepwise EPD could be seen as a cost effective way of trying out to work with EPDs. But it does not go the whole way of engaging the interested parties in the process, i.e. step B, and making a detailed life cycle assessment, which often means involving the suppliers, i.e. step C.

Table 15.2 outlines the activities of the case study companies and the results achieved in relation to using the Stepwise EPDs for market communication.

As shown in Table 15.2, in terms of new orders, customers' responses on the Stepwise EPDs were disappointing. Only Etac AB experienced that the Stepwise EPD actually qualified them to take part in a bid (concerning building products) that they would otherwise have been excluded from. Some companies (Mercatus, Melitek, Etac, Polisport, and Cruzinox) mention general strengthening of the image and likely long-term effects on sales thanks to the Stepwise EPD.

The main motivation for the SMEs to participate in the project was an anticipation of a demand from some key client or client group. In only one case, Melitek, this anticipated demand was confirmed.

In terms of identifying different ways of using the EPD in marketing of product and/or company, project targets are exceeded in quantitative terms. Even though some positive aspects of EPD as a marketing tool are highlighted above, all of the companies also experienced EPD as a difficult marketing tool. Too technical, does not tell you if the product is environmentally good or bad, incomprehensible terminology, perceived lack of some well-known environmental problems, nothing to compare with and not enough graphics were some typical customer reactions.

Using the underlying LCA as a platform for in-depth communication with selected parties in the supply chain showed more promise.

Table 15.2 Market communications

Company	Customer	Used ways of communicating	Customer response
Huskvarna Prototyper AB	BT, producer of forklift trucks.	Eco-design workshop with client BT	Good response from BT designers, but no interest in EPD
	Automotive industry	Posted on website	
		E-mail to customer's personnel in charge of environment issues	No noted response
			No response yet
Mercatus engineering AB	Power stations burning waste or bio fuels	Given to clients during sales meetings with the words: "We have done an LCA. We know how our equipment impact on the environment".	Appreciation but no interest in any details of the EPD. Not triggering any further discussions
	Process industry		No response yet
		Sent with bids for public contracts including the UN	Strengthening company image
			No response yet
		CEO included in general presentation of company.	
		Environmental part of environmental technology verification	
Etac AB	Retailers of building products	Sent with bids for public contracts	Bids in the building sector required an EPD
		Meeting with public procurement	
	Public procurers	Body Westma	EPD can be used for establishing relevant product requirements in public purchasing
		Customer service	No recorded response
Konto Ltd	Furniture industry	Questionnaire to customers	One out of seven customers thought that EPD could be positive for business. The rest were more negative.
		Given to clients in sales meetings information about Stepwise EPD in printed materials about Konto	No order yet due to the EPD but it contributes to the customers having a more positive picture of Konto

(continued)

Table 15.2 (continued)

Company	Customer	Used ways of communicating	Customer response
Melitek A/S	Pharmaceutical companies	Several discussions and work sessions with one key customer	Very positive response. Cooperation continues.
		Presentation to 5-10 other customers	Impressed. Reinforces Melitek's green image
Danish Company B	Public institutions such as nursing homes, hospitals, day care centres, and municipalities	Presenting the EDP to environmental and technical managers and discussing the need for environmental documentation of products.	The EDP is too technical, not understandable and not credible. Cost, not environment is important when purchasing rehab aids.
		Presenting the EDP to public procurers and asking for a written evaluation of the EDP as a tool for green public procurement.	The EDP is of no use in this context; it is too technical. Environmental issues are not important compared to issues like price.
		Meetings with therapists responsible for selecting rehab aids focusing on how to select rehab aids based on environmental merits.	Lot of support for the idea of going green, but the information has to be brief and present real advantages for customers and users. Cost, not environment is important.
		An open questionnaire to visitors to the Scandinavian rehab exhibition	The environmental information must be brief, relevant and present real advantages for the customer. Instructions or labels on the product are preferred.
Danish Company C	Mainly private property owners in densely populated rural areas, who buy from local plumbers, who buy from wholesalers	Presenting the EDP to wholesalers and local plumbers and discussing the need for environmental documentation of products.	Too technical, not understandable, not credible.
		Presenting the EDP to municipal environmental authorities responsible for regulating the wastewater treatment locally and discussing the use of the	Does not answer the questions relevant for the kind of political and technical decisions related to local wastewater treatment policy.
			Not relevant – how does it work, and what is the price?

(continued)

Table 15.2 (continued)

Company	Customer	Used ways of communicating	Customer response
Informal meetings with visitors to an environmental exhibition.			
Cruzinox Lda	Retailers, often large commercial chains	Posted on website together with explanatory information.	No recorded response
		Sent by e-mail to around 100 customers together with explanatory information.	Few answers. Little interest. Some concern with cost increase due to EPD.
			Suggestion to put Stepwise EPD logo on the packaging. Two foreign clients showed commercial interest.
		Presented at phone meetings.	
		Presented at national and international business fairs	No recorded response.
Polisport Lda	Retailers, often large commercial chains	Sent by e-mail to clients together with explanatory information.	Customer feedback generally positive to EPDs and suggesting more graphics and more comprehensible language.
		Posted on website together with explanatory information.	
		Presented at business fairs.	
		Wrote about doing a Stepwise EPD in 2006/2007-year catalogue and in newsletter	Polisport believes that an EPD can differentiate the product from the rest of the market. This would be advantageous in markets with high environmental concerns.
Concretope Lda	Private and public building companies	Posted on website together with explanatory information.	No recorded response.
		Sent with bids.	Positive response.
		Sent by e-mail to around 26 clients together with questionnaire.	Eight answers. The main conclusion was that the Stepwise EPD was not so useful for selecting a product because it was found difficult to understand. Nevertheless six of eight respondents found it useful for acquiring information about the environmental profile of a product (ready-mixed concrete).

Meetings or work sessions with clients focusing on the underlying LCA rather than the EPD met with more appreciation. LCA as such stimulates cooperation in the supply chain as one need to ask ones suppliers about environmental data and the end-users about how they use the product. Vattenfall, who took part in the project as a mentor, has experienced that just by asking suppliers regularly about certain data can induce environmental improvements. An EPD, which uses generic data for upstream processes, does not automatically engage with the suppliers and end-users like a normal LCA (or full EPD) does.

15.4 Environmental Management Systems, ISO 14000 Standards

The resource-based view of the firm (Barney 1991) holds that firm characteristics such as strategy, structure, and core capability affect firms' innovation activities (Fagerberg et al. 2005). Based on this, Hart (1995) develops a concept of green capabilitie that is a firm's knowledge of environmental issues relevant to its business and procedures implemented to act and react on these issues. Russo and Fouts (1997) and Sharma and Vredenburg (1998) further elaborate and empirically corroborate this concept. Regarding environmental innovation, many studies look into organizational capabilities, particularly environmental management systems (EMS).

The assumption is that (certified) EMS such as ISO 14001 facilitate environmental innovations directly by introducing environmental goals and management structures as well as programs to achieve them (Coglianese and Nash 2001) and indirectly by inducing organizational learning and providing critical environmental information (Melnyk et al. 2003). Gonzalez-Benito and Gonzalez-Benito (2006) point out that the popularity and visibility of EMS certification offers potential for opportunistic (mis-) use to reduce stakeholder pressure without actually improving any environmentally relevant activities. Empirically, a positive impact of EMS in general on environmental product innovation activity was found in recent studies by Rehfeld et al. (2007) and Wagner (2008) whereas Rennings et al. (2006) showed evidence for the stimulating effect of EMS induced learning processes.

Stoller (1995) suggests that ISO 14000 will consist of several standards classified into six categories: (1) environment management system; (2) environmental auditing; (3) environmental performance evaluation; (4) life cycle assessment; (5) environmental labeling; and (6) environmental aspects in product standards. A brief discussion of each of these categories follows.

The ISO 14000 Environmental Management System (EMS) standard (ISO standards 14001, 14004) provides basic requirements for firms implementing an environmental management system. The standard defines an environmental management system as "that part of the overall management system which includes organization structure, planning activities, responsibilities, practices, procedures, processes, and resources for developing, implementing, achieving, reviewing, and maintaining the environmental policy" (Tibor and Feldman 1996). Environmental policy describes a firm's intentions, values, culture, and principles with respect to environmental performance. ISO 14000 parallels the approach TQM takes toward quality management, where a quality management system is seen as a philosophical business orientation. In this way the ISO 14000 standards help to define the corporation's environmental policy. ISO 14000 requires that the firm declare in a policy statement a strong commitment to comply with environmental regulations and minimize pollution.

Many firms have attempted to seek an effective environmental management system. These have led to implementation and development of the ISO 14001 standard for assessing environmental management processes. Today, all over the world, many firms are seeking ISO 14001 certification. According to statistics published by ISO, by the end of 2006, 129,199 certificates have been issued in 140 countries, an increase of 18,037 certificates since the end of 2005 when the total number was 21, 225 in 138 countries (The ISO survey 2006).

While Europe has 44.05% of regional share expressed in 2006, Far East countries have 41.24% share. The other regional shares are respectively North America 5.94%, Africa/West Asia 3.74%, Central and South America 3.37, and Australia and New Zealand 1.66%. The top 10 countries for growth in ISO 14001 certification were China (6159), Italy (2745), Spain (2505), Germany (975), Korea (938), Sweden (729), Romania (702), Turkey (505), and Switzerland (503) by 2006. In Turkey, while 918 firms were registered at the end of 2005, the number of certifications increased to 1423 at the end of 2006. This is an increase of 64.5% in 1 year in the number of ISO 14001 certificates in Turkey. This growth is rather striking.

According to statistics published by ISO, the top five industrial sectors for ISO 14001 certifications are electrical and optical equipment (9423), construction (9095), basic metal and fabricated metal products (7521), chemicals, chemical products and fibers (5041), and machinery and equipment (4554), respectively. The share of construction certificates in industrial sectors is quite high.

While 4660 firms were registered at the end of 2005, the number of certifications rose significantly to 9095 at the end of 2006. In this sector, the share all over the world has increased by 51.2% in 1 year (The ISO survey 2006). The more and more increasing interest of construction firms to obtain the ISO 14001 certificate depends on benefits associated with it (Turk 2009).

Government regulations usually require companies to reduce or eliminate the pollution (Morrow and Rondinelli 2002). Since the 1980s, governments and industry associations have significantly increased their promotion and reliance on voluntary environmental policies as a mean of encouraging firms to establish management and operational practices that reduce pollution and increase material and energy efficiencies. The term 'voluntary policy' includes a wide range of programes that employ explicit or implied regulatory and market incentives to obtain commitments from polluters in service and manufacturing industries to reduce the environmental damage for which they are responsible. The prevalence of voluntary policies and programes is representative of a broader shift toward more flexible instruments and away from standards-based regulation (Mori and Welch 2008). The first environmental management standard, BS7750, was prepared in 1992. In 1993, the Eco-management Audit Scheme (EMAS), prepared by the European Union, started to be applied. Following the BS7750 and EMAS, various countries developed their own EMS (Kein et al. 1999). Later, ISO 14001 environmental management system standard was introduced in 1996.

ISO 14001 provides guidelines by firms or organizations design and implement an EMS that identifies the organization's environmental policy, the environmental aspects of its operations, legal and other requirements, a set of clearly defined objectives and targets for environmental improvement and a set of environmental management programs (Jackson 1997). The ISO 14001 is a set of guidelines by which a facility can establish or strengthen its environmental policy, identify environmental aspects of its operations, define environmental objectives and targets, implement a program to attain environmental performance goals, monitor and measure effectiveness, correct deficiencies and problems, and review its management systems to promote continuous improvement.

Murat Turk (2009), using a structured questionnaire survey, investigated whether there is any dependence or relation between construction firms characteristics and having ISO 14001 certification and any difference in the perceptions related to ISO 14001 by considering both firm characteristics and two different groups as certified and non-certified firms. Additionally, they examine the perceived benefits of having ISO 14001 for certified construction firms.

Table 15.3 presents the behavior of firms toward environmental management measured in terms of whether the firm received complaints, how the firm responded and whether the firm had been punished due to violation of environmental protection regulations. Of the respondents of all certified firms, 71.4% (20) declared that they occasionally received complaints about the impact of their construction work on the environment. 75% (30) of the non-certified firms stated that they occasionally received complaints.

While, of the respondents of all certified firms, 64% (14) agreed that they would operate necessary procedures about these complaints, 56% (14) of surveyed non-certified firms declared the same. The majority of the respondents (97%) claimed that they had not been punished due to violation of environmental protection regulations.

89.3% (25) of the surveyed certified firms, stated that ISO 14001 was suitable for construction firms. Again 92.9% (26) of those firms declared that the ISO 14001 certificate had a positive impact within the Turkish construction sector. 100% (28) of the surveyed certified firms accepted that its importance would increase in the near future. According to 75% (21) of the certified firms, the ISO 14001 certificate should be mandatory within the Turkish construction sector (Table 15.4).

Table 15.3 Behavior toward environmental management (source: Turk 2009)

Item	All firms		Certificated firms		Non-certificated firms	
	Number	Percentage(%)	Number	Percentage(%)	Number	Percentage(%)
Has your company ever received any complaints related to environment?						
Always	2	3.0	2	7.1	–	–
Occasionally	50	73.5	20	71.4	30	75
No	15	22.1	5	17.9	10	25
Others	1	1.4	1	3.6	–	–
How has your company responded to these complaints?						
Complaints are taken seriously and care given	18	39.3	8	36	10	40
Necessary procedures operated	28	58.9	14	64	14	56
Ignoring the complaints	1	1.8	–	–	1	4
Whether firm had been punished due to violation of environmental protection regulations						
Yes	2	3	1	3.6	1	0
No	65	97	27	96.4	38	100
Whether firms employs staff for EM						
Yes	26	38.2	16	57.1	10	25
No	42	61.8	12	42.9	30	75

Table 15.4 Implementing ISO 14001 in construction industry (source: Turk 2004)

Item	All firms		Certificated firms		Non-certificated firms	
	Number	percentage(%)	Number	percentage(%)	Number	percentage(%)
ISO 14001 certification is an appropriate tool for the construction industry						
Yes	53	80.3	25	89.3	28	73.7
No	13	19.7	3	10.7	10	26.3
ISO 14001 certification has a positive impact on the Turkish construction sector						
Yes	54	83.1	26	92.9	28	75.7
No	11	16.9	2	7.1	9	24.3
ISO 14001 certification has to be mandatory for the Turkish construction industry						
Yes	48	72.7	21	75	27	71.1
No	18	27.3	7	25	11	28.9
The importance of ISO 14001 certification will increase for the Turkish construction sector soon						
Yes	63	95.5	28	100	35	92.1
No	3	4.5	0	0	3	7.9

Reasons for obtaining ISO 14001 of certified firms have been analyzed using t statistical testing (Table 15.5). According to the results, the most significant reason for the firms to obtain ISO 14001 certification is its easy access opportunity into the international market. In particular, the adaptation process to access the European Union and the extensive use of ISO 9000 QMS and ISO 14001 EMS and standardization within construction sectors of member countries of the European

Table 15.5 Reasons for seeking ISO 14001 certification, benefits gained from ISO 14001 registration and difficulties encountered during certification process and implementation, and t-test results

Items		Mean	Stand. Dev.	t-statistic	p-value
Reasons for seeking ISO 14001 certification	Easy access to international markets	1.6296	0.4921	17.207	0.000
	Desire of firm to develop its EMS	1.4074	0.6939	10.539	0.000
	For obligations in tender specifications	1.1481	1.0267	5.811	0.000
	Desire of firm to change and development	1.1154	0.9089	6.257	0.000
	Common opinion that ISO 14000 EMS to be mandatory in the near future	0.8519	1.1995	3.690	0.001
	Clients request ISO 14001	0.6296	1.2449	0.6296	0.014
	Competitors have ISO 14001	-0.4815	1.3118	-1.907	0.068
Benefits gained from ISO 14001 registration	Improves environmental awareness of company	1.4697	0.7888	15.137	0.000
	Improves standardization in environmental management	1.2879	0.8729	11.986	0.000
	Decreases adverse impacts on environment	1.2576	0.8649	11.813	0.000
	Provides sustainable development in environment	1.2424	0.8604	11.731	0.000
	Enhances company's image	1.1364	0.8751	10.550	0.000
	Decreases complaints against the company about environmental problems	0.9848	1.0889	7.352	0.000
	Increases self-confidence of the company	0.8939	0.9943	7.304	0.000
	Enlarges market share	0.8741	0.9989	7.241	0.000
	Improves client satisfaction	0.8636	1.0653	6.586	0.000
	Gives more stringent control over subcontractors	0.5152	1.0988	3.809	0.000
	Increases social recognition of the company	0.5303	1.1925	3.613	0.001
Difficulties encountered during ISO 14001 certification process and implementation	Firm management is not open to research and criticism	1.1429	1.0079	6.000	0.000
	Long period of certification process	1.000	0.9428	5.612	0.000
	Increased amount of paperwork	0.5926	1.0099	3.049	0.005
	High implementation costs	0.4290	1.3801	0.548	0.588
	Construction firm's size limits ability to get ISO 14001	0.4286	1.3174	1.721	0.097
	Lack of information regarding the certificate	0.3214	1.2188	1.396	0.174
	Difficulty in understanding the terminology of EMS	0.2857	1.1819	1.279	0.212
	Need of reorganization of firm in terms of management	0.2857	1.3569	1.114	0.275
	Lack of qualified personnel in the firm	0.1786	1.1564	0.817	0.421
	Failure to provide control of the sub-contractor	-0035	1.1701	-0.162	0.873
	Lack of client support	-0.0071	1.2150	-0.311	0.758
	Lack of government support	-0.2857	1.3569	-1.114	0.275

Union have accelerated the urgency to seek ISO 14001 certification in the construction sector in Turkey.

One of the reasons for the certified firms to obtain certificates is the requirement of tender specifications to have the certificate. As can be understood from the results of the analysis, there is a similarity between the reasons for obtaining ISO 9001 certificates and the reasons for obtaining ISO 14001 certificates.

According to the results of analysis, although there is not any difference in perceptions on ISO 14001 certification in terms of firm characteristics and being as certified and non-certified and their both positive opinions about ISO 14001 certification. There is a relation between firms characteristics and having ISO 14001 certification. ISO 14001 certification contributes to construction firms not only in terms of environmental benefits but also with corporate management and marketing effects, thus verifying that the ISO 14001 has a positive impact on the Turkish construction sector.

15.5 Environmental Value Chain

"Value chain" refers to all the activities and services that bring a product (or a service) from conception to end use in a particular industry—from input supply to production, processing, wholesale, and finally, retail. It is so called because value is being added to the product or service at each step. Taking a "value chain approach" to economic development means addressing the major constraints and opportunities faced by businesses at multiple levels of the value chain.

Consider the typical, vaunted "value chain": Suppliers extract raw materials from nature, the manufacturing sector turns them into products, and retailers sell them to consumers, who throw away the packaging, and often eventually the products, as waste. In each link, companies and consumers—consciously or not—despoil the environment. Nonetheless, businesses generally concentrate solely on their own place in the chain, making sure to add at least enough value to generate a profit while using whatever materials and energy are needed to create that value. For their part, consumers are by and large just as cavalier about the consequences of their actions.

One possible solution is what I call a "value loop" (Hardin Tibbs 2010). Under this approach, the beginning and end of the value chain are linked together so that materials, products, and waste can flow among suppliers, manufacturers, and customers in a sustained cycle. The goal: to promote technologies and business models that have minimum impact on nature throughout the loop—or that incorporate it in a beneficial way.

The forward half of the loop—from raw materials to manufactured product to trash—is already in place. The challenge now is to create the return half of the loop, collecting waste material and reprocessing it into new "raw" material. This requires a sequence of steps, including "product take-back," product de-manufacture (breaking down an item into its basic elements), and materials reprocessing. The business challenge of the return path is to create value with each of these steps—just as value is created during the forward half. This could be accomplished through regulation-based or through innovation-based business models.

A recent example of the former has already been introduced in Europe, in the form of so-called product take-back legislation, which mandates the recycling of consumer products at the end of their useful life and requires that new products contain a minimum percentage of recycled materials. The European Union's End-of-Life Vehicles Directive of 2000, for instance, requires automakers to pay for the cost of taking back and recycling old cars, starting with vehicles sold in 2001, while setting a date of 2015 for 85% of the metal in cars to be recycled and banning the use of hazardous heavy metals.

As a result of this directive, European car manufacturers such as BMW and Volkswagen are beginning to develop environment-friendly strategies that include targets for recycled content, the redesign of cars to make them easier to de-manufacture, and the use of materials that are easier to reprocess into a reusable form, thus keeping down the costs of the return loop and making it economically viable for third-party reprocessors to take part in the value loop.

As critical as legislation will be in promoting the creation of value loops, companies need not wait until laws are passed; instead, they could devise innovation-based business models that protect natural resources. For example, as existing environmental regulations continue to drive up the cost of routine waste disposal, recycling businesses will be able to charge higher fees for picking up recyclable trash. This disposal fee can subsidize both the cost of collection and the expense of breaking down the material so it can be reused. This would make the price of recycled goods competitive with non-recycled products.

Mining companies could rethink their role in the value chain by "mining" the scrap flows in urban areas—such as cars, trucks, and other steel-rich equipment—and cutting out the recycling middlemen. This notion has already placed steel mini-mills in such scrap-rich locations as the suburbs of Sydney, Australia, where steel maker OneSteel has established a new plant devoted to just this kind of reclamation. Mining companies might also begin to sell reusable metals like "cyclic copper," which they would lease to product manufacturers that would include it as an environmental feature in their products. When the product is not useful anymore, the manufacturers would take it back, retrieve, and extract the cyclic copper, and return it to the mining company for recycling, putting it back in the value loop.

The energy required to maintain value loops must not be ignored, however, and to get the most out of this new closed value chain we need to both develop ways to use less energy and perfect more easily renewable sources of energy—hydrogen cells, for instance.

Box AB describes the application of EVCA to product end-of-life treatment programs proposed by Rose et al. (2000). The analysis examines the information, money, and product flows between players.

Table 15.6 General relationship between players

Actors	Government	Consumers	Producers	Recyclers
Government	Political decision making	Votes ! Taxes $	Lobbying Taxes $	Lobbying Taxes $
Consumers	Representation, services	Feelings Fads	Products Services	Used Products Services
Producers	Regulation, representation, services	Complaints Product, Pre-owned $	Competition	Material $ Recycled material
Recyclers	Regulation, representation, services	Products	Products Recycling cost $	Competition

Box AB: Environmental Value Chain Analysis Applied to End-of-Life Systems

In searching for a more direct route of managing and communicating value, Environmental Value Chain Analysis is able to reduce complexity and cost of building programs for eco-efficient treatment of products at the end-of-life. The graphical representation of the EVCA provides clear demonstration of the actors participating.

EVCA can be used to understand the relationships between the players: producers, consumers, government, and recyclers. The lines of jurisdiction, roles and responsibility are blurred within the groups themselves. A generic view of the relationships between government, producers, consumers, and recyclers is shown in the following Table 15.6.

The following figure demonstrates an external EVCA applied generally to the relationships among producers, consumers, recyclers, and government. EVCA can also be applied internally, as shown in the examination of Philips Consumer Electronics (Box AC). This diagram demonstrates the various types of flows between the players, focusing on information, product, and money.

Complaints, sadly the most frequent, comes from dissatisfaction with communication, service, product or information provided. Appropriate exchange of information leads to mutual benefit of the participants. Top-down information is a limited transfer of information from a 'high' member of the chain to lower members of the chain (depending on the 'power' in the chain). Top-down usually offers information as the higher member requires. Feedback is the response of lower members to these requirements or to other requests. It is necessary to know the individual

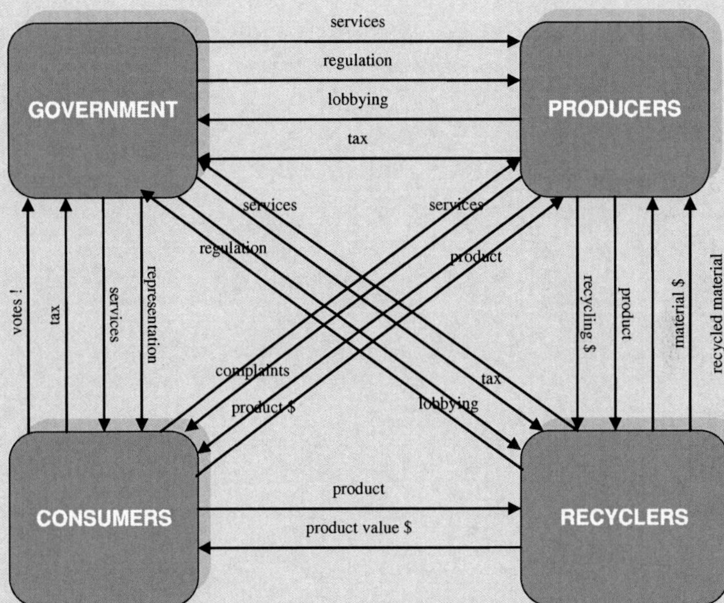

Fig. 15.3 Example of external environmental value chain analysis

goals of the players regarding the end-of-life treatments of products—whether to minimize waste, make profits, improve customer, and shareholder perception or to abide by regulation. It is also helpful to understand the goals and objectives of the other members.

Awareness of the, often conflicting, goals and objectives allows for the appropriate values to be placed on the EVCA diagram. EVCA looks different depending on the reasons for initiating the end-of-life system and end-of-life strategy. An end-of-life system focused on the end-of-life strategy—product reuse—looks different from one aiming to achieve high percentages of material recycling—product recycling through disassembly and without disassembly, leads to different relationships as well. The results of Environmental Value Chain Analysis (EVCA) allow decision makers to identify the critical success factors of such programs upfront (Fig. 15.3)

The case studies reveal that, apart from technical and economical factors, organization factors, and external perceptions play a major role in the success of Ecodesign activities Rose et al. (2000).

HP printers

Hewlett Packard sells a wide variety of information technology equipment from computers, servers to printers (Fig. 15.4).

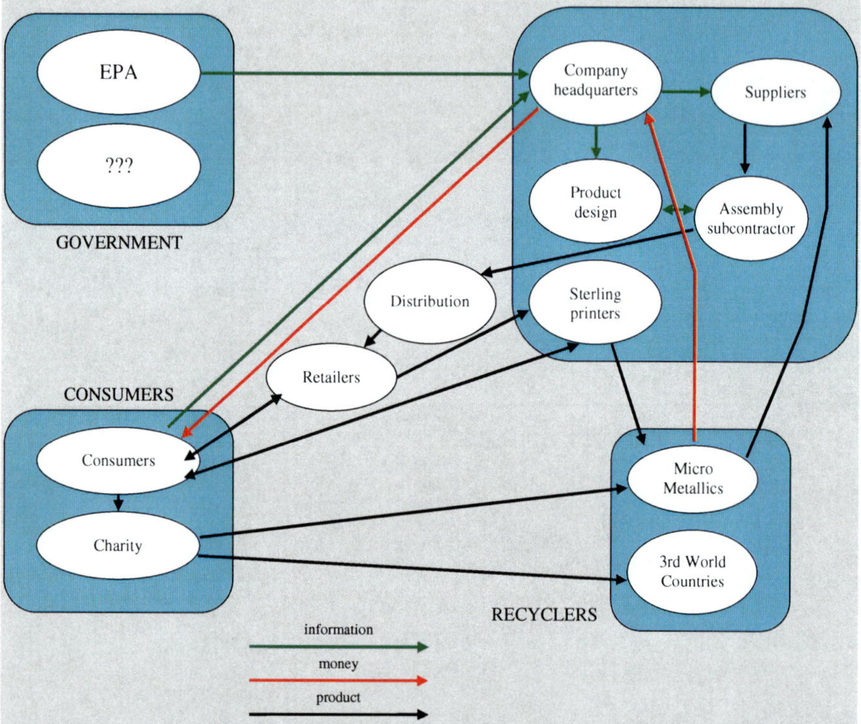

Fig. 15.4 HP value chain

The Hewlett Packard and Micro Metallics partnership between an original equipment manufacturer and a recycling company recovers service parts and recycles useful materials from end-of-life products. HP realizes great savings to the company by collecting products to harvest for service parts. As of yet, they have had limited success in reuse, or reselling their products on the market. To provide incentives for consumers to return their product, HP has started a program, Trade-in Trade-Up, that gives trade-in value to apply toward the purchase of new products.

General Electric

General Electric provides aircraft engines for military and commercial applications. They also include an extensive service contract through the life of the product, in which service is performed by another organization in the company. The long life of the product as well as the high value of materials at the final disposition allow for high reuse, refurbishment, and recycling of this product. The incentives are inherent to the product characteristics. The largest challenge for GE is staying informed on changing regulations regarding fuel efficiency and usage conditions (Fig. 15.5).

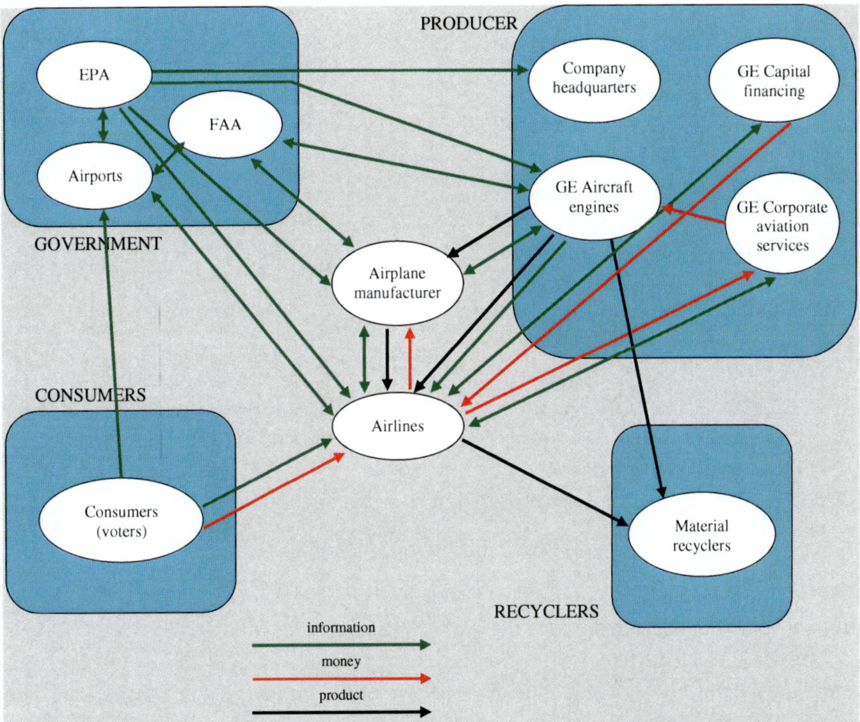

Fig. 15.5 GE value chain

The examination of the product end-of-life systems shows weak lines of communication, demonstrating relationships that may be neglected. In some cases, there are redundant mechanisms for collecting products, leading to confusion from consumers and increasing end-of-life costs. Where products have perceived end-of-life value, through materials or application, the incentives for product return exist naturally. The use of EVCA shows the incentives, such as service to the consumer or trade-in value. On the other hand, if consumers are burdened with additional fees, such as the Dutch take back system, there are many disincentives for consumers.

Secondly, recognizing that fee systems must be decided and agreed on by all participants in the end-of-life systems is necessary. For example, fee systems that place heavy burdens on one group will not be successful. Incentives are more successful, through trade-in or rebate systems, and will encourage more participation.

The existing fee structure charges producers equally, even if some products are cheaper to recycle. Therefore, there is no reward for companies that have products which have undergone extensive redesign to reduce the end-of-life treatment cost. Additionally, the department charged for the end-of-life costs does not always transfer those charges to the design

Table 15.7 Developments in environmental design at Philips Electronics

	1992–1996	1996–2000	Now
Management Involvement	+	+ +	+ + +
Program	Policy and interpretation	Environmental opportunity	Eco Vision
Orientation	Defensive	Cost oriented	Proactive
Approach	Procedures	Technical	Business
Tools	Mandatory rules	Manual	Managing processes

departments. Therefore, there is no incentive from a design perspective to improve the end-of-life treatment costs. If the financial mechanism involved in these end-of-life systems does not involve the design departments, then the objective to encourage eco-design is not achieved.

Box AC shows the Environmental Value Chain Analysis (EVCA) carried out at Philips Consumer Electronics. This case study was proposed by Ishii and Stevels (2000). EVCA helped to transform Philips Consumer Electronics from an environmentally defensive organization in 1994 into a proactive one at present. The Green TV project and the current Eco Vision program serve as illustrative examples of EVCA. The analysis enables us to formulate the lessons to be learned from the past and also in which way current programs can be further enhanced.

Box AC: From Defensive to Proactive

Table 15.7 shows the evolution of approaches to Design for Environment (DFE) at Philips Consumer Electronics.

The Defensive Period (1992–1996).

There was a policy statement and a detailed interpretation of it, however, the actual implementation was left to the divisions and business groups. This resulted in basically a defensive attitude: efforts were focussed on legal and regulatory compliance and preventing bad environmental press.

They achieved these objectives by applying mandatory environmental design rules to be checked at product release. Subjects to be considered included (and still include): banned substances, packaging rules, marking and labelling, customer information, and batteries. Although the orientation of this program was chiefly internal, particularly on product development technicalities, its effects were wider. The effort created environmental

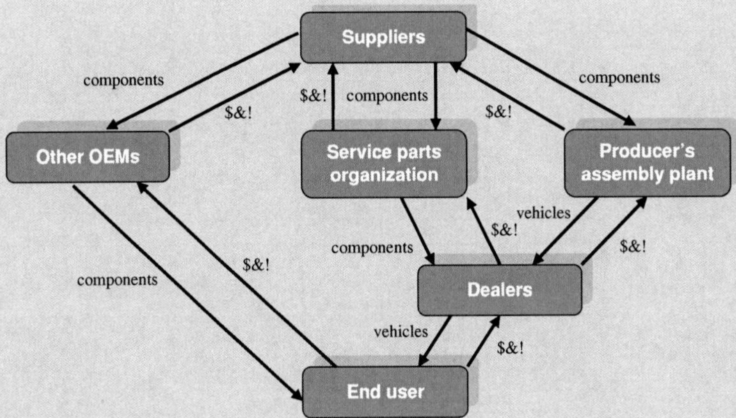

Fig. 15.6 Automotive components CVCA

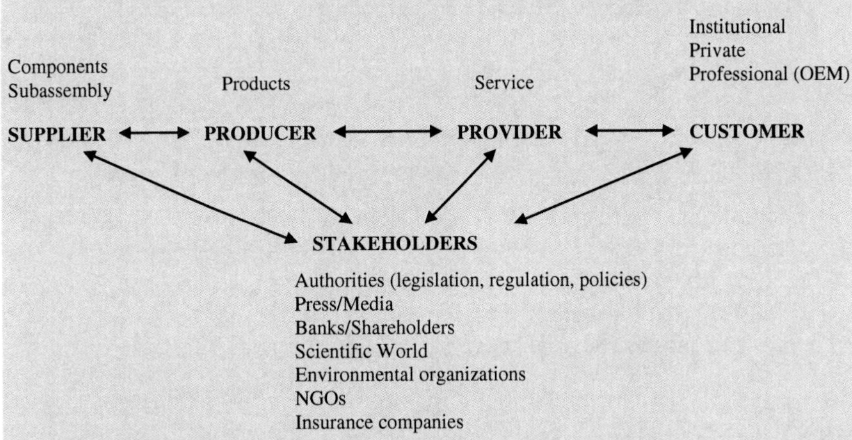

Fig. 15.7 General form of an environmental value chain

awareness and, while the scope was limited by present standards, the outside world saw the company a first mover taking initiatives on the environment.

The Environmental Opportunity Program (1996–2000).

This program showed stronger management involvement. The aim was to strengthen further the environmental organization and to achieve cost reductions through an environmental approach. The major goal was that all factories should have an environmental management system in place per IS0 14001, should achieve 25% energy reduction in all operations, and 15% packaging reduction. Furthermore, the program called for creation of internal and external network and active participation in legislation and regulation discussions. Life cycle design and supplier requirements were

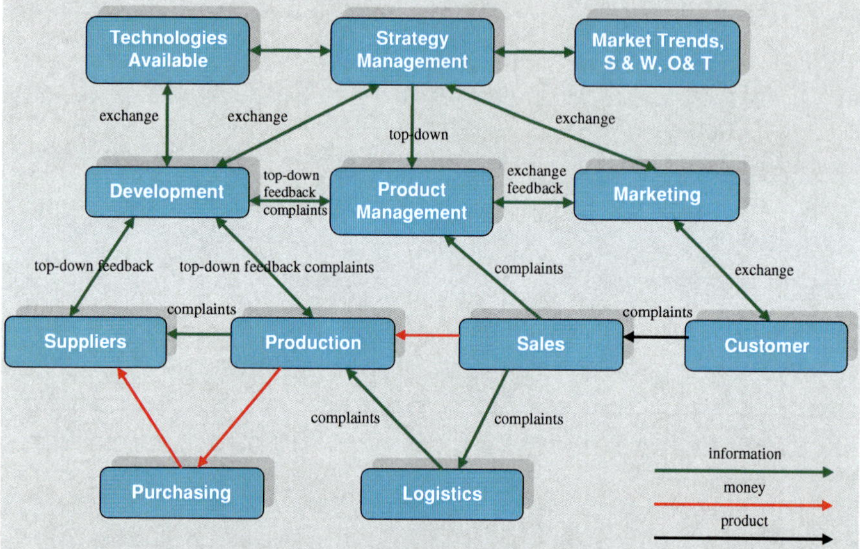

Fig. 15.8 The internal Environmental Value Chain

mentioned as well, but these were still on a voluntary basis for divisions and business units. The big step forward was that this program,

- forced business groups to confront systematically various environmental concerns,
- brought clear cost saving through energy reduction and reducing packaging that were visible for the organization, and
- enhanced substantially the internal profile of environmental programs.

In spite of its voluntary character, the practice of life cycle design (Eco Design) started to takeoff. The program produced an environmental design manual that provided guidance to these processes. Although the program scored a big success, two elements were substantially lacking:

- *Creativity*. Both the ISO 14001 standard and a design manual (providing information, recommendations, and establishing "rules") are both fairly formal and static. As such they do not challenge people to unleash their creativity.
- *External orientation*. Since the program was to a large extent organizationally and technically oriented, the programs did not cater downstream (to customers and other stakeholders who should be the environmental beneficiaries) and upstream (to suppliers who should contribute proactively).

A clear communication of results is just as important as obtaining the results themselves. Top achievements are possible only when the effort involves total value chain (from supplier up to end-of-life processors).

Table 15.8 An internal Issue Correlation Matrix (ICM) for environmental and related business issues

	Environmental department	Strategy management	Product management	Development	Purchasing	Production	Marketing	Sales	Totals
Environmental strategy	X	X							2
Specification (Functionality)	X		X	X	X		X		5
Quality	X				X	X	X	X	5
Environmental score	X								1
Project management	X		X	X		X			4
Environmental communication/ competition		X	X				X	X	4
Production technology		X	X	X					3
Cost		X	X		X	X	X	X	6
Investment		X				X			2
Time to market			X	X			X	X	4
Suppliers				X	X				2
Logistics					X	X		X	3

The current Eco Vision Program (now).

The cornerstone of the Eco Vision program is the communication of top achievements as embodied in Green Flagship products to customers and other stakeholders. These achievements come primarily from management of the cross-functional processes around creation, production and marketing, and sales of these products.

The Eco Vision program has been introduced top-down in the organization that is through initiative and commitment of the president and chief executive officer of the company.

Environmental Value Chain Analysis (EVCA) has been developed from Customer Value Chain Analysis (CVCA). The method seeks to identify pertinent customer and other stakeholders' interests, their value perceptions and the relationship between these parties in green product or process development projects.

Basic stem involved in CVCA

CVCA facilitates the first steps product definition and feeds into other structured methodologies such as Quality Function Deployment. CVCA should be a team effort involving multi-functional teams and top level management. Figure 15.6 shows a typical example of a customer value chain graph for an automotive interior components (panel), manufactured by a supplier and assembled into a vehicle by a product integrator.

The basic steps involved in generating such a graph and analyzing them are as follows:

1. List the pertinent parties involved: stakeholders, customers, partners, regulatory bodies, etc.
2. Identify the relationship among these parties by defining the flow of the following:

   Money or funds (indicate the flow by $)
   Stuff: machines, materials, services, or information (use appropriate icon)
   Complaints, regulatory influences, votes, etc. (!)
3. Analyze the resulting CVC "Graph" to address the following questions:

   Who are the customers that are critical to the project?
   Trace the $ and ! from your own position
   What are the value propositions of these parties?
   Look at the input/out of $, !, and other icons.
   How are they going to make money?
   Use this information to generate the Voice of Customers (VOCs).
   Use the flow of !, particularly complaints to identify negative VOCs.
4. Feed the information into Product Definition Assessment (PDA)

   The CVC Graph should facilitate the completion of the PDA checklists
   Use CVC results to identify partners and management needs
   Do not be afraid to cancel the project if the value proposition is weak

5. Use the CVCA results down stream in the

Development **Process** (Flow Down)
Use Quality Function Deployment to flow down the VOCs
Use Failure Modes and Effects Analysis on negative VOCs and generate robust designs.
Adapting CVCA to Environmental Value Chain

Now let us adapt the CVCA to environmental values. A general form of an Environmental Value Chain looks as follows (Fig. 15.7):

The upper part of the graph represents the main stream of values between suppliers, producers, providers, and customers. Stakeholder interactions with the players in the main stream mostly involve one of this information flow as defined above.

One must also address the value chains internal to organizations. For manufacturing companies, Fig. 15.8 shows this chain focuses around the main stream between suppliers, producers, providers, and customers.

The interaction between internal stakeholders is mostly focused on information flows. Note that individual internal stakeholders communicate with different stakeholders outside the company. This fact shows that significant cross functional exchange is needed to form a unified position with respect to the outside world. In order to make this interaction more transparent, an Issue Correlation Matrix (ICM) has been developed. The rows represent the environmental and the business issues.

The columns represent the various departments involved are listed. For each department, crosses indicate which environmental and related business issues rank highest from their perspective. The number of crosses per department is "normalized," i.e., the numbers rank the department priorities with a score of 1 to 5 (Table 15.8).

One can see that, during this period, only the specification and quality items have high correlation among the departments. At the time, the Environmental Department did not adequately address issues such as the competition, communication, cost issues, and time to market. The Eco Vision Program substantially resolved this deficiency. As will be shown, the company must still address the supplier and logistics.

References

Acs ZJ, Audretsch DB (1988) Innovation in large and small firms—an empirical analysis. Am Econ Rev 78(4):678–690

Antil JH (1984) Socially responsible consumers: profile and implications for public policy. J Macromarketing 1:18–39

Barbier E (1987) The concept of sustainable economic development. Environ Conserv 14: 101–110

Barney J (1991) Firm resources and sustained competitive advantage. J Manag 17(1):99–120

Baylis R, Connell L, Flynn A (1998) Company size, environmental regulation and ecological modernization: further analysis at the level of the firm. Business Strategy Environ 7(5): 285–296

Cleff T, Rennings K (1999) Determinants of environmental product and process innovation. Eur Environ 9(5):191–201

Coglianese C, Nash J (2001) In: Coglianese C, Nash J (Eds.) Regulating from the inside—Can environmental management systems achieve policy goals? Resources for the future, Washington D.C (2001)

Drumwright ME (1994) Socially responsible organizational buying: Environmental concern as a non-economic buying criterion. J Mark 58(3):1–19

Ellen PS, Wiener JL, Cobb-Walgren C (1991) The role of perceived consumer effectiveness in motivating environmentally conscious behaviors. J Public Policy Mark 10:102–117

Engels S (2008) Determinants of environmental innovation in the Swiss and German food and beverages industry—What role does environmental regulation play? ETH Zurich, Zurich, Switzerland

Esty DC, Winston AS (2006) Green to gold: How smart companies use environmental strategy to innovate create value, and build competitive advantage. Yale University Press, New Haven

Fagerberg J, Mowery DC, Nelson RR (eds.) (2005) The Oxford Handbook of Innovation. Oxford University Press, Oxford

Gonzalez-Benito J, Gonzalez-Benito O (2006) A review of determinant factors of environmental proactivity. Business Strategy Environ 15(2):87–102

Greening DW, Gray B (1994) Testing a model of organizational response to social and political issues. Acad Manag J 37(3):467–498

Gronroos C (1990) Service management and marketing. Lexington Books, Lexington

Hardin Tibbs (2010) http://www.socialedge.org/features/resources/human-technologies/scenario-planning/hardin-tibbs-towards-normative-scenarios

Hart SL (1995) A natural-resource-based view of the firm. Acad Manag Rev 20:986–1014

Hart SL (2005) Capitalism at the crossroads: The unlimited business opportunities in solving the world's most difficult problems. Wharton School Publishing, Philadelphia

Hawken P, Lovins A, Lovins H (1999) A road map of green capitalism. Harvard Business Review, May–June

Hitchens DMWN, Birnie E, Thompson W, Triebswetter U, Bertossi P, Messori L (2000) Environmental regulation and competitive advantage: a study of packaging waste in the european supply chain. Edward Elgar, Cheltenham, UK

Hunt S, Morgan RM (1995) The comparative advantage theory of competition. J Mark 59:1–15

Ishii K, Stevels A (2000) Environmental value chain analysis: a tool for product definition in Eco Design. IEEE International Symposium on Electronics

Jackson SL (1997) The ISO 14001 implementation guide: creating an integrated management system. Wiley, New York

Kein ATT, Ofori G, Briffett C (1999) ISO 14000: its relevance to the construction industry of Singapore and its potential as the next industry milestone. Constr Manag Econ 17:449–461

Kotler P (2003) Marketing management, 11th edn. Upper Saddle River. Prentice Hall

Lash J, Wellington F (2007) Competitive advantage on a warming climate. Harvard Bus Rev 85(3):94–103

Lewis M (2008) Under the needle: TV repair shop pulling the plug. Seattle postintelligencer. February 14th. http://seattlepi.nwsource.com/local/351347_needle15.html. Accessed 2 Apr 2008

Lovins AB, Lovins LH, Hawken P (1999) A road map for natural capitalism. Harvard Business Review 77:145–158

Melnyk SA, Sroufe RP, Calantone R (2003) Assessing the impact of environmental management systems on corporate and environmental performance. J Oper Manag 21(3):329–351

Menon A, Menon A (1997) Enviropreneurial marketing strategy: the emergence of corporate environmentalism as market strategy. J Mark 61:51–67

Mori Y, Welch EW (2008) The ISO 14001 environmental management standard in Japan: results from a national survey of facilities in four industries. J Environ Plan Manag 51(3):421–445

Morrow D, Rondinelli D (2002) Adopting corporate environmental management systems: motivations and results of ISO 14001 and EMAS certification. Euro Manag J 20(2):159–171

Narver JC, Slater SF (1990) The effect of a market orientation on business profitability. J Mark 54:20–35

Rehfeld K-M, Rennings K, Ziegler A (2007) Integrated product policy and environmental product innovations: an empirical analysis. Ecol Econ 61(1):91–100

Rennings K, Ziegler A, Ankele K, Hoffmann E (2006) The influence of different characteristics of the EU environmental management and auditing scheme on technical environmental innovations and economic performance. Ecol Econ 57(1):45–59

Robotis A, Bhattacharya S, Van Wassenhove LN (2005) The effects of remanufacturing on procurement decision for resellers in secondary markets. Eur J Oper Res 163:688–695

Rose CM, Stevels A, Ishii K (2000) A new approach to end-of-life design advisor. In: IEEE International Symposium for Electronics and the Environment Conference

Russo MV, Fouts PA (1997) A resource-based perspective on corporate environmental performance and profitability. Acad Manag J 40(3):534–559

Savitz AW, Weber K (2006) The triple bottom line: How today's best-run companies are achieving economic, social and environmental success—and how you can too. Jossey-Bass, San Francisco

Seijas-Nogareda J (2007) Determinants of Environmental Innovation in the German and Swiss Chemical Industry—With Special Consideration of Environmental Regulation. ETH Zurich, Zurich

Sharma A, Iyer GR, Mehrotra A, Krishnan R (2010) Sustainability and business-to-business marketing: a framework and implications. Ind Mark Manag 39:330–341

Sharma S, Vredenburg H (1998) Proactive corporate environmental strategy and the development of competitively valuable organizational capabilities. Strateg Manag J 19(8):729–753

Sharma A, LaPlaca P (2005) Marketing in the emerging era of build-to-order manufacturing. Ind Mark Manage 34:476–486

Sheth JN, Sisodia R, Sharma A (2000) The antecedents and consequences of customer-centric marketing. J Acad Mark Sci 28:55–66

Shrivastava P (1995) Environmental technologies and competitive advantage. Strategic Manag J 16(3):183e200

Sisodia R, Wolfe D, Sheth J (2007) Firms of endearment: How world-class companies profit from passion and purpose. Wharton School Publishing, Philadelphia

Stoller FL (1995) Correction symbols: Developing editing skills. In: White RV (Ed.) New ways in teaching writing, TESOL. Alexandria, pp 139–141

The ISO survey of ISO 9001 and ISO 14001 certificates. International Organisation for Standardization (2006) ISO International Organization for Standardization CD-ROM, Geneva

Tibor T, Feldman I (1996) ISO 14000: A guide to the new environmental standards. Irwin Professional Publishing, Chicago, IL

Turk AM (2009) The benefits associated with ISO 14001 certification for construction firms: Turkish case. J Clean Prod 17:559–569

Wagner M (2008) Empirical influence of environmental management on innovation: evidence from Europe. Ecol Econ 66(2–3):392–402

Winsemius P, Guntram U (1992) Responding to the environmental challenge. Business Horizons 35:12–20

Zackrisson M, Rocha C, Christiansen K, Jarnehammar A (2008) Stepwise environmental product declarations: ten SME case studies. J Clean Prod 16:1872–1886

Zimmerman A (2006, August 21) Wal-Mart sees profit in green: Some recycling initiatives are helping retailer's bottom line. Wall Str J, B3

Printed by Publishers' Graphics LLC